SUSTAINABLE
HORTICULTURE
Volume 2:

Food, Health, and Nutrition

Innovations in Horticultural Science

SUSTAINABLE HORTICULTURE
Volume 2:

Food, Health, and Nutrition

Edited by
Debashis Mandal, PhD
Amritesh C. Shukla, DPhil, DSc
Mohammed Wasim Siddiqui, PhD

Apple Academic Press Inc.
3333 Mistwell Crescent
Oakville, ON L6L 0A2
Canada

Apple Academic Press Inc.
9 Spinnaker Way
Waretown, NJ 08758
USA

© 2019 by Apple Academic Press, Inc.

First issued in paperback 2021

Exclusive worldwide distribution by CRC Press, a member of Taylor & Francis Group
No claim to original U.S. Government works

ISBN-13: 978-1-77463-125-6 (pbk)
ISBN-13: 978-1-77188-647-5 (hbk)

Library and Archives Canada Cataloguing in Publication

Sustainable horticulture / edited by Debashis Mandal, PhD, Amritesh C. Shukla, DPhil, DSc, Mohammed Wasim Siddiqui, PhD.

(Innovations in horticultural science)
Includes bibliographical references and indexes.
Contents: Volume 2. Food, health, and nutrition.
Issued in print and electronic formats.
ISBN 978-1-77188-647-5 (v. 2 : hardcover).--ISBN 978-1-315-14799-4 (v. 2 : PDF)

1. Sustainable horticulture. I. Mandal, Debashis, editor II. Shukla, Amritesh C., editor III. Siddiqui, Mohammed Wasim, editor IV. Series: Innovations in horticultural science

SB319.95.S87 2018	635.028'6	C2018-903327-4	C2018-903328-2

CIP data on file with US Library of Congress

Apple Academic Press also publishes its books in a variety of electronic formats. Some content that appears in print may not be available in electronic format. For information about Apple Academic Press products, visit our website at **www.appleacademicpress.com** and the CRC Press website at **www.crcpress.com**

CONTENTS

INNOVATIONS IN HORTICULTURAL SCIENCE

Editor-in-Chief:

Dr. Mohammed Wasim Siddiqui Assistant Professor-cum- Scientist
Bihar Agricultural University | www.bausabour.ac.in
Department of Food Science and Post-Harvest Technology
Sabour | Bhagalpur | Bihar | P. O. Box 813210 | INDIA
Contacts: (91) 9835502897
Email: wasim_serene@yahoo.com | wasim@appleacademicpress.com

The horticulture sector is considered as the most dynamic and sustainable segment of agriculture all over the world. It covers pre- and postharvest management of a wide spectrum of crops, including fruits and nuts, vegetables (including potatoes), flowering and aromatic plants, tuber crops, mushrooms, spices, plantation crops, edible bamboos etc. Shifting food pattern in wake of increasing income and health awareness of the populace has transformed horticulture into a vibrant commercial venture for the farming community all over the world.

It is a well-established fact that horticulture is one of the best options for improving the productivity of land, ensuring nutritional security for mankind and for sustaining the livelihood of the farming community worldwide. The world's populace is projected to be 9 billion by the year 2030, and the largest increase will be confined to the developing countries, where chronic food shortages and malnutrition already persist. This projected increase of population will certainly reduce the per capita availability of natural resources and may hinder the equilibrium and sustainability of agricultural systems due to overexploitation of natural resources, which will ultimately lead to more poverty, starvation, malnutrition, and higher food prices. The judicious utilization of natural resources is thus needed and must be addressed immediately.

Climate change is emerging as a major threat to the agriculture throughout the world as well. Surface temperatures of the earth have risen significantly over the past century, and the impact is most significant on agriculture. The rise in temperature enhances the rate of respiration, reduces cropping periods, advances ripening, and hastens crop maturity, which adversely affects crop productivity. Several climatic extremes such as droughts, floods, tropical cyclones, heavy precipitation events, hot extremes, and heat waves cause a negative impact on agriculture and are mainly caused and triggered by climate change.

In order to optimize the use of resources, hi-tech interventions like precision farming, which comprises temporal and spatial management of resources in horticulture, is essentially required. Infusion of technology for an efficient utilization of resources is intended for deriving higher crop productivity per unit of inputs. This would be possible only through deployment of modern hi-tech applications and precision farming methods. For improvement in crop production and returns to farmers, these technologies have to be widely spread and adopted. Considering the above-mentioned challenges of horticulturist and their expected role in ensuring food and nutritional security to mankind, a compilation of hi-tech cultivation techniques and postharvest management of horticultural crops is needed.

This book series, Innovations in Horticultural Science, is designed to address the need for advance knowledge for horticulture researchers and students. Moreover, the major advancements and developments in this subject area to be covered in this series would be beneficial to mankind.

BOOKS IN THE SERIES

- **Spices: Agrotechniques for Quality Produce**
  Amit Baran Sharangi, PhD, S. Datta, PhD, and Prahlad Deb, PhD

- **Sustainable Horticulture, Volume 1: Diversity, Production, and Crop Improvement**
  Editors: Debashis Mandal, PhD, Amritesh C. Shukla, PhD, and Mohammed Wasim Siddiqui, PhD

- **Sustainable Horticulture, Volume 2: Food, Health, and Nutrition**
  Editors: Debashis Mandal, PhD, Amritesh C. Shukla, PhD, and Mohammed Wasim Siddiqui, PhD

- **Underexploited Spice Crops: Present Status, Agrotechnology, and Future Research Directions**
  Amit Baran Sharangi, PhD, Pemba H. Bhutia,
  Akkabathula Chandini Raj, and Majjiga Sreenivas

ABOUT THE EDITORS

Debashis Mandal, PhD

Assistant Professor, Department of Horticulture, Aromatic and Medicinal Plants, Mizoram University, Aizawl, India

Debashis Mandal, PhD, is an Assistant Professor in the Department of Horticulture, Aromatic and Medicinal Plants at Mizoram University in Aizawl, India. Dr. Mandal started his academic career as Assistant Professor at Sikkim University, India. He had postdoctoral research experience as a project scientist at the Precision Farming Development Centre, Indian Institute of Technology, Kharagpur, India. He has done research on sustainable hill farming for the past seven years and has published 25 research papers in reputed journals and three book chapters and two books. He is a working member of the global working group on Lychee and Other Sapindaceae Crops, International Society for Horticultural Science (ISHS), Belgium, and as Editor-in-Lead (Hort.) for the *International Journal of Bio-resource & Stress Management*. In addition, he is a consultant horticulturist to the Department of Horticulture and Agriculture (Research and Extension), Government of Mizoram, India, where he works on various research projects. He has visited many countries for his work, including Thailand, China, Nepal, Bhutan, Vietnam, South Africa, and South Korea.

Amritesh C. Shukla, DPhil, DSc

Professor, Department of Botany, University of Lucknow, India

Amritesh C. Shukla, DPhil, DSc, is currently a Professor in the Department of Botany, University of Lucknow, India. He was formerly a Professor in the Department of Horticulture, Aromatic and Medicinal Plants, Mizoram University, Aizawl, India. He began his academic career as Associate Professor at the same university. He has more than 20 years of experience on natural products and drug development and standardized various scientific methods, viz., MSGIT, MDKT, MS-97, NCCLS (2002–2003).

He has developed six commercial herbal formulations and holds four patents. He has published 75 research papers and 15 book chapters. He has also authored six books and has handled many externally funded research projects. He is a Fellow of five national and international scientific societies and also a Visiting Professor at the University of British Columbia, Canada, and the University of Mauritius. In addition, Dr. Shukla is working as an Editor, Associate Editor, and editorial board member of many internationally reputed journals, including the *American Journal of Food Technology, Journal of Pharmacology & Toxicology, Research Journal of Medicinal Plants*, etc. He has also been a visiting scientist at universities in Australia, Germany, China, and USA.

Mohammed Wasim Siddiqui, PhD

Assistant Professor and Scientist in the Department of Food Science and Post-Harvest Technology, Bihar Agricultural University, Sabour, India

Mohammed Wasim Siddiqui, PhD, is an Assistant Professor and Scientist in the Department of Food Science and Post-Harvest Technology, Bihar Agricultural University, Sabour, India, and author or co-author of 31 peer-reviewed research articles, 26 book chapters, two manuals, and 18 conference papers. He has 11 edited and an authored books to his credit, published by Elsevier, CRC Press, Springer, and Apple Academic Press. Dr. Siddiqui has established the international peer-reviewed *Journal of Postharvest Technology.* He is Editor-in-Chief of two book series (Postharvest Biology and Technology and Innovations in Horticultural Science), published by Apple Academic Press. Dr. Siddiqui is also a Senior Acquisitions Editor in Apple Academic Press, New Jersey, USA, for Horticultural Science. He has been serving as an editorial board member and active reviewer of several international journals.

Dr. Siddiqui has received several grants and respected awards for his research work by a number of organizations. He is an active member of the organizing committees of several national and international seminars, conferences, and summits. He is one of key members in establishing the World Food Preservation Center (WFPC), LLC, USA, and is currently an active associate and supporter.

Dr. Siddiqui acquired a BSc (Agriculture) degree from Jawaharlal Nehru Krishi Vishwa Vidyalaya, Jabalpur, India. He received his MSc (Horticulture) and PhD(Horticulture) degrees from Bidhan Chandra Krishi Viswavidyalaya, Mohanpur, Nadia, India, with specialization in Post-harvest Technology. He is a member of Core Research Group at the Bihar Agricultural University (BAU), which provides appropriate direction and assistance in prioritizing research.

LIST OF CONTRIBUTORS

Pinaki Acharyya
Department of Horticulture, University of Calcutta, 51/2, Ballygunge Circular Road, Kolkata–700019, India

Sibdas Baskey
Regional Research Station (Hill Zone), North Bengal Agriculture University, Kalimpong, Darjeeling, West Bengal – 734301, India

Ajanta Birah
ICAR-National Research Centre for Integrated Pest Management, New Delhi–110012, India

Sukanta Biswas
Department of Horticulture, University of Calcutta, 51/2, Ballygunge Circular Road, Kolkata–700019, India

R. Chacoory
Crop Department, Food and Agricultural Research and Extension Institute, Reduit, Mauritius

Pradip K. Chatterjee
Thermal Engineering Division, CSIR Central Mechanical Engineering Research Institute, Ministry of Science and Technology, Government of India, M. G. Avenue, Durgapur – 713209, India

C. Chattopadhyay
Uttar Banga Krishi Viswavidyalaya, Pundibari, Coochbehar–736165, West Bengal, India

Rekha Chaurasia
Division of Post Harvest Management, ICAR–CISH Lucknow 226106, India

M. Chinlampianga
Department of Horticulture, Aromatic and Medicinal Plants, Mizoram University, Aizwal – 796004, India

Akoijam Ranjita Devi
Department of Spices and Plantation Crops, Faculty of Horticulture, Bidhan Chandra Krishi Viswavidyalaya, Mohanpur–741252, Nadia, West Bengal, India

Anupam Dikshit
Biological Product Lab, Department of Botany, University of Allahabad, Allahabad–211004, India

Manisha Dubey
Department of Plant Pathology, Punjab Agricultural University, Ludhiana–141004, India

R. K. Dubey
Department of Floriculture and Landscape, Punjab Agricultural University, Ludhiana–141004, India

Neelima Garg
Central Institute for Subtropical Horticulture, Rehmankhera, P.O. Kakori, Lucknow–226101, India

Murray B. Isman
Faculty of Land and Food Systems, University of British Columbia, Vancouver V6T1Z4, Canada

S. K. Jindal
Department of Vegetable Science, Punjab Agricultural University, Ludhiana–141004, India

Bharati Killadi
Division of Post Harvest Management, ICAR–CISH Lucknow 226106, India

Awadhesh Kumar
Department of Horticulture, Aromatic and Medicinal Plants, Mizoram University, Aizawl–796004, India

Sanjay Kumar
Central Institute for Subtropical Horticulture, Rehmankhera, P.O. Kakori, Lucknow–226101, India

K. Lalhmingsangi
Department of Forestry, School of Earth Sciences and Natural Resource Management, Mizoram University, Aizawl–796004, India

B. Lalramhlimi
Department of Vegetable Crops, Faculty of Agriculture, Bidhan Chandra Krishi Viswavidyalaya, Mohanpur–741252, Nadia, West Bengal, India

R. Lawmzuali
Department of Chemistry, School of Physical Sciences, Mizoram University, Aizawl–796004, India

J. Lenka
Division of Post Harvest Management, ICAR–CISH, Rehmankhera, P.O. Kakori, Lucknow–226 101, India

A. R. Logesh
Department of Horticulture, Aromatic and Medicinal Plants, Mizoram University, Aizwal – 796004, India

Jayoti Majumder
Bidhan Chandra Krishi Viswavidyalaya (BCKV), Kalyani, West Bengal, India

Moumita Malakar
Department of Horticulture, University of Calcutta, 51/2, Ballygunge Circular Road, Kolkata–700019, India

T. P. Mall
Postgraduate Department of Botany, Kisan PG College, Bahraich, U.P., India

Bikash Mandal
Advanced Centre of Plant Virology, Division of Plant Pathology, Indian Agricultural Research Institute, New Delhi–110012, India

D. Mandal
Department of Horticulture, Aromatic & Medicinal Plants, Mizoram University, Aizawl-796001, India.

R. Mandi
Department of Agricultural Entomology, Bidhan Chandra Krishi Viswavidyalaya, Mohanpur, Nadia, 741252, West Bengal, India

Bernadette Montanari
Department of Geography, University of Urbana Champaign, USA [Social Dimensions of Environmental Policy (SDEP)]

Roop Soodha Munbodh
Food and Agricultural Research and Extension Institute (FAREI), Reduit, Mauritius

Khaidem Nirja
Department of Agricultural Extension, Faculty of Agriculture,
Bidhan Chandra Krishi Viswavidyalaya, Mohanpur–741252, Nadia, West Bengal, India

Sellam Perinban
Directorate of Floricultural Research, Indian Agricultural Research Institute, Pusa Campus,
New Delhi–110012, India

A. Pramanik
Department of Agricultural Entomology, Bidhan Chandra Krishi Viswavidyalaya, Mohanpur, Nadia,
741252, West Bengal, India

Puja Rai
Bidhan Chandra Krishi Viswavidyalaya (BCKV), Kalyani, West Bengal, India

N. Ramburn
Crop Department, Food and Agricultural Research and Extension Institute, Reduit, Mauritius

Zuliana Razali
Institute of Biological Sciences, Faculty of Science and the Centre for Research in Biotechnology for Agriculture (CEBAR), University of Malaya, 50603, Kuala Lumpur, Malaysia

U. K. Sahoo
Department of Forestry, School of Earth Sciences and Natural Resource Management,
Mizoram University, Aizawl–796004, India

Vicknesha Santhirasegaram
Institute of Biological Sciences, Faculty of Science and the Centre for Research in Biotechnology for Agriculture (CEBAR), University of Malaya, 50603, Kuala Lumpur, Malaysia

Rita Seffrin
Faculty of Land and Food Systems, University of British Columbia, Vancouver V6T1Z4, Canada

Amritesh C. Shukla
Department of Botany, University of Lucknow, Lucknow–226007, India

D. K. Shukla
Division of Post Harvest Management, ICAR–CISH Lucknow 226106, India

Babita Singh
Directorate of Floricultural Research, Indian Agricultural Research Institute, Pusa Campus,
New Delhi–110012, India

N. Mohondas Singh
Department of Chemistry, School of Physical Sciences, Mizoram University, Aizawl–796004, India

Preeti Singh
Central Institute for Subtropical Horticulture, Rehmankhera, P.O. Kakori, Lucknow–226101, India

Hasvinder Kaur Baldev Singh
Institute of Biological Sciences, Faculty of Science and the Centre for Research in Biotechnology for Agriculture (CEBAR), University of Malaya, 50603, Kuala Lumpur, Malaysia

K. Birla Singh
Department of Zoology, Pachhunga University College, Aizawl–796001, India

Chandran Somasundram
Institute of Biological Sciences and Centre for Research in Biotechnology for Agriculture (CEBAR), Faculty of Science, University of Malaya, 50603 Kuala Lumpur, Malaysia

George Srzednicki
School of Chemical Engineering, Food Science and Technology, University of New South Wales, Sydney 2052, Australia

S. Subramaniam
Crop Department, Food and Agricultural Research and Extension Institute, Reduit, Mauritius

T. S. Thind
Department of Plant Pathology, Punjab Agricultural University, Ludhiana–141004, India

R. B. Tiwari
Division Post Harvest Technology, IIHR, Bangalore–560089, India

S. C. Tripathi
Postgraduate Department of Botany, Kisan PG College, Bahraich, U.P., India

Biplab Tudu
Regional Research Station (Hill Zone), North Bengal Agriculture University, Kalimpong, Darjeeling, West Bengal – 734301, India

D. K. Upreti
Lichenology Laboratory, Plant Biodiversity, Systematics and Herbarium Division, CSIR-National Botanical Research Institute, Lucknow – 226001, Uttar Pradesh, India

Kaushlesh K. Yadav
Central Institute for Subtropical Horticulture, Rehmankhera, P.O. Kakori, Lucknow–227107, India

Ramesh S. Yadav
Department of Plant Pathology, Sardar Vallabhbhai Patel University of Agriculture & Technology, Meerut–250110, India

LIST OF ABBREVIATIONS

AAE	ascorbic acid equivalent
ACPV	Advanced Center of Plant Virology
ANS	average number of spores
AOA	aminooxy acetic acid
APSA	All Purpose Suspension Agent
ASA	acetyl salicylic acid
BA	benzyl adenine
CASE	custard apple seed
CBCS	Central Biological Control Stations
CD	critical difference
CE	catechin equivalent
CMV	cucumber mosaic virus
CP	coat protein
CPPS	Central Plant Protection Stations
CRBD	completely randomized block design
CSS	Central Surveillance Stations
DAT	days after treatment
DMRT	Duncan's multiple range test
DMSO	dimethyl sulfoxide solvent
DPPH	diphenyl-1-picrylhydrazyl
EIL	economic injury level
ELISA	enzyme-linked immunosorbent assay
EOs	essential oils
EPA	entry point activities
ETL	economic threshold
FAO	Food and Agriculture Organization
FFSs	Farmers' Field Schools
FGI	fungal growth inhibition
GAE	gallic acid equivalent
GAP	good agricultural practices
GBNV	groundnut bud necrosis virus

GC	gas chromatography
GoI	Government of India
HACCP	hazard analysis critical control point
HPLC	high performance liquid chromatography
HSD	honestly significant difference
HTST	high temperature short time
IA	intensive agriculture
IARI	Indian Agricultural Research Institute
ICAR	Indian Council of Medical Research
ICT	Information Communication Technology
IDS	International Drying Symposium
IDSS	integrated decision support system
IFW	initial fresh weight
IgG	immunoglobulin G
IPM	integrated pest management
ISR	induced systemic resistance
IVM	initial volume of medium
IVPD	in vitro multienzyme protein digestibility
JFM	Joint Forest Management
K	potassium
LFA	lateral flow assay
LPS	lipopolysaccharides
MAb	monoclonal antibody
MAPs	medicinal and aromatic plants
MARDI	Malaysian Agricultural Research and Development Institute
MBC	minimum bactericidal concentration
MCCs	minimum cidal concentration
MEC	minimum effective concentration
MIC	minimum inhibitory concentration
MKT	minimum killing time
MPL	maximum permissible limit
MPTs	multipurpose tree species
MR	moisture ratio
MRLs	maximum residue limit
MS	mass spectrometry

MSGIT	modified spore germination inhibition technique
MT	metric tons
NBRI	National Botanical Research Institute
NCIPM	National Research Centre for Integrated Pest Management
NE	north-east
NGO	Non Governmental Organization
NPV	nucleopolyhedrovirus
NSKE	neem seed kernel extract
NTFPs	non timber forest products
OPD	out patient department
PAb	polyclonal antibody
PCR	polymerase chain reaction
PDI	percentage disease index
PEF	pulsed electric field
PR	pathogenesis related
PRA	participatory rural appraisal
PRSV	papaya ringspot virus
PT	persistent toxicity
RBD	randomized block design
RDA	recommended dietary allowance
RFW	relative fresh weight
RSM	red spider mite
scF	single chain variable fragment
SD	standard deviation
SDA	sabourad dextrose agar medium
SEM	standard error of mean
SG	solid gain
SHGs	self help groups
TA	titratable acidity
TLC	thin layer chromatography
TSS	total soluble solids
UV-C	ultraviolet-c
VFDC	Village Forest Development Committee
VMF	volume of microscopic field
VOCs	volatile organic compounds

VSUR	vase solution uptake rate
WHO	World Health Organization
WL	water loss
WR	weight reduction
YMA	Young Mizo Association
YPDA	yeast potato dextrose agar

PREFACE

Global food demand is expected to be doubled by 2050, while the production environment and natural resources are continuously shrinking and deteriorating. Horticulture, a major sector of agriculture, needs to take part in enhancing crop production and productivity in parity with agricultural crops to meet the emerging food demand. There are projections that demand for food grains would increase from 192 million tonnes in 2000 to 345 million tonnes in 2030. Hence, in the next 15 years, production of food grains needs to be increased at the rate of 5.5 million tonnes annually. The demand for high-value commodities (such as horticulture, dairy, livestock and fish) is increasing faster than food grains—for most of the high-value food commodities demand is expected to increase by more than 100% from 2000 to 2030. These commodities are all perishable and require, different infrastructure for handling, value-addition, processing and marketing.

Asia, the major crop-producing continent, thought to be the global super power because of its increasing skilled and energetic human resources and faster developing technological and economic growth, has to play a major role to meet the projected global food demand. However, the developing Asiatic countries are more or less facing the common problem of diminution of cultivable land with massive rapid urbanization. Apart from that, the contemporary production limitations—depleting land fertility, unequal cross-subsidy and, more predominantly, vagaries of climate change—put forth a tough task to perform.

India, the second most important Asiatic food grain producer, is facing a more grim production situation. The average size of land holding declined to 1.32 ha in 2000–01, from 2.30 ha in 1970–71, and absolute number of operational holdings increased from about 70 million to 121 million. If this trend continues, the average size of holding in India would be mere 0.68 ha in 2020, and would be further reduced to a low of 0.32 ha in 2030. This is a very complex and serious problem, when the share of agriculture, including horticulture, in gross domestic product is declining, average size of landholding is contracting and fragmenting, and numbers of operational holdings are increasing. In addition, annually, India is losing nearly 0.8 million tonnes of nitrogen, 1.8 million tonnes of phosphorus and 26.3 million tonnes of potassium—the deteriorating quality and health of soil is something to be checked. Problems are further aggravated by imbalanced application of nutrients (especially nitrogen, phosphorus and potash) and excessive mining of micronutrients, leading to deficiency of macro- and micro-nutrients in the soils. Similarly, the water-table is lowering steeply in most of the irrigated areas, and water quality is also deteriorating, due to leaching of salts and other pollutants.

Amidst this situation, fulfilling the target with a sustainable model of crop production is really an enormous endeavor. There is an earnest need to develop promising technologies and management options to raise productivity to meet the growing food demand in this situation of deteriorating production environment at the lowest cost; and to develop appropriate technologies, to create required infrastructures, and to evolve institutional arrangements for production, postharvest and marketing of high-value and perishable commodities and their value-added products. Improved agro-techniques, quality planting materials, improved varieties, climate resilient production models, involvement of information technology and biotechnology, improved postharvest handling-storage, and marketing are the key issues on which to focus on to bring about the desired metamorphosis in global horticulture. However, to achieve the goal with a sustainable production environment (i.e., sustainability in terms of economy-ecology-society), is the greatest challenge to meet.

The International Symposium on Sustainable Horticulture, organized by the Department of Horticulture, Aromatic and Medicinal Plants, Mizoram University, Aizawl, India, from 14–16th March, 2016, provided a platform for the exchange of ideas and research experience and discussed several facets of sustainable horticulture with the thematic areas such as management of genetic resources and biodiversity conservation; production technology of horticultural crops; crop improvement and biotechnology; plant protection in horticultural crops; postharvest technology and value addition; trade, marketing, entrepreneurship development and extension; and horticulture for food, health and nutrition.

In this regard, this research compendium has been categorically divided to form two volumes; *Sustainable Horticulture, Volume 1: Diversity, Production, and Crop Improvement*, and *Sustainable Horticulture, Volume 2: Food, Health, and Nutrition*. The first volume outlines the contemporary trends in sustainable horticulture research, in particular: crop diversity, species variability and conservation strategies, production technology, tree architecture management, plant propagation, and nutrition management, organic farming, new dynamics in breeding, and marketing of horticulture crops. The second volume depicts the research trends in sustainable horticulture comprising postharvest management and processed food production from horticulture crops, crop protection and plant health management, and horticulture for human health and nutrition.

We extend our sincere thanks to the contributors, reviewers, and Apple Academic Press for their efforts and contributions. It is hoped that these book volumes will be useful for students, teachers, scientists, extension workers, and researchers in horticulture and allied disciplines.

—**Debashis Mandal**
Amritesh C. Shukla
Wasim Siddiqui

PART I

POSTHARVEST MANAGEMENT AND PROCESSED FOOD

CHAPTER 1

EMERGING FRUIT JUICE PROCESSING TECHNOLOGIES: QUALITY IMPROVEMENT

CHANDRAN SOMASUNDRAM, ZULIANA RAZALI, and VICKNESHA SANTHIRASEGARAM

Institute of Biological Sciences and Centre for Research in Biotechnology for Agriculture (CEBAR), Faculty of Science, University of Malaya, 50603 Kuala Lumpur, Malaysia, Tel: +60379674423; Fax: +60379674178, E-mail: chandran@um.edu.my

CONTENTS

ABSTRACT

Fruit juice has the highest acceptability among other beverages, generally due to its natural taste as well as its nutritional value. The presence of various phytochemicals in fruit juice is related to various health-promoting properties such as protection against several chronic human diseases such as cancer, cardiovascular diseases, and diabetes. However, the number of outbreaks and cases of illness caused by consumption of contaminated juices, especially unpasteurized juices, has increased over the last decade. Currently, conventional thermal treatment is the preferred technology to inactivate microorganisms and enzymes causing spoilage, thus prolonging the shelf-life of juice. Because of the relatively high temperatures generally needed to inactivate food poisoning- and spoilage-causing microorganisms, thermal treatment can adversely affect the quality of food products, by reducing their nutritional value and altering sensory attributes such as color and flavor. The growing interest for fresh-like products has promoted the effort for the development of innovative nonthermal food preservation methods. Nonthermal processing techniques have been explored for their efficacy in extending shelf-life and enhancing the safety of fresh juice while preserving organoleptic and nutritional qualities. As consumers continue to seek food products with improved nutritional value and functionality, juice producers have the opportunity of improving product marketability through the use of these novel technologies that provide better retention of phytonutrients. Therefore, there is a need for nonthermal processing techniques to be tested on a pilot scale, so that these methods can be developed as an alternative to thermal pasteurization.

1.1   INTRODUCTION

According to Business Insights (2010), the global market for juices valued about US$ 93 billion in 2014. The key driver for the growth of fruit juice market is the increase in awareness among consumers on preventive healthcare and wellness benefits. Natural fruit juices are susceptible to spoilage, mainly due to their intrinsic properties such as pH, water activity, redox potential, and nutrients (Odumeru, 2012).

Fruit juice deterioration is mostly caused by enzymatic, chemical, and microbial reactions. Enzymes in fruit juices such as polyphenol oxidase and peroxidase may react with oxygen, thus contributing to juice browning and off-flavor (Bharate and Bharate, 2014). The causal agents of microbial spoilage of fruit juices are bacteria, yeast, and molds. Yeast and molds are the main spoilage agents due to the low pH of fruit juices. According to the

Centre for Disease Control and Prevention (1996), one of the current food-borne disease outbreaks has been linked to pathogens such as *Escherichia coli* O157:H7, where the emphasis was on unpasteurized juices.

Spoilage of fruit juice and related products as a result of microbial growth may contribute to physical and chemical changes in food products. These alterations include unacceptable flavor and odor, changes in color and turbidity, gas production, and formation of slime. Usually, growth of microorganisms to high numbers is necessary before spoilage becomes noticeable. Hence, it is important to control the growth of spoilage organisms in order to inhibit microbial spoilage (Odumeru, 2012).

The increase in outbreaks and cases of illness related to consumption of unpasteurized juices have urged the development of a more effective food safety control program, known as the hazard analysis and critical control point (HACCP) program. HACCP is a systematic approach to identify, assess, and control microbiological, chemical, and physical hazards of public health concern (Odumeru, 2012). Currently, there are a number of fruit juice preservation technologies for controlling microbial growth and survival. These preservation methods must be evaluated to avoid significant organoleptic changes in food products (Bates et al., 2001).

This paper will provide an overview of emerging thermal and non-thermal processing methods. This information is necessary to improve the progress of positive implementation of novel processing methods in the fruit juice industry.

1.2 FRUIT JUICE PROCESSING

The main objective of fruit juice processing is to prevent microbiological spoilage while assuring safety and maintaining quality characteristics. Fruit juice processing technologies can be divided into two groups, namely thermal and nonthermal processing.

1.2.1 THERMAL PROCESSING

Conventional thermal treatment is the preferred technology to inactivate microorganisms and enzymes causing spoilage, thus prolonging the shelf-life of juice. Traditional thermal processing depends on the generation of heat outside the product to be heated and its transfer into the product via convection and conduction mechanisms (Pereira and Vicente, 2010). Pasteurization is an example of thermal treatment that is commonly practiced in the food industry. The most common pasteurization method for fruit juice is high temperature short time (HTST) or also known as flash pasteurization (David et al., 1996).

According to Nagy et al. (1993), HTST treatment for fruit juices range from 90°C to 95°C for 15 to 60 seconds to assure at least 5-log reduction in microbial count. The time and temperature variables for pasteurization of juice depend on the type of juice, initial microbial count, pH, water activity, and thermal inactivation kinetics of microorganisms present in juice. Hence, pasteurization conditions should be selected appropriately to avoid overprocessing. However, underprocessing may not completely inactivate microbial growth, thus resulting in juice spoilage (Rawson et al., 2011).

Because of the relatively high temperatures generally needed to inactivate food poisoning- and spoilage-causing microorganisms, conventional pasteurization can adversely affect the quality of food products by reducing their nutritional value or altering sensory attributes such as color and flavor (Rawson et al., 2011). Some studies on thermally treated fruit juices such as orange (Cortes et al., 2008) and strawberry (Aguilo-Aguayo et al., 2009) reported significant loss of quality and degradation of bioactive compounds such as ascorbic acid. In addition, Rattanathanalerk et al. (2005) reported significant color degradation in thermal-treated pineapple juice (at 85°C and 95°C for 60 seconds).

1.2.2 NON-THERMAL PROCESSING

The growing interest for fresh-like products has promoted the effort for developing innovative nonthermal food preservation methods. Nonthermal processing techniques have been explored for their efficacy to extend shelf-life and enhance the safety of fresh juice while preserving organoleptic and nutritional qualities. In addition, these preservation methods are considered to be more energy efficient and provide better retention of quality when compared to conventional thermal processing. Some of the nonthermal processing methods extensively studied for juice preservation include ultrasound, ultraviolet light irradiation, and pulsed electric field (Morris et al., 2007).

1.2.2.1   Ultrasound

Power ultrasound (10–1000 W/cm^2) is used to alter food properties, either physically or chemically, such as by disrupting cells and inactivating enzymes. Ultrasonic processing equipment includes ultrasonic bath and probe system (Carcel et al., 2012). The mechanism of action for ultrasonic processing or sonication is explained in three different approaches, which include cavitation, localized heating, and formation of free radicals. When high power ultrasound at low frequencies (20–100 kHz) propagates in a liquid, cavitation (formation and collapse of bubbles) occurs. As a result, there is elevation of localized pressure (up to 500 MPa) and temperature (up to 5000°C). These "tiny hotspots" provide the energy to alter the properties of food product either physically or chemically. Accordingly, these cavitation bubbles induce microstreaming and shear stress, resulting in the disintegration of the microbial cells. Besides that, cavitation causes intracellular micromechanical shock that disrupts the functional components of the cell, thus inactivating enzymes (O'Donnell et al., 2010; Abid et al., 2013).

Sonication is a potential technology to achieve the US FDA condition of a 5-log reduction of foodborne pathogens in fruit juices. Several studies using ultrasonic processing on fruit juice reported minimal effect on the degradation of quality parameters and improved functionalities, such as in kasturi lime (Bhat et al., 2011a), apple (Abid et al., 2013), and carrot juice (Jabbar et al., 2014). Besides that, Rawson et al. (2011) reported that sonication provides better retention of bioactive compounds. In addition, Tiwari et al. (2009) reported that ultrasonic processing (25 kHz for 2 min) improves the cloud value and stability of orange juice during storage. However, significant color degradation was observed in orange juice subjected to sonication (Tiwari et al., 2009).

1.2.2.2   Ultraviolet-C (UV-C) Light

Ultraviolet-C (UV-C) light is a part of the electromagnetic spectrum with wavelengths between 200 to 280 nm and exhibits germicidal properties as it inactivates bacterial and viral microorganisms. Although UV-C radiation technology is considered as an effective method for food preservation, consumers' misconception about this process have delayed many of its potential applications in the food industry. Actually, UV-C radiation is a physical treatment that does not result in chemical residues. Hence, the consumption of UV-C-treated food products is not harmful to humans.

One of the limitations of the application of UV in juice processing is associated with the high absorbance coefficients of juice. According to Koutchma et al. (2004), UV-C penetration largely depends on the presence of dissolved organic solutes (suspended solids) and colored compounds that act as a barrier, thus exhibiting UV-C attenuation effects. Hence, an appropriate UV reactor should be designed to reduce the interference of UV absorbance and improve microbial inactivation efficiency. The reactor design should include a narrow laminar flow or conditions with high turbulence, where juices are mixed resulting in all parts of the juice being exposed to the UV light source (Koutchma et al., 2004).

The US FDA criterion of a 5-log reduction of the chosen pathogen in fruit juices can be achieved by UV-C radiation. Several studies using short-wave UV-C light treatment on fruit juices reported better retention of nutritional and quality attributes, such as in starfruit (Bhat et al., 2011b) and orange juice (Pala and Toklucu, 2013). Besides that, Bhat et al. (2011b) reported that UV-C processing (for 30 and 60 minutes) induces a significant increase in polyphenol and flavonoid content of starfruit juice. However, there is an increasing trend in browning degree and color changes corresponding to increased UV-C treatment time, as previously reported by Bhat et al. (2011b).

1.2.2.3   Pulsed Electric Field (PEF)

Pulsed electric field (PEF) treatment applies short pulses (1 to 100 microseconds) with high voltage (10 to 50 kV/cm) to liquid products in a continuous system. A simple PEF system consists of a high voltage power supply, a pulse generator, treatment chamber, and a switch to discharge energy to electrodes.

In addition, there is a cooling system to balance moderate temperature rise during treatment (Morris et al., 2007; Lopez-Gomez et al., 2009). The effectiveness of PEF processing is dependent on variables such as pulse width, electric field strength, flow rate, treatment temperature, and time of exposure.

Some studies reported that juices treated with lower intensity pulse fields and shorter pulse width exhibited higher retention of ascorbic acid, such as observed in orange (Elez-Martinez and Martin-Belloso, 2007). In addition, an enhancement of antioxidant capacity was observed in PEF-processed juices due to increased extraction yield of secondary metabolites and generation of free radicals (Rawson et al., 2011).

1.3 CONCLUSION

The value for fruit juices has been increasing in the global market due to their health benefits. Conventional thermal pasteurization is the preferred technology used to achieve microbial inactivation and extend the shelf-life of juices. Lately however, consumers' demand for a new preservation technology that retains freshness and at the same time ensures food safety has resulted in growing interest for nonthermal processing methods. Therefore, there is a need for other nonthermal processing techniques to be tested on a pilot scale, so that these methods can be developed as an alternative to thermal pasteurization. In future, the combination of nonthermal processing methods as a hurdle concept would be a new trend of preservation of fruit juices that improves the microbiological quality and safety with minimal impact on the quality of the food product.

KEYWORDS

- **fruit juice**
- **non-thermal processing**
- **quality**
- **safety**
- **thermal processing**

REFERENCES

Abid, M., Jabbar, S., Wu, T., Hashim, M. M., Hu, B., Lei, S., Zhang, X., & Zeng, X., (2013). Effect of ultrasound on different quality parameters of apple juice. *Ultrasonics Sonochemistry, 20,* 1182–1187.

Aguilo-Aguayo, I., Oms-Oliu, G., Soliva-Fortuny, R., & Martin-Belloso, O., (2009). Changes in quality attributes throughout storage of strawberry juice processed by high-intensity pulsed electric fields or heat treatments. *LWT- Food Science and Technology, 42,* 813–818.

Bates, R. P., Morris, J. R., & Crandall, P. G., (2001). Principles and practices of small and medium scale fruit juice processing: Food and Agriculture Organization (FAO), *Agricultural Services Bulletin Issue, 146,* FAO, Rome.

Bharate, S. S., & Bharate, S. B., (2014). Non-enzymatic browning in citrus juice: Chemical markers, their detection and ways to improve product quality. *Journal of Food Science and Technology*, 51(10):2271-2288. DOI: 10.1007/s13197–012–0718–8.

Business Insights (2010). Innovations in fruit and vegetable juices: Emerging opportunities in premiumization, sustainability and positive health. Available online: <http://www.scribd. com/doc/50026681/Innovations-in-Fruit-and-Vegetable-Juices>.

Carcel, J. A., Garcia-Perez, J. V., Benedito, J., & Mulet, A., (2012). Food process innovation through new technologies: Use of ultrasound. *Journal of Food Engineering, 110,* 200–207.

Centers for Disease Control and Prevention (CDC), (1996). Outbreak of Escherichia coli O157:H7 infections associated with drinking unpasteurised commercial apple juice: British Columbia, California, Colorado, and Washington. *Morbidity and Mortality Weekly Report, 45,* 975.

David, J. R. D., (1996). Principles of thermal processing and optimization. In: David, J. R. D., Graves, R. H., & Carlson, V. R., (eds.), *Aseptic Processing and Packaging of Food: A Food Industry Perspective.* CRC Press LLC. Florida, pp. 14.

Elez-Martinez, P., & Martin-Belloso, O., (2007). Effects of high intensity pulsed electric field processing conditions on vitamin C and antioxidant capacity of orange juice and gazpacho, a cold vegetable soup. *Food Chemistry, 102,* 201–209.

Jabbar, S., Abid, M., Hu, B., Wu, T., Hashim, M. M., Lei, S., Zhu, X., & Zeng, X., (2014). Quality of carrot juice as influenced by blanching and sonication treatments. *LWT-Food Science and Technology, 55*(1), 16–21.

Koutchma, T., Keller, S., Chirtel, S., & Parisi, B., (2004). Ultraviolet disinfection of juice products in laminar and turbulent flow reactors. *Innovative Food Science and Emerging Technologies, 5,* 179–189.

Lopez-Gomez, A., Fernandez, P. S., Palop, A., Periago, P. M., Martinez-Lopez, A., Marin-Iniesta, F., & Barbosa-Canovas, G. V., (2009). Food safety engineering: An emergent perspective. *Food Engineering Reviews, 1,* 84–104.

Morris, C., Brody, A. L., & Wicker, L., (2007). Non-thermal food processing/preservation technologies: A review with packaging implications. *Packaging Technology and Science, 20,* 275–286.

Nagy, S., Chen, C. S., & Shaw, P. E., (1993). *Fruit Juice Processing Technology.* Ag. Science, Florida.

O'Donnell, C. P., Tiwari, B. K., Bourke, P., & Cullen, P. J., (2010). Effect of ultrasonic processing on food enzymes of industrial importance. *Trends in Food Science and Technology, 21*, 358–367.

Odumeru, J. A., (2012). Microbial safety of food and food products. In: Simpson, B. K., Nollet, L. M. L., Toldra, F., Benjakul, S., Paliyath, G., & Hui, Y. H., (eds.), *Food Biochemistry and Food Processing, 2nd edition.* John Wiley & Sons, UK, pp. 787–797.

Pala, C. U., & Toklucu, A. K., (2013). Microbial, physicochemical and sensory properties of UV-C processed orange juice and its microbial stability during refrigerated storage. LWT - *Food Science and Technology, 50*, 426–431.

Pereira, R. N., & Vicente, A. A., (2010). Environmental impact of novel thermal and non-thermal technologies in food processing. *Food Research International, 43*, 1936–1943.

Rattanathanalerk, M., Chiewchan, N., & Srichumpoung, W., (2005). Effect of thermal processing on the quality loss of pineapple juice. *Journal of Food Engineering, 66*, 259–265.

Rawson, A., Patras, A., Tiwari, B. K., Noci, F., Koutchma, T., & Brunton, N., (2011). Effect of thermal and non-thermal processing technologies on the bioactive content of exotic fruits and their products: Review of recent advances. *Food Research International, 44*, 1875–1887.

Tiwari, B. K. O.,'Donnell, C. P., Muthukumarappan, K., & Cullen, P. J., (2009). Ascorbic acid degradation kinetics of sonicated orange juice during storage and comparison with thermally pasteurised juice. *LWT-Food Science and Technology, 42*, 700–704.

CHAPTER 2

DEVELOPMENT OF POSTHARVEST PROCESSING TECHNOLOGY FOR GINGER, TURMERIC, AND CHILLI IN MIZORAM

PRADIP K. CHATTERJEE[1] and GEORGE SRZEDNICKI[2]

[1]*Thermal Engineering Division, CSIR Central Mechanical Engineering Research Institute, Ministry of Science and Technology, Government of India, M. G. Avenue, Durgapur – 713209, India*

[2]*School of Chemical Engineering, Food Science and Technology, University of New South Wales, Sydney 2052, Australia, E-mail: g.srzednicki@unsw.edu.au*

CONTENTS

ABSTRACT

Ginger (*Zingiber officinale* Roscoe) and turmeric (*Curcuma longa*) belong to Zingiberaceae family. They are known for their pungent and aromatic flavor and for their medicinal properties. Both crops are often grown by smallholders in mountain areas on rich former forest soils without the need for fertilizers and pesticides. They can be consumed fresh or dried. Both of them are major cash crops in Mizoram. Given the lack of proper postharvest processing and storage, a significant amount of the crop perishes, and the remainder is sold in the local market at very low price. The traditional sun drying is not effective due to insufficient availability of sunshine. Therefore, a complete postharvest processing package consisting of a rotary drum washer, a slicing unit, a dryer, and a grinder was developed and implemented by Central Mechanical Engineering Research Institute (CMERI) in collaboration with a local NGO Community Development Action and Reflection (CDAR). Freshly harvested ginger/turmeric is first washed, sliced, and then dried in a cabinet dryer. Proper washing of ginger/turmeric is essential for the quality of the dried product as fresh rhizomes coming from the field are covered with mud. The mud is difficult to remove by hand wash due to the critical shape of the rhizomes. Dried product is ground into a fine powder and stored.

The project resulted in the following achievements:

- Establishment of two centers for postharvest processing and research.
- Processing technology for ginger and turmeric transferred (two patents and two copyrights registered).
- Training of rural youths on ginger/turmeric processing.
- Employment of local youths.
- Enhancement of farmers' economy.

2.1   INTRODUCTION

Ginger (*Zingiber officinale* Roscoe) and turmeric (*Curcuma longa*), like cardamom and galangal, are members of Zingiberaceae family. They are known for their pungent and aromatic flavor and for their medicinal properties. Chili is the fruit of plants from the genus *Capsicum*, members of the nightshade family Solanaceae. In Mizoram, ginger, turmeric, and chili are generally grown by smallholders in mountain areas on rich former forest soils without the need for fertilizers and pesticides.

Various bioactive compounds have been identified in their rhizomes, and their content affects the price of the dried product. The bioactive compounds in ginger rhizome (*Z. officinale* Roscoe) are gingerols (polyphenolic compounds including 10-gingerol, 8-gingerol, 6-gingerol, and its derivatives and also shogaols that are produced by the dehydration reaction of gingerols under high temperature). They are present in the rhizomes of ginger, and their extracts have been proved to have high antioxidant activity (Stoilova et al., 2007; Sakulnarmrat et al., 2015) and anti-inflammatory effect (Dugasani et al., 2010). New phenolics have also been identified, which have antioxidant and anti-inflammatory properties. Turmeric (*C. longa*) is known for its bright yellow color and pharmacological properties due to curcumin, a phenolic compound. Curcumin is the component of turmeric responsible for its color and all its medicinal properties. Its structure has been identified as diferuloylmethane. Curcumin and its two related demethoxy compounds, namely demethoxycurcumin and bisdemethoxycurcumin, are known as curcuminoids. These components have been identified as antioxidants. Cyclocurcumin is a newly identified curcuminoid isolated from the fraction of turmeric found to be active as a nematicide. In chili, the substances that give chili peppers their intensity when ingested or applied topically are capsaicin and several related chemicals, collectively called capsaicinoids that belong to the group of amides. Although all the three crops can be consumed fresh, the distance from the major markets or industrial clients makes it essential for them to be preserved, mostly by dehydration. Given the economic importance of the bioactive compounds used in pharmaceutical and food industry, it is important to maximize the retention of bioactive compounds in the dried products.

Ginger and turmeric are among the major cash crops in northeastern India and particularly in Mizoram. They are generally harvested in October–November. So far, there has been no proper postharvest processing for ginger and turmeric rhizomes. Open sun drying is a widespread postharvest practice, but given that the sunshine is erratic, this process is not very effective in this part of India. The relative humidity is generally high during the nighttime in the mountain areas, and the drying process is slow, often leading to fungal contamination. Moreover, the dried product also gets contaminated by dust and insects (Loha et al., 2008).

Therefore, CSIR-Central Mechanical Engineering Research Institute (CMERI) in collaboration with a local NGO, Community Development Action and Reflection (CDAR), set up a project to establish and implement rural R&D centers at Mizoram, which will continuously cater to the technology need of the people of Mizoram in the area of postharvest processing and enhance the economy of the local farmers. It was expected that as a result of the project, Mizoram farmers would get a better access to the market, finance, and education about their crops. The initial project (RSP-0011), started in 2007, received Rs. 3 crore (INR 30 million) funding for 5 years from the Indian Government. The primary goal was to develop a rapid, safe, and controllable drying system. The forced convection hot air drying is an efficient method to produce a uniform, hygienic, and attractively colored product. Therefore, a forced convection cabinet dryer was developed to address such issues (Loha et al., 2012). The project is currently focusing on processing of ginger and turmeric but will consider expanding the technology to process other cash crops such as chili at a later stage.

2.2   EQUIPMENT DEVELOPMENT

A complete postharvest processing package consisting of a rotary drum washer, a slicing unit, a dryer, and a grinder was developed and implemented as shown in Figure 2.1.

2.2.1   ROTARY DRUM WASHER

Freshly harvested ginger/turmeric rhizomes from the field are first washed and cut into slices before being subjected to the drying process. Proper washing of ginger/turmeric is very important to maintain the quality of the dried product because the ginger/turmeric rhizomes coming from the field

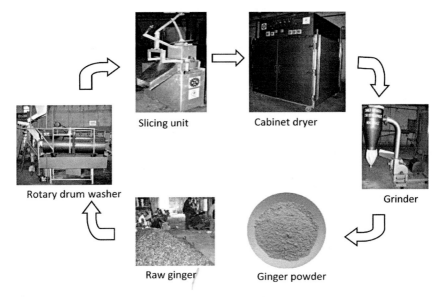

Slicing unit   Cabinet dryer

Rotary drum washer   Grinder

Raw ginger   Ginger powder

FIGURE 2.1   Complete postharvest processing system for ginger/turmeric.

contain a lot of mud on it. It is very difficult to remove the mud by hand washing due to the particular shape of the rhizomes and also because hand wash is not practicable for large-scale application. Water consumption for washing is another important aspect due to the scarcity of water in the eastern Indian states. Therefore, a continuous rotary drum washing unit of 500 kg/h capacity was developed as shown in Figure 2.2, where the waste water is filtered and re-circulated.

2.2.2   SLICING UNIT

After washing, the material is cut into slices before being introduced into the dryer. The material is sliced to achieve faster and uniform drying by providing a larger surface area for heat and mass transfer. Therefore, a 50 kg/h capacity slicing unit (SU-50 KPH) was developed as shown in Figure 2.3. The slicing unit consists of three blades, which can be adjusted to get different slice thickness according to the requirement.

FIGURE 2.2   Rotary drum washer with filtration and recirculation of water.

FIGURE 2.3   Slicing unit.

2.2.3   CABINET DRYER

Ginger/turmeric slices should be dried from an initial moisture content
of 85% to 90% to a final moisture content of less than 10% to preserve

them for long time. Therefore, a 50 kg/batch cabinet dryer was developed as shown in Figure 2.4. The dryer has four chambers, and each chamber houses six perforated stainless steel trays. Air is heated by an electrical heater placed at the bottom of the drying chamber. Air is forced to move in a round-about path inside the chamber to increase the heat and mass transfer rate before being expelled by an exhaust fan at the top. Figure 2.5 shows an experimental 10 kg/batch dryer used for studies of drying kinetics and design optimization. To evaluate the thin layer drying characteristics of ginger slices, the 10 kg/batch dryer was slightly modified. Top four trays were removed and a perforated sample tray was suspended between two trays and attached to an electronic balance placed at the top of the drying chamber in order to monitor the weight loss during drying.

2.2.4 GRINDER

The dried material should be grinded into powder for the end use. Therefore, a grinder of 50 kg/h capacity was developed as shown in Figure 2.6.

FIGURE 2.4 Cabinet dryer (50 kg/batch capacity).

FIGURE 2.5   Cabinet dryer (10 kg/batch capacity).

FIGURE 2.6   Grinder for dry ginger/turmeric chips.

2.3 STUDIES CONDUCTED IN THE EXPERIMENTAL DRYER

A series of experiments were conducted in the experimental dryer (10 kg/batch capacity).

The first set of experiments was related to the effects of drying air temperature on the drying rates and total drying time required to reach the equilibrium moisture content. The second set of experiments dealt with the effects of moisture content of the slices on thermal conductivity.

2.3.1 EFFECT OF DRYING AIR TEMPERATURE ON THE DRYING RATES

Experiments were conducted at 45°C, 50°C, 55°C, and 60°C drying air temperature and a constant air velocity of 1.3 m/s. Moisture contents of ginger slices were monitored online using an electronic balance until a constant weight of the product was reached. The moisture content was expressed as moisture ratio (MR) (see Eq. (1)).

$$MR = \frac{\left(w_t - w_e\right)}{w_i - w_e} = a \, \exp\left(-kt\right) \tag{1}$$

where W_i is the initial moisture content, W_e the equilibrium moisture content, W_t the moisture content at time t, and a and k are temperature-dependent constants.

The variation in MR is plotted against the drying time for four different drying air temperatures of 45°C, 50°C, 55°C, and 60°C as shown in Figure 2.7. It is evident from the figure that the MR decreases exponentially with the drying time. With an increase in drying air temperature, the drying time decreases considerably. The drying time required to reach equilibrium moisture content starting with an initial moisture content of around 88–87% (w.b.) to a final moisture content of around 6–7% (w.b.) are 8.5, 7.5, 6, and 4.5 h with drying air temperatures of 45°C, 50°C, 55°C, and 60°C, respectively.

The experimental data were fitted to various mathematical models. The model developed by Midilli et al. (2002) gave the highest R^2 value

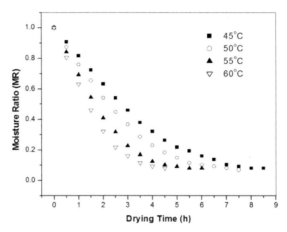

FIGURE 2.7   MR of ginger slices at different air temperatures and air velocity of 1.3 m/s.

and lowest RMSE value for all the temperatures. R^2 values and root mean square error (RMSE) values for the Midilli et al. (2002) model varied from 0.9993 to 0.9996 and 0.007072 to 0.009637, respectively.

2.3.2   EFFECT OF MOISTURE CONTENT ON THERMAL CONDUCTIVITY

The experimental thermal conductivity variation with moisture content (% w.b.) is shown in Figure 2.8. The thermal conductivity decreases with the reduction in moisture content. It varies from 0.571 to 0.358 W/m.K with the reduction in moisture content from 80% to 40% (w.b.) at room temperature of 24°C. The moisture content has the most significant effect on thermal conductivity, whereas the effect of temperature is negligible.

As the moisture content has the most significant effect on thermal conductivity, a correlation was developed to calculate the thermal conductivity of sliced ginger (see Eq. (2)).

$$\kappa = 3.098 \times 10^{-006} \, (M^3) - 0.0004412 \, (M^2) + 0.02294 \, (M) - 0.02775 \quad (2)$$

The correlation predicts the value within 1.5% accuracy for a moisture range of 80% to 40% (w.b.) and a temperature of 24°C.

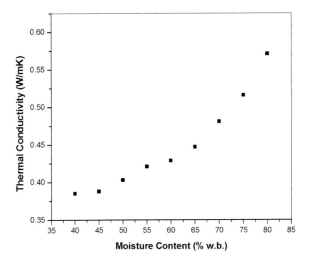

FIGURE 2.8   Variation in thermal conductivity of sliced ginger with moisture content at 24°C.

2.4   PROJECT ACHIEVEMENTS

The project Development of Postharvest Processing Technology for Ginger, Turmeric, Chili, and its Implementation for augmenting Regional Economy of Mizoram (RSP-0011) produced the following achievements:

- Establishment of two centers for postharvest processing and research.
- Cabinet dryer for ginger and turmeric (Reference No. 0090NF2009) patented and copyrighted (Reference No: IPMG/Copyright/2009).
- Rotary drum washer for ginger and turmeric (Reference No. 0151NF2009) patented and copyrighted (Reference No: IPMG/Copyright/2008/05).
- The know-how of the technology of (1) cabinet dryer, (2) rotary drum washer, and (3) slicing unit was transferred on a nonexclusive basis with one-time license fee and royalty on the sold items using the technologies.
- Processing technology for ginger & turmeric transferred.
- Training of rural youths on ginger/turmeric processing.

- Employment of local youths.
- Enhancement of farmers' economy.

Moreover, after the establishment of the postharvest processing system and market linkages, more than 6000 farmer families (30,000 people approx.) in Mizoram benefitted from the technology by getting good value of their organic products. About 14,110 hectares of land (see Table 2.1) and a number of self-help groups (SHGs) in Mizoram have been brought under the project. The technology has also led to employment generation in the states in the form of workers engaged in the processing units; 35–40 tribal youths are earning their daily livelihood in these centers. More than 10000 man-days have been generated in Mizoram. The use of technology has reduced the drudgery of the farmers in processing their produce.

2.5 CONCLUSIONS

The project RSP-0011, carried out by the CSIR-CMERI in collaboration with a local NGO, CDAR, has introduced the modern processing technology for ginger and turmeric in Mizoram. The project team developed a complete system including a rotary washing unit, a slicing unit, a cabinet dryer, and a grinder. The project has not only developed the processing technology (including two patents and copyrights) but also transferred this technology to the users. A total of 68 villages were included and more than 6000 farmers were trained and are now capable of adding value to their cash crops and to significantly improve their revenue.

TABLE 2.1 Statistics Regarding the RSP-0011 Project in Mizoram (Aizawl District)

Block Name	No. of Villages	No. of Farmers	Area (hectares)
1. Tlangnuam	18	1452	3212
2. Thingsulthliah	17	1713	3430
3. Aibawk	15	1172	2731
4. Phullen	7	653	1749
5. Darlawn	11	1288	2988
Total	68	6278	14110

KEYWORDS

- ginger
- grinding
- income generation
- slicing
- training
- turmeric

REFERENCES

Dugasani, S., Pichika, M. R., Nadarajah, V. D., Balijepalli, M. K., Tandra, S., & Korlakunta, J. N., (2010). Comparative antioxidant and anti-inflammatory effects of [6]-gingerol, [8]-gingerol, [10]-gingerol and [6]-shogaol. *Journal of Ethnopharmacology, 127*(2), 515–520.

Loha, C., Choudhury, B., & Chatterjee, P. K., (2008). Development of tray dryer for processing ginger for the rural people in Mizoram. *Proceedings of the 16th International Drying Symposium (IDS 2008),* Hyderabad, 1358–1362.

Loha, C., Das, R., Choudhury, B., & Chatterjee, P. K., (2012). Evaluation of air drying characteristics of sliced ginger (*Zingiber officinale*) in a forced convective cabinet dryer and thermal conductivity measurement. *J., Food Process Technol. 3,* 130. doi:10.4172/2157–7110.1000160.

Midilli, A., Kucuk, H., & Yapar, Z., (2002). A new model for single-layer drying. *Drying Technology, 20,* 1503–1513.

Sakulnarmrat, K., Srzednicki, G., & Konczak, I., (2015). Antioxidant, enzyme inhibitory and antiproliferative activity of polyphenolic-rich fraction of commercial dry ginger powder. *International Journal of Food Science & Technology* (in print). doi:10.1111/ijfs.12889.

Stoilova, I., Krastanov, A., Stoyanova, A., Denev, P., & Gargova, S., (2007). Antioxidant activity of a ginger extract (*Zingiber officinale*). *Food Chemistry, 102*(3), 764–770.

CHAPTER 3

WASTE PRODUCTS OF HORTICULTURAL CROPS: AN ALTERNATIVE FOR DEVELOPING ENTREPRENEURSHIP

AMRITESH C. SHUKLA[1] and D. MANDAL[2]

[1]Department of Botany, University of Lucknow, Lucknow–226007, India, E-mail: amriteshcshukla@gmail.com

[2]Department of Horticulture, Aromatic & Medicinal Plants, Mizoram University, Aizawl-796001, India.

CONTENTS

ABSTRACT

Antimicrobial activity of the secondary metabolites extracted from the leafy wastes of some species of *Curcuma*, viz., *Curcuma angustifolia, C. aromatica, C. domestica,* and *C. zedoaria,* was screened against three common dermatophytic fungi causing ringworm infection in human beings. The volatile oil of *C. domestica* Valet (Family Zingiberaceae) was found to be the strongest toxicant against the test fungi. The minimum inhibitory concentration (MIC) of the oil was 1.6 μL/mL against *Epidermophyton floccosum* and 1.4 μL/mL against *Microsporum gypseum* and *Trichophyton rubrum*; however, it was fungicidal at 1.6 μL/mL against *M. gypseum* and *T. rubrum* and 2.0 μL/mL against *E. floccosum*. The efficacy contains heavy doses of inoculums (25 discs of 5 mm each.). The minimum killing time (MKT) of the oil was 30 s against *E. floccosum* and *M. gypseum* and 20 s against *T. rubrum*, while, its minimum fungicidal concentration (MFCs) required 6.30 h against *E. floccosum* and *M. gypseum* and 5.30 h against *T. rubrum*. The oil efficacy was thermostable up to 100°C and for 36 months of storage, with the maximum unit taken into consideration. Moreover, the oil of *C. domestica* did not exhibit any adverse effect on mammalian skin up to 5% conc. The clinical trial of the oil in the form of ointment (at 1% v/v conc) to topical testing on patients attending outpatient department (OPD) of mLN Medical College, Allahabad, is still in progress.

3.1 INTRODUCTION

The World Health Organization (WHO) estimated that 80% of the population of developing countries rely on traditional medicine, mostly plant drugs, for their primary healthcare needs. Medicinal plants being natural, non-narcotic, having no side effects, safe, and cost-effective offer preventive and curative therapies that could be useful in achieving the goal of "Health for all" in a cost-effective manner. Demand for medicinal plants is increasing in both developing and developed countries, but 90% malarial la harvested from wild sources without applying scientific management hence many species are under threat to become extinct.

In fact, traditional herbal remedies have led the scientists to the development of numerous modern drugs. At this point, the discovery of reserpine from *Rauvolfia serpentine* can be cited as an example of how a plant utilized by the indigenous people eventually becomes the source of one of the most important pharmaceuticals of the world.

Keeping these views in mind, in the present investigations, a scientific attempt has been made to explore the possibilities of *Curcuma* spp., as a protecting measure against ringworm infection in human beings.

3.2 MATERIALS AND METHODS

3.2.1 IN VITRO INVESTIGATION

3.2.1.1 Extraction and Isolation of Essential Oil

The essential oil(s) were extracted separately from the wasted leaves of *Curcuma angustifolia, C. aromatica, C. domestica,* and *C. zedoaria* (Family- Zingiberaceae) by hydro distillation using Clevenger's apparatus (Clevenger, 1928). A clear light yellow colored oily layer was obtained on the top of the aqueous distillate, which was then separated and dried over anhydrous sodium sulfate. The oils thus obtained were subjected to various antimicrobial investigations.

3.2.1.2 In vitro Antimicrobial Investigations of the Essential Oil(s)

The minimum effective concentration (MEC) of the oil against some common human pathogenic fungi, namely, *Epidermophyton floccosum* Hartz, *Microsporum gypseum* (Bodin) Guiart et Grigorakis, and *Trichophyton rubrum* Castellani, was determined using the technique of Shahi et al. (2001), with a slight modification. Two sets were maintained: one for the treatment set and another for the control. The treatment set at different concentrations of the oil was prepared by mixing the required quantity of the oil samples in acetone (2% of the total quantity of the medium) and then added in pre-sterilized Sabouraud dextrose agar (SDA) medium. In the control set, sterilized water (in place of the oil) and acetone were used in the medium in appropriate amounts. The fungistatic/fungicidal (MSC/MCC) action of the oil was tested by aseptically reinoculating the fungi in culture tubes containing Sabouraud dextrose broth (Tables 3.1–3.3).

The data recorded were the mean of triplicates, repeated twice. The percentage of fungal growth inhibition (FGI) was calculated according to the formula:

$$FGI(\%) = \frac{Dc\text{-}Dt}{Dc} \times 100$$

TABLE 3.1   Minimum Effective Concentration of Oil of Four Different spp. of Curcuma against Some Common Human Pathogenic Fungi

Curcuma spp.	Human Pathogenic Fungi		
	Epidermophyton floccosum	*Microsporum gypsum*	*Trichophyton rubrum*
Curcuma angustifolia	2.6 μL/mL	2.2 μL/mL	2.4 μL/mL
C. aromatica	1.8 μL/mL	1.6 μL/mL	1.8 μL/mL
C. domestica	1.6 μL/mL	1.4 μL/mL	1.4 μL/mL
C. zedoaria	2.2 μL/mL	1.8 μL/mL	2.0 μL/mL

TABLE 3.2   Minimum Effective Concentration of the Oil Extracted from the Waste Leaves of Curcuma Domestica Against Test Fungi

Concentration (μL/mL)	Human Pathogenic Fungi		
	Epidermophyton floccosum	*Microsporum gypsum*	*Trichophyton rubrum*
2.0	100[C]	100[C]	100[C]
1.8	100[s]	100[C]	100[C]
1.6	100[s]	100[s]	100[C]
1.4	88	60	100[s]
1.2	60	—	80
1.0	—	—	60

[c] indicates cidal and [s] indicates static.

where Dc indicates colony diameter in the control set and Dt indicates colony diameter in the treatment sets.

3.2.1.3   Effect of Inoculum Density

The effect of inoculum density on the minimum cidal concentration (MCC) of the oil against the test fungi was determined using the method of Shukla et al. (2001). Mycelial discs of 5 mm diameter of 7-day oil cultures were inoculated in culture tubes containing oil at their respective MCCs. In the control set, sterilized water was used in place of the oil and

TABLE 3.3 Detailed in-Vitro Investigations of Curcuma Domestica Against the Test Fungi

Properties studied	Observations		
	Epidermophy-ton floccosum	Microsporum gyp-sum	Trichophyton rubrum
Minimum Inhibitory Concentration			
MEC (µL/mL)	1.6 µL/mL	1.4 µL/mL	1.4 µL/mL
MFC (µL/mL)	2.0 µL/mL	1.6 µL/mL	1.6 µL/mL
Minimum Killing Time			
Pure oil	30 sec	30 sec	20 sec
MFC	6.30 hrs	6.30 hrs	5.30 hrs
Inoculum Density (25 disc, 5 mm diam)	No Growth	No Growth	No Growth
Thermostability (up to 100°C)	No Growth	No Growth	No Growth
Effect of Storage (36 months)	No Growth	No Growth	No Growth

*MEC indicates minimum effective concentration.; MFC indicates minimum fungicidal concentration.

run simultaneously. The numbers of mycelial discs in the treatment as well as control sets were increased progressively up to 25 discs, in multiply of five. Observations were recorded up to the 7th day of incubation. Absence of mycelial growth in the treatment sets up to 7th day exhibited the oil potential against heavy doses of inoculums (Table 3.3).

3.2.1.4 Effect of Some Physical Factors

Effect of some physical factors, viz., temperature (40°C, 60°C, and 80°C) and autoclaving (up to 15 lb/sq inch pressure for 30 min) on the efficacy of the oil at MCC was also determined following the method of Shukla et al. (2001) and Shahi et al. (2001). Samples of oil in small vials, each containing 1 mL, were exposed to 40°C, 60°C, and 80°C temperatures in a hot water bath. Further, the oil's efficacy was tested against the test fungi at their respective MCCs (Table 3.3).

3.2.1.5 Minimum Killing Time

The minimum killing time (MKT) of the pure oil and their respective MCCs of *C. domestica* against the test fungi was determined using the method of Shahi et al. (1999) (Table 3.4).

3.2.1.6 Fungitoxic Spectrum

The fungitoxic spectrum of the oil at lethal and hyperlethal concentration (i.e., 2 and 4 µL/mL) was determined against some common human pathogenic fungi, viz., *Microsporum auddouinii* Gruby, *M. canis* Bodin,

TABLE 3.4   Minimum Killing Time of the Oil Extracted from the Waste Leaves of *Curcuma Domestica* Against Test Fungi

Mycelial Growth Inhibition (%)						
Minimum Killing Time (MKT)	Epidermophyton floccosum		Microsporum gypseum		Trichophyton rubrum	
	P.O.	MFC	P.O.	MFC	P.O.	M.F.C.
7.0	100	100	100	100	100	100
6.30	100	100	100	100	100	100
6.0	100	60	100	80	100	100
5.30	100	—	100	—	100	100
5.0	100		100		100	80
2.30	100		100		100	—
2.0	100		100		100	
1.30	100		100		100	
1.00	100		100		100	
30 min	100		100		100	
15 min	100		100		100	
5 min	100		100		100	
60 sec	100		100		100	
30 sec	100		100		100	
20 sec	90		80		100	
10 sec	60	—	70	—	88	—

*P.O. indicates pure oil; MFC indicates minimum fungicidal concentration.

M. nanum Fuentes, *Trichophyton mentagrophytes* (Robin) Blanchard, *T. tonsurans* Malmstem, and *T. violaceum* Bodin. This was done using the method of Shahi et al. (2001) (Table 3.5).

The oil's efficacy was also tested against some plant pathogenic fungi, viz., *Aspergillus parasiticus* Speare, *Cladosporium cladosporioides* (Fresenius) de Vries, *Curvularia lunata* (Wakker) Boedijin, *Colletotrichum capsici* (Syd.) Butler and Bisby, *C. falcatum* Went, *Fusarium oxysporum* Schlecht, *F. udum* de vries, *Helminthosporium maydis* Nisikado & Miyakel, *H. oryzae* Breda de Haan, *Penicillium implicatum* Biourge, and *P. minioluteum* Dierckx by using the technique of Shukla et al. (2001) (Table 3.5).

TABLE 3.5 Fungitoxic Spectrum of the Oil of *Curcuma Domestica* Against Some Common Pathogenic Fungi

Fungi Tested	Lethal Concentration (2.0 μL/mL)	Hyper Lethal Concentration (4.0 μL/mL)
Human Pathogens		
Microsporum auddouinii	100[s]	100[c]
M. canis	100[s]	100[c]
M. nanum	100[c]	100[c]
Trichophyton mentagrophytes	100[c]	100[c]
T. tonsurans	100[c]	100[c]
T. violaceum	100[c]	100[c]
Plant Pathogens		
Aspergillus parasiticus	100[s]	100[c]
Cladosporium cladosporioides	100[c]	100[c]
Curvularia lunata	100[c]	100[c]
Colletotrichum capsici	100[c]	100[c]
C. falcatum	100[c]	100[c]
Fusarium oxysporum	100[c]	100[c]
F. udum	100[c]	100[c]
Helminthosporium maydis	100[c]	100[c]
H. oryzae	100[c]	100[c]
Penicillium implicatum	100[c]	100[c]
P. minio-luteum	100[c]	100[c]

[s] indicates static; [c] indicates cidal in nature.

3.2.1.7 Comparison with Some Synthetic Fungicides:

The comparative efficacy of oil of *C. domestica* with some synthetic anti-fungal drugs was carried out by comparing MECs. This was done using the method of Shahi et al. (1999) (Tables 3.6 and 3.7).

All the experiments were repeated twice, and each contained three replicates; the data presented in the tables are the mean values.

3.3 RESULTS

On comparing the MECs of the oils of *Curcuma angustifolia, C. aromatica, C. domestica,* and *C. zedoaria* against the test fungi, the MEC of the oil of *C. domestica* was found most effective (Table 3.1).

The MEC of *C. domestica* oil was 1.4 µL/mL against *M. gypseum* and *T. rubrum* and 1.6 µL/mL against *E. floccosum;* however, it was fungicidal at 1.6 µL/mL against *M. gypseum* and *T. rubrum* and at 2.0 µL/mL against *E. floccosum* (Table 3.2).

The oil's efficacy contains heavy doses of inoculums (i.e. up to 25 discs, each of 5 mm), thermostable up to 80°C and also persisted after autoclaving at 15 lb/ sq inch pressure for 30 min (Table 3.3).

The pure oil kills the test fungi within 30 s; however, its MCC ranges from 5.30 to 6.30 h to kill all the fungi (Table 3.4).

Fungitoxic spectrum of the oil at lethal and hyperlethal concentration (i.e., 2 µL/mL and 4 µL/mL), against some common pathogenic fungi reveals that the oil has a broad fungicidal spectrum (Table 3.5).

TABLE 3.6 Comparative MECs of the Oil Extracted From the Waste Leaves of *Curcuma domestica* With Some Synthetic Antifungal Drugs

Oil & Trade Name of Anti-fungal Drugs	Active Ingredients	Minimum Effective Concentration (µL/mL)		
		Epidermophyton floccosum	*Microsporum gypseum*	*Trichophyton-rubrum*
Curcuma domestica	Essential oil	1.6	1.4	1.4
Dactrine	Miconazole nitrate	6.0	6.0	6.0
Nizaral	Ketoconazole	6.0	0.5	5.0
Tenaderm	Tolnaftate	2.0	1.5	0.8

TABLE 3.7 Comparative Efficacy of the Oil Extracted from the Waste Leaves of *Curcuma Domestica* With Some Synthetic Antifungal Drugs

Antimycotic Drugs	Drugs %	Cost (Rs.) Ointment/g	Cost (Rs.) Lotion/mL	Adverse Effects	Expiry Duration (months)	Environmental impact
C. domestica	1%v/v	0.90	0.70	No adverse effects	24-36	Renewable, biodegradable, non-residual toxicity.
Dactrine	2%w/w	2.80	—	Occasionally produced gastrointestinal side effects viz., nausea, vomiting, diarrhea	35	Non-renewable, non-biodegradable and residual toxicity
Nizaral	2%w/w	3.75	3.17	Adverse reaction observed were mainly burning, irritation. Drug may block testosterone synthesis	24	——do——
Tenaderm	1%w/v	1.06	1.30	Adverse effects were fever, nausea, vomiting, diarrhoea & skin rash, rarely produced irritation	24	——do——
Batrafine	1%w/v	1.50	1.60	——do——	24	——do——

Furthermore, comparison of the MECs of the oil with those of some synthetic antifungal drugs revealed that the MECs of the oil was higher than those of Dactrine, Nizaral, and Tenaderm (Tables 3.6 and 3.7).

3.4 CONCLUSION

The preliminary in vitro investigations of the extracted oil from the wastes of *Curcuma* spp. against some common human pathogenic fungi *E. floccosum, M. gypseum,* and *T. rubrum* revealed that on the basis of the detailed in vivo and multicenter clinical trials, *Curcuma domestica* can not only be an effective antimicrobial agent against dermatophytes but can also be an alternative for developing entrepreneurship in pharmaceutical industries.

ACKNOWLEDGMENTS

Authors are thankful to Prof. Anupam Dikshit, Biological Product Laboratory, Department of Botany, University of Allahabad, for providing some research facilities during this research work. Authors are also thankful to Prof. A. K. Bajaj, Former Head of Department of Dermatology, MLN Medical College, Allahabad, and to Dr. Uma Banerjee, Division of Microbiology All India Institute of Medical Sciences, New Delhi, for providing the fungal cultures and their identification. Author are also highly thankful to the authorities of the Mizoram University, Aizawl, as well as University of Lucknow, for providing various kinds of support as and when required during compilation of the work.

KEYWORDS

- **antimicrobial** activity
- **dermatophtes**
- **herbal drug**
- **medicinal plants**
- **minimum inhibitory concentration**

REFERENCES

Anonymous, (1948–88). *Wealth of India: Raw Materials*, vol. I-XI. Publications & Information Directorate, CSIR, New Delhi.

Chopro, R. N., Nayar, S. L., & Chopra, I. C., (1956). *Glossary of Indian Medicinal Plants.* Publications & Information Directorate, CSIR, New Delhi.

Clevenger, J. F., (1928). Apparatus for the determination of volatile oil. *J. Am. Pharm. Assoc., 17,* 346.

Ganesan, S., Venkateshan, G., & Banumathy, N., (2006). Medicinal plants used by ethnic group Thottianaickans of Semmalai hills (reserved forest), Tiruchirappali district, Tamil Nadu. Indian Journal of Traditional Knowledge, *5,* 253–258.

Garber, R. H., & Houston, B. R., (1959). An inhibitor of *Verticillium alboatrum* in cotton seed. *Phytopathology, 49,* 449–450.

Grover, G. S., & Rao, J. T., (1978). In vitro antimicrobial studies of the essential oil of *Eugenia jambolana. Indian Drugs, 15,* 143–144.

Grover, R. K., & Moore, J. D., (1962). Toxicometric studies of fungicides against brown rot organisms *Sclerotinia fructicola* and *S. laxa. Phytopathology, 52,* 876–880.

Jadhav, D., (2006). Ethnomedicinal plants used by *Bhil* tribe of Bibdod, Madhya Pradesh. Indian Journal of Traditional Knowledge, *5,* 268–270.

Jain, S. K., (1991). *Dictionary of Indian Folk Medicine and Ethanobotany, (eds).* Deep Publications, New Delhi, India.

Kirtikar, K. R., & Basu, B. D., (1935). *Indian Medicinal Plants.* International Book Distributers, Dehradun, vol. *1–4.*

Lalramnghinglova, H., (2003). *Ethno-Medicinal Plants of Mizoram,* Bishen Singh, Mahendra Pal Singh, Dehradun.

Lalramnghinglova, J. H., (1996). Ethnobotany of Mizoram-a preliminary survey. *J. Econ. Taxon. Bot. Add. Ser., 12,* 439–450.

Langenau, I. E. E., (1948). The examination and analysis of essential oils, synthetics and isolates. Guenther, (eds). *The Essential Oil.* Robert, E., Krieger Publishing Co. Huntington, New York.

Nehrash, A. K., (1961). The antimicrobial properties of cultivated radish. The antimicrobial activity of extracts and essential oils from cultivated & wild radish. *J., Mircrobial, 23,* 32–37.

Pandey, D. K., Tripathi, N. N., Tripathi, R. D., & Dixit, S. N., (1982). Fungitoxic and phytotoxic properties of the essential oil of *Caesulia axillaris,* Roxb. (Compositae). *Angew. Botanik, 56,* 259–267.

Pandey, M. C., Sharma, J. R., & Dikshit, A., (1996). Antifungal evaluation of the essential oil of *Cymbopogon pendulus* (Nees ex Steud.) Wats. cv. Praman. *Flavor and Fragrance Journal, 11,* 257–260.

Rao, J. T., (1976). Antifungal activity of essential oil of *Curcuma aromatica. Ind. J., Pharm., 38,* 53–54.

Shahi, S. K., Patra, M., Shukla, A. C., & Dikshit, A., (2001). Botanical drug for therapy against fungal infections in human beings. *Nat. Acad. Sci. Letters, 24*(5–12), 73–78.

Shahi, S. K., Shukla, A. C., Bajaj, A. K., Banerjee, U., Rimek, D., Medgely, G., & Diksit, A., (2000). Broad spectrum herbal therapy against superficial fungal infections. *Skin Pharmacol. Appl. Skin Physiol., 13,* 60–64.

Shahi, S. K., Shukla, A. C., Bajaj, A. K., Medgely, G., & Diksit, A., (1999). Broad spectrum antimycotic drug for the control of fungal infection in human beings. *Current Science, 76*(6), 836–839.

Shahi, S. K., Shukla, A. C., Dikshit, A., & Uperti, D. K., (2001). Broad-spectrum antifungal properties of the lichen, *Heterodermia leucomela*. *The Lichenologist*, UK, *33*(2), 177–179.

Shahi, S. K., Shukla, A. C., Dikshit, S., & Diksit, A., (1997). Modified spore germination inhibition technique for evaluation of candidate fungitoxicant (*Eucalyptus* spp.) Dehne, H. W., et al., (eds.). *Proc. of the 4th Int. Symposium on Diagnosis and Identification of Plant Pathogens*, Kluwer Academic Publishers, Netherlands, pp. 257–263.

Sharma, G. P., Jain, N. K., & Garg, B. D., (1978). Antifungal activity of some essential oil. *Indian Drugs, 16*, 21–33.

Sharma, P. K., Chauhan, N. S., & Lal, B., (2005). Studies on plant associated indigenous knowledge among the *Malanis* of Kullu district, Himachal Pradesh. Indian Journal of Traditional Knowledge, *4*, 409–411.

Shukla, A. C., (1998). Fungitoxic studies of some aromatic plants against storage fungi. D. Phil Thesis, University of Allahabad, India.

Shukla, A. C., Shahi, S. K., & Dikshit, A., (2000). Epicarp of *Citrus sinensis*: a potential source of natural pesticide. *Indian Phytopathology, 53*(3), 318–322.

Shukla, A. C., Shahi, S. K., Dikshit, A., & Saksena, V. C., (2001). Plant product as a fumigant for the management of stored product pests. *Proc. Int. Conf. on Controlled Atmosphere and Fumigation in Stored Products*, USA (Eds. Donahaye, E. J., Navarro, S., & Leesch, J. G.), pp. 125–132.

Singh, A. K., Dikshit, A., & Dixit, S. N., (1983). Antifungal studies of *Peperomia pellucida*. *Beitr. Biol. Pflanzen, 58*, 357–368.

Singh, A. K., Tripathi, S. C., & Dixit, S. N., (1978). Fungitoxicity on volatile fractions of some angiospermic plants. *Proc. Sym. Environmental Science and Human Welfare, 13*.

Singh, R., Shukla, A. C., & Prasad, L., (2007). Antifungal screening of some higher plants against storage fungi. *Progressive Agriculture, 7*(1–2), 128–131.

Slavenas, J., & Razinskaite, D., (1962). Some studies of phytocidal substances and Juniper oil from common Juniper Leit. *TSR. Moks. Acad. Darbal. Ser. C., 2*, 63–64.

WCS, (1980). *World Conservation Strategy*. International union of conservation of Nature and Natural Resources.

Wellman, R. H., (1967). Commercial development of fungicides. Hollen, et al., (ed.). *Plant Pathology Problem and Progress*. Indian University Press, Allahabad, 1908–1958.

CHAPTER 4

OSMOTIC DEHYDRATION OF MANGO VARIETIES ALPHONSO AND TOTAPURI

J. LENKA[1] and R. B. TIWARI[2]

[1]*Division of Post Harvest Management, ICAR–CISH, Rehmankhera, P.O. Kakori, Lucknow–226 101, India*

[2]*Division Post Harvest Technology, IIHR, Bangalore–560089, India*

CONTENTS

ABSTRACT

Osmotic dehydration is one of the new preservation methods for value addition to fruits. It is preferred over others due to better retention property of vitamin and minerals, color, flavor, and taste. It is less energy intensive than air or vacuum drying process because partial dehydration takes place at low or ambient temperature. The study on osmotic dehydration of mango varieties Alphonso and Totapuri was done. The treatments were Alphonso slices treated with 55 °Brix (T_1) and 70 °Brix (T_2) and Totapuri slices 40 °Brix (T_3), 40 °Brix (Used) (T_4), 45 °Brix (Used)

(T_5), 50 °Brix (T_6), and 60 °Brix (T_7) for 24 h. The conditions used in the dehydration process were syrup/fruit ratio of 2:1 (v/w) in room temperature. Higher weight loss and solid gain (SG) were observed in Alphonso slices (T_2) for 24 h than in Totapuri slices. Solid gain, water loss (WL), and weight reduction (WR) ranged from 6.60% to 7.70%, 34.50% to 44.40%, and 27.96% to 36.70% in Alphonso and from 7.55% to 13.00%, 27.15% to 48.89%, and 19.60% to 36.10% in Totapuri, respectively. It was also observed that an increase in duration of osmosis and syrup concentration increased weight loss, moisture loss, and SG in slices of both the varieties. Final dried yield ranged from 25.34% to 28.52% in Totapuri and was highest in Alphonso slices (T_2) at 29.46%. The sensory acceptance of osmo-dehydrated fruits was determined for the attributes of aroma, flavor, texture, and overall acceptance using a hedonic scale. Highest sensory score (84.08) was found in Alphonso slices made using 70 °Brix (T_2). Osmotic dehydration process also significantly affected other parameters and values were moisture content (10.07 to 15.45%), a_w (0.631 to 0.684) and Total solids (83.97 to 86.77%). Osmo-dehydrated slices were found to be microbially safe for direct consumption after 2 months of storage at room temperature. The process of osmotic dehydration could be applied in rural areas for value addition to fruits by entrepreneurs and at home scale, as it is a simple process and can contribute to sustainable horticulture.

Mango (*Mangifera indica* L.) is one of the most important commercial crops worldwide in terms of production, marketing, and consumption. India, China, and Mexico are the main producers of mango. Unripe mangoes are rich in vitamin C; the ripe fruits are rich in provitamin A and contain moderate levels of vitamin C and are an excellent source of fiber. All mango varieties represent a potential source of natural antioxidants. A vast diversity of products can be prepared from fresh mango. However, mangoes are extremely perishable like other farm produce, especially fruits and vegetables. Besides the fresh fruit, the range of processed products includes juices, nectars, concentrates, jam, fruits bars, flakes, chutney, and dried fruit (Schieber et al., 2000; Berardini et al., 2005). Dried mango is the commonly preserved form of the fruit in Asia, and it has also become increasingly popular in Europe (Tedjo et al., 2002). However, conventionally dried mango fruit has undesirable tough texture, poor color, and cooked flavor with a loss of nutritive value, which reduce its economic importance (Durance et al., 1999).

India is the second largest producer of fruits after China. In India, the area of mango cultivation is 2516,000 ha and productivity is 7.3 MT/ha (NHB, 2014). Fruit and vegetable losses in the developing countries are considerably high. In India, postharvest losses of fruits and vegetables are estimated to be more than 25%. Osmotic dehydration is an important pretreatment that involves the immersion of the fruit in concentrated solutions where both partial dehydration of water from the fruit and solid uptake are obtained. Mass transfer rates during osmotic dehydration are influenced by several factors including temperature, concentration of osmotic medium, size and geometry of the samples, sample to solution ratio, and the level of agitation of the solution (Torreggiani, 1993; Raoult-Wack, 1994). Osmotic dehydration technique is an effective method for the preservation of fruits that has gained more attention due to its potential application in the food processing industry. The main advantages of osmotic dehydration include better color, texture, flavor, nutrient retention, and prevention of microbial spoilage. The product obtained by osmotic dehydration is more stable during storage due to low water activity imparted by solute gain and WL (Tiwari, 2005).

Osmotic dehydration has been reported to be useful prior to drying to produce a variety of shelf-stable dried products with improved quality attributes such as color, texture, and aroma (Heng et al., 1990). Osmotic dehydration is a complementary treatment in the processing of dehydrated foods, as it presents some advantages such as minimizing heat damage to the color and flavor, inhibiting enzymatic browning, and reducing energy costs (Torres et al., 2006, 2007). The technique aims to dehydrate food products by immersing them in a hypertonic solution. Water is removed due to the difference of osmotic potential between the food and the osmotic solution, thus reducing the water activity of the food and consequently the water availability for chemical and biological deterioration.

Osmotic concentration is the process of water removal from fruits because the cell membranes are semi-permeable and allow water to pass through them more rapidly than sugar. During osmosis, small quantity of fruit acid is removed along with water. It is a dynamic process, in which water and acid are removed at first and then move slowly, while sugar penetration is very slight at first but increases with the time. The characteristics of the product can be varied by controlling temperature, sugar syrup concentration, concentration of osmosis solution, time of osmosis, etc., to make the osmotic concentration process faster. The present study on osmotic dehydration of mango varieties Alphonso and Totapuri has the following objectives: to assess the effect of osmotic pre-treatment, viz., concentration of sugar syrup, on WL, SG, WR, yield, and physiochemical and quality parameters and to evaluate sensory quality of osmo-dehydrated Alphonso and Totapuri mango slices (Figure 4.1).

4.1 MATERIALS AND METHODS

The materials used in present investigation on osmotic dehydration of mango are Alphonso and Totapuri. The experiment was conducted at the Processing Laboratory of Division of Post Harvest Technology, Indian Institute of Horticultural Research (IIHR), Bengaluru. Fruits of Alphonso and Totapuri varieties were harvested from IIHR farms. Fruits were allowed to ripe at room temperature, and hard ripe fruits

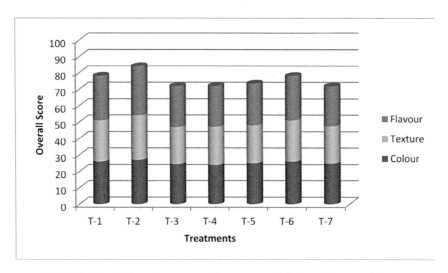

FIGURE 4.1 Effect of sensory quality on osmo-dehydrated mango slices.

were selected for the experiment. Fresh fruits with uniform shape and size free from insect damage and diseases were selected for osmo-dehydrated products. The selected mango fruits were weighed and peeled. The edible fruit portion was cut into slices. Prepared slices were again weighed to record the yield recovery of fresh slices for osmotic dehydration. Prepared Alphonso mango slices of 1 kg each were dipped in 55 and 70 °Brix fresh sugar syrup solution and Totapuri slices were treated with 40 °Brix fresh and used syrup, 45 °Brix used syrup, and 50 and 60 °Brix fresh sugar syrup in the ratio of 1:2 of fruit to syrup and allowed to continue osmosis for 24 h at room temperature (20–30°C). Changes in syrup Brix was observed periodically. During the process of osmosis, water flows out of the fruit pieces to the syrup and a fraction of solute moves into the fruit slices. At the end of the treatment, the fruit slices were taken out of the osmotic solution, and water adhering to the surface of the slices was removed. These osmosed mango slices were weighed to know the extent of water removal from the slices by osmosis.

4.1.1 PRE-TREATMENTS DETAILS

- T_1: Alphonso Slices 55 °Brix for 24 h
- T_2: Alphonso Slices 70 °Brix for 24 h
- T_3: Totapuri Slices 40 °Brix for 24 h
- T_4: Totapuri Slices 40 °Brix (Used) for 24 h
- T_5: Totapuri Slices 45 °Brix (Used) for 24 h
- T_6: Totapuri Slices 50 °Brix for 24 h
- T_7: Totapuri Slices 60 °Brix for 24 h

The following characteristics of fresh and osmo-dehydrated mango slices were recorded:

$$\textbf{Weight loss}\,(\%) = \frac{\text{initial weight} - \text{weight at time}}{\text{initial weight}} \times 100$$

$$\textbf{Solid gain (\%)} = \text{Moisture loss (\%)} - \text{Weight loss (\%)}$$

$$\text{Yield}(\%) = \frac{\text{Weight of prepared fruit pieces}}{\text{Weight of fresh fruit}} \times 100$$

Moisture content of fresh pulp, osmosed slice, and osmotically dehydrated samples was determined on percentage basis. Five grams of sample was taken in a pre-weighed China dish and kept in a hot air oven for overnight, and the weight was then recorded using an electronic balance. Moisture content was determined on fresh weight basis. Total solids were calculated by subtracting moisture content from 100.

$$\text{Moisture content}(\%) = \frac{\text{Moisture loss}}{\text{Sample weight}} \times 100$$

4.2 RESULTS AND DISCUSSION

Data given in Table 4.1 indicate that various osmotic treatments significantly affected the different physical parameters of mango slices in both the varieties.

4.2.1 MOISTURE CONTENT

Moisture content of osmo-dehydrated mango slices ranged from 10.07% to 16.03%. The maximum moisture content (16.03%) of osmo-dehydrated products was recorded in the treatment of Totapuri slices at 40 °Brix used syrup for 24 h (T_4), followed by 15.45% in Alphonso mango slices at 55 °brix for 24 h (T_1). Treatments T_2, T_7, and T_5 were found to be statistically at par with each other (Table 4.1).

4.2.2 TOTAL SOLID CONTENT

Total soluble solid (TSS °brix) in fresh mango slices was 16 to 17 °brix in Alphonso and 13 to 13.5 °brix in Totapuri. Total solid of osmo-dehydrated mango slices ranged from 83.97% to 89.93%. In Alphonso, total highest

TABLE 4.1 Effect of Different Osmotic Treatments on Physical Composition, Yield and Mass Transfer Kinetics in Osmotically Dehydrated Alphonso and Totapuri Mango Slices

Treatments	Osmosed-mango slices				Osmo-dehydrated samples			
	Moisture (%)	Total Solid (%)	Yield (%)	Water Activity (a_w)	Water Loss (%)	Solid gain (%)	Weight Reduction (%)
T_1 Alphonso slices at 55°Brix for 24 h	15.45	84.55	27.81	0.643	34.50	6.60	27.96
T_2 Alphonso slices at 70° Brix for 24 h	14.93	85.07	29.46	0.642	44.40	7.70	36.70
T_3 Totapuri slices at 40°B for 24 h	13.23	86.77	27.98	0.64	32.93	13.00	19.93
T_4 Totapuri slices at 40°Brix (Used) for 24 h	16.03	83.97	27.62	0.655	34.55	12.68	21.87
T_5 Totapuri slices at 45°B (Used) for 24 h	14.18	85.82	27.03	0.631	40.71	11.11	29.60
T_6 Totapuri slices at 50°B for 24 h	10.07	89.93	25.34	0.684	27.15	7.55	19.60
T_7 Totapuri slices at 60°B for 24 h	14.80	85.20	28.52	0.648	48.89	12.79	36.10
Sem ±	0.02	0.02	0.22	0.008	0.15	0.01	0.27
CD at 5%	0.43	0.43	1.49	0.001	1.05	0.09	1.81

solid (85.07%) was recorded in Alphonso at 70 °brix (T_2) and lowest in Alphonso at 55 °brix (T_1) In Totapuri slices, maximum total solid was recorded in Totapuri slices treated at 50 °brix (T_6) and lowest in Totapuri slices at 40 °brix used syrup (T_4) and the values ranged from 83.97% to 86.77%.

4.2.3 YIELD OF OSMOTICALLY DEHYDRATED MANGO SLICE

There was a statistically significant difference among the various osmotic treatments of final dried yield of mango slices, and it ranged from 25.34% to 29.46%. Highest yield (29.46%) was recorded in Alphonso at 70 °brix (T_2) because of maximum SG followed by Totapuri slices (28.52%) at 60 °brix (T_7). Treatments T_3, T_1, and T_5 were found to be statistically at par with each other and lowest in Totapuri slices at 50 °brix (T_6) due to minimum SG among the Totapuri mango slices. WL in osmosed mango slices ranged from 27.15 to 48.89. It also showed that the drying rate was better in concentrated syrup due to the increased osmotic pressure in the sugar syrup at higher concentrations, which increased the driving force available for water transport. These results are in conformity with the findings of Thippanna (2005) for banana.

4.2.4 WATER LOSS AND SOLID GAIN

Maximum WL of 48.89% was recorded in Totapuri slices at 60 °brix for 24 h (T_7) followed by Alphonso (44.4%) slices treated with 70 °brix for 24 h (T_2), while minimum WL of 27.15% was found in Totapuri slices at 50 °brix (T_6) . Highest SG of 13.00% was recorded in Totapuri slices at 40 °brix for 24 h (T_3), followed by 12.79% in Totapuri slices at 60 °brix for 24 h (T_7). These findings are also in conformity with observations made by other workers in case of mango (Varany-Anond et al., 2000), pineapple (Rahaman and Lamb, 1990). Sharma et al. (2003) reported that during osmotic dehydration, WL is always favored over solid uptake that leads to mass loss of pear fruit.

4.2.5 WEIGHT REDUCTION DURING OSMOSIS

Weight reduction in osmosed mango slices ranged from 19.60% to 36.70%. Highest WR (36.70%) was recorded in Alphonso slices treated with 70 °brix for 24 h (T_2), followed by 36.10% in Totapuri slices at 60 °brix for 24 h (T_7), while minimum WR (19.60%) was found in Totapuri slices at 50 °brix for 24 h (T_6). Further, it was also observed that increase in syrup concentration led to higher WR in osmosed mango slices. Variation in weight loss during osmotic dehydration among the varieties of apricot was observed by Sharma et al. (2004). Lombard et al. (2008) studied the effect of osmotic dehydration on mass fluxes (WL, SG, and WR).

4.2.6 WATER ACTIVITY (AW) IN DEHYDRATED SAMPLE

Water activity in osmo-dehydrated mango slices ranged from 0.640 to 0.684. Minimum water activity was recorded in Totapuri slices at 40 °brix for 24 h (T_3) followed by Alphonso at 70 °brix for 24 h (T_2), while maximum water activity was found in Totapuri slices at 50 °brix for 24 h (T_6) treatment. Similar results were reported by Pointing (1973). The product obtained by osmotic dehydration is more stable during storage due to low water activity imparted by solutes gain and WL (Tiwari, 2005). They also found that all these parameters depend on the concentration of syrup and syrup to fruit ratio.

4.2.7 EFFECT OF VARIOUS OSMOTIC TREATMENTS ON ACIDITY, ASCORBIC ACID, CAROTENOIDS, AND NON-ENZYMATIC BROWNING IN OSMO-DEHYDRATED MANGO SLICES

Data given in Table 4.2 indicate that various osmotic treatments significantly affected acidity, ascorbic acid, carotenoids and non-enzymatic browning (NEB) in osmo-dehydrated mango slices. Acidity content in fresh Alphonso and Totapuri slices was 1.45% and 0.90%, respectively. Acidity content in osmo-dried mango slices ranged from 0.60 to 1.79%. Highest acidity was recorded in Totapuri slices at 45 °brix used

TABLE 4.2 Effect of Different Osmotic Treatments on Chemical and Sensory Quality Parameters of Osmotically Dehydrated Alphonso and Totapuri Mango Slices

Treatments	Total Acidity (%)	Ascorbic Acid (mg/100 g)	Carotenoids (mg/100 g)	NEB (OD 440 nm)
T_1 Alphonso slices at 55°Brix for 24 h	0.77	123.03	13.60	0.167
T_2 Alphonso slices at 70° Brix for 24 h	1.54	96.43	20.25	0.196
T_3 Totapuri slices at 40°B for 24 h	0.90	36.93	2.35	0.110
T_4 Totapuri slices at 40°Brix (Used) for 24 h	1.54	24.90	3.67	0.100
T_5 Totapuri slices at 45°B (Used) for 24 h	1.79	53.95	4.82	0.090
T_6 Totapuri slices at 50°B for 24 h	0.90	41.50	6.22	0.129
T_7 Totapuri slices at 60°B for 24 h	0.60	52.29	5.67	0.100
Sem ±	0.01	0.24	0.02	0.001
CD at 5%	0.08	1.59	0.17	0.008

for 24 h (T_5), followed by Totapuri slices at 40 °brix used for 24 h (T_3) and Alphonso at 70 °brix for 24 h (T_2). Lowest acidity was recorded in Totapuri slices at 60 °brix fresh for 24 h (T_7). Ascorbic acid content in osmo-air dried samples ranged from 24.90 to 123 mg per 100 g. Highest ascorbic acid was recorded in Alphonso slices treated with 55 °brix for 24 h (T_1) followed by Alphonso at 70 °brix for 24 h (T_2) while lowest was recorded in Totapuri slices at 40 °brix used for 24 h (T_4). Carotenoid content in osmo-dried mango slices ranged from 2.35 to 20.25 mg/100 g. Highest carotenoid content (20.25 mg/100 g) was recorded in Alphonso slices treated with 70 °brix sugar syrup for 24 h (T_2) followed by Alphonso at 55 °brix for 24 h (T_1), while lowest carotene content (2.35 mg/100 g) was recorded in Totapuri slices at 40 °brix fresh for 24 h (T_3) (Figure 4.2). Non-enzymatic browning (NEB OD at 440 nm) in osmo-air dried mango slices ranged from 0.090 to 0.196. Lowest NEB value (0.100 OD at 440 nm) was observed in Totapuri slices treatment with 45 °brix used for 24 h (T_5) followed by Totapuri slices at 60 °brix fresh for 24 h (T_7), while maximum NEB value (0.196 OD at 440 nm) was recorded in the treatment Alphonso at 70 °brix for 24 h (T_2). Torregianni et al. (1986) reported on sugar content, color, acidity, vitamin C, pH, and organoleptic distinctiveness on osmo-dehydrated cherry.

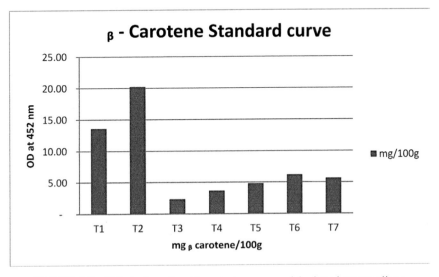

FIGURE 4.2 Effect of carotenoid contents on osmo-dehydrated mango slices.

4.2.8 COLOR, TEXTURE, FLAVOR OF OSMO-DRIED MANGO SLICES

Significant differences were recorded due to different treatments for color score in osmo-dehydrated mango slices. Sensory score for color in osmo-dehydrated mango slices ranged from 24 to 27.08, which was in acceptable range (Table 4.3, Plate 1). The highest score recorded in Alphonso at 70 °brix for 24 h (T_2) followed by Alphonso at 55 °brix for 24 h (T_1) and Totapuri slices at 50 °brix for 24 h (T_6), and lowest score was found in Totapuri slices at 40 °brix used for 24 h (T_4). Sensory score for texture in osmo-dehydrated mango slices ranged from 22.50 to 27.00. The highest score recorded was in Alphonso at 70 °brix for 24 h (T_2), followed by

TABLE 4.3 Effect of Different Osmotic Treatments on Sensory Quality Parameters of Osmotically Dehydrated Alphonso and Totapuri Mango Slices

Treatments		Sensory Score			
		Colour (30)	Texture (30)	Flavour (40)	Total (100)
T_1	Alphonso slices at 55°Brix for 24 h	26.00	25.00	27.08	78.08
T_2	Alphonso slices at 70° Brix for 24 h	27.08	27.00	30.00	84.08
T_3	Totapuri slices at 40°B for 24 h	24.50	22.50	24.91	71.91
T_4	Totapuri slices at 40°Brix Used for 24 h	24.00	23.20	24.80	72.00
T_5	Totapuri slices at 45°B Used for 24 h	25.00	23.00	25.41	73.41
T_6	Totapuri slices at 50°B for 24 h	26.00	25.00	27.00	78.00
T_7	Totapuri slices at 60°B for 24 h	24.80	22.65	24.30	71.75
	Sem ±	0.31	0.33	0.27	0.78
	CD at 5%	2.11	2.23	1.82	5.26

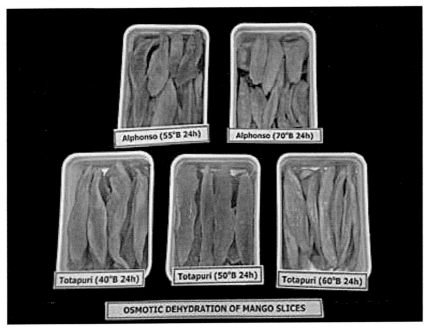

PLATE 1 **(See color insert.)** Osmotic dehydration of mango slices.

Alphonso at 55 °brix for 24 h (T_1) and Totapuri slices at 50 °brix for 24 h (T_6) while lowest score was found in Totapuri slices at 40 °brix for 24 h (T_3). Sensory score for flavor in osmo-dehydrated mango slices ranged from 24.30 to 30.00 (Figure 4.1). The highest score was recorded in Alphonso at 70 °brix for 24 h (T_2), followed by Alphonso at 55 °brix for 24 h (T_1) and Totapuri slices at 50 °brix for 24 h (T_6), while lowest score was found in Totapuri slices at 60 °brix fresh for 24 h (T_7). Overall, sensory score in osmo-dehydrated mango slices ranged from 71.75 to 84.08. The highest score 84.08 was recorded in Alphonso slices treated with 70 °brix sugar syrup for 24 h (T_2). Treatments T_1 and T_6 were statistically at par with each other, while lowest score was found in Totapuri slices at 60 °brix for 24 h (T_7). These results are in conformity with observations made by other workers Heng et al. (1990); Torres et al. (2006, 2007, 2008) improved quality attributes such as color, texture and aroma by inhibiting enzymatic browning by osmotic dehydration. Ramanuja and Jayaraman (1980) prepared intermediate moisture banana with better flavor and appearance with good storability.

4.2.9 MICROBIAL QUALITY IN OSMO-DRIED MANGO SLICES

Microbial quality in the samples was evaluated after 2 months of storage at room temperature in the Microbiology Laboratory of post-harvest technology (PHT) Division at IIHR. Different media were used for yeast, mold, lactic acid bacteria, *E. coli*, and total count. Total plate count was recorded in 13×10^2, 4×10^4, and Nil $\times 10^6$ dilution in Totapuri slices at 60 °brix fresh for 24 h (T_7), and all other treatments were recorded nil in different dilutions. *E. coli*, yeast, and mold counts were negligible in osmo-dehydrated mango slices, and they were found to be microbially safe. Therefore, these slices were safe for direct consumption. Similar results were reported by Ramanuja and Jayaraman (1980) and Kandekar et al. (2005).

4.3   CONCLUSION

It was concluded that various osmotic treatments had significant effect on product quality. Osmotic dehydration improved acceptability in mango samples. Variation in syrup concentration significantly affected WL, SG, WR, and final osmotically dehydrated yield. Varietal variation was also noticed. The overall sensory score in osmo-dehydrated mango slices ranged from 71.75 to 84.08. The highest score of 84.08 was noted in Alphonso slices treated with 70 °brix sugar syrup for 24 h (T_2), which was nonsignificantly followed by Totapuri slices treated with 50 °brix sugar syrup for 24 h (T_6).

KEYWORDS

- mango
- moisture loss
- organoleptic quality
- osmotic dehydration
- preservation

REFERENCES

Berardini, N., Knodler, M., Schieber, A., & Carle, R., (2005). Utilization of mango peels as a source of pectin and polyphenolics. *Innovative Food Science and Emerging Technologies, 6*, 442–452.

Durance, T. D., Wang, J. H., & Meyer, R. S., (1999). *Processing for Drying Mango and Pineapples.* US Patent No. 5962057.

Heng, W., Guilbert, S., & Cug, J. L., (1990). Osmotic dehydration of papaya: influence of process variables on the quality. *Science Des. Aliments, 10*, 831–848.

Khandekar, S. V., Chavan, U. D., & Chavan, J. K., (2005). Preservation of pulp and preparation of toffee from fig fruit. *Beverage and Food World, 32*, 55–56.

Lombard, G. E., Oliveira, J. C., Fito, P., & Andres, A., (2008). Osmotic dehydration of pineapple as a pre-treatment for further drying. *J. Food Eng., 85*, 277–284.

National Horticulture Board Database (2015). National Horticulture Board, Government of India, Gurugram, Haryana. (http://www.nhb.gov.in).

Ponting, J. D., (1973). Osmotic dehydration of fruits-recent modifications and applications. *Process Biochem., 8*, 18–20.

Rahman, M. S., & Lamb, J. (1990). Osmotic dehydration of pineapple. *J. Food Sci. Technol. (India), 27*(3), 150–152.

Ramarjuna, M. N., & Jayaraman, K. S., (1980). Studies on the preparation and storage stability of intermediate banana. *J. Food Sci. Technol, 17*, 183.

Raoult-Wack, A. L., (1994). Recent advances in the osmotic dehydration of fruits. *Trends in Food Sci. Technol., 5*, 255–260.

Sharma, H. K., Pandey, H., & Kumar, P., (2003). Osmotic dehydration of sliced pears. *J. Agric. Eng., 40*(1), 65–68.

Sharma, K. D., Kunen, R., & Kaushal, B. B. L., (2004). Mass transfers characteristics of yield and quality of five varieties of osmotically dehydrated apricot. *J. Food Sci. Technol., 41*, 264–275.

Tedjo, W., Taiwo, K. A., Eshtiaghi, M. N., & Knorr, D., (2002). Comparison of pretreatment methods on water and solid diffusion kinetics of osmotically dehydrated mangos. *Journal of Food Engineering, 53*, 133–142.

Thippanna, K. S., (2005). Studies on osmotic dehydration of banana (Musa *spp.*) fruits. *M., Sc. (Hort.) Thesis*, University of Agricultural Sciences, Bangalore.

Tiwari, R. B., (2005). Application of osmo–air dehydration for processing of tropical fruits in rural areas. *Indian Food Industry, 24*, 62–69.

Torreggiani, D., (1993). Osmotic dehydration in fruit and vegetable processing. *Food Res. Intl., 26*, 59–68.

Torreggiani, D., Giagiacamo, R., Bertolo, G., & Abbo, E., (1986). Research on the osmotic dehydration of fruits-I. Suitability of cherry varieties. *Ind. Conserv., 61*(2), 101–107.

Torres, J. D., Castello, M. L., Escriche, I., & Chiralt, A., (2008). Quality characteristics, respiration rates, and microbial stability of osmotically treated mango tissue (*Mangifera indica* L.) with or without calcium lactate. *Food Sci. Technol. Int., 14*(4), 355–365.

Torres, J. D., Talens, P., Carot, J. M., Chiralt, A., & Escriche, I., (2007). Volatile profile of mango (Mangifera indica L.), as affected by osmotic dehydration. *Food Chem., 101,* 219–228.

Torres, J. D., Talens, P., Escriche, I., & Chiralt, A., (2006). Influence of process conditions on mechanical properties of osmotically dehydrated mango. *J. Food Engg., 74,* 240–246.

Varney, W., Wongkrajang, K., Warunee, V. A., & Wongkrajan, K., (2000). Effects of some parameters on the osmotic dehydration of mango Cv. Kaew. *Thai. J. Agri. Sci., 33,* 123–135.

CHAPTER 5

BREADNUT: INNOVATIVE PRODUCTS FOR THE AGRO FOOD SECTOR

S. SUBRAMANIAM, N. RAMBURN, and R. CHACOORY

Crop Department, Food and Agricultural Research and Extension Institute, Reduit, Mauritius, Tel.: +230 6708249, E-mail: fruit@farei.mu, smyovana@gmail.com

CONTENTS

Breadnut (*Artocarpus camansi*) is an untapped crop in Mauritius. Breadnut grows with minimum inputs and presents high potential for food security. Breadnut seeds are considered to be nutritious, being a good source of protein, namely essential amino acids, minerals, and carbohydrates. This study aimed at evaluating the potential of breadnut as (i) a cooked nut; and (ii) processed products in view of providing new healthy innovative products for the local population and for the export market targeting celiac patients in particular. The work confirmed that locally available immature and mature green breadnut fruits can be consumed as a vegetable, while the mature seeds from fully ripened fruits can be used as a staple, a vegetable, or a snack with high acceptability. Proximate analysis showed that breadnut is low in fat and is gluten free. Breadnut seeds can be preserved in brine or can be frozen for subsequent use as a vegetable. A protocol was also developed for the production of breadnut flour. A 100-g portion of breadnut flour contains 74 g carbohydrates, 9.74 g protein, 2.27 g ash, 1.65 g fat, and minerals like potassium (390 mg), phosphorous (260 mg), calcium (110 mg), and magnesium (70 mg). The breadnut flour was found to be suitable for enrichment of wheat flour and can be used in different forms for the preparation of various nutritious dishes like porridge, stews, and milk-based drink with high sensory value, thereby indicating the potential of an unexploited crop for value addition.

5.1   INTRODUCTION

Breadnut (*Artocarpus camansis Blanco*, Family: Moraceae), also known as *chataigne*, *castana*, *kamansi*, and *rima*, is a tropical tree nut distinct from breadfruit with its spiny texture and numerous seeds embedded in the fruit pulp. It is an underutilized crop with significant potential as a nutritious food source (Ragone, 2006). Immature fruits are bright green, firm with white pulp and soft immature seeds. Mature fruits are light green, firm with a light yellow pulp and mature seeds enclosed within a brown thick, hard endocarp and an inner thin brown testa. Most fruits are harvested at this stage as the seeds are at their best eating quality. Ripe fruits are yellow brown with a soft pulp and mature seeds; ripe fruits should be immediately harvested as any delay leads to declination of seed eating quality due to its high perishable nature (Robrts-Nkrumah, 2015). Breadnut seeds are considered to be nutritious: a 100-g portion of dried seeds contain 13.3–20 g protein, 6.2–12.8 g fat, 76.2 g carbohydrates, 2.5–3.9 g fiber, and essential amino acids (Quijano et al., 1979; Negro de Bravo et al., 1983); they are lower in fat content compared to almond, brazil nut, cashew, and macadamia (Moreira et al., 1998).

Immature breadnut fruits are used in Trinidad and Tobago and Guyana as a curried dish. Seeds from mature fruits are consumed after boiling, frying, or roasting and used for flour production in some Caribbean and Latin American countries (Williams and Badrie, 2005; Ordonez, 2011; Roberts-Nkrumah, 2015).

Breadnut is an untapped crop in Mauritius; it is not widely distributed, and it is mainly used for the production of seedlings that are used as rootstock for the propagation of breadfruit locally. Breadnut trees, which require minimum inputs, can be introduced in agro-forestry in Mauritius and can have high potential as a food and nutrition security crop. It can also be a new source of produce and products for celiac patients particularly as breadnut seeds are gluten free.

This study aimed at evaluating the potential of breadnut as (i) a cooked nut and (ii) processed products in view of providing healthy innovative products for the local population and for the export market through low-cost processing methods and creating new commercial opportunities in Mauritius.

5.2 MATERIAL AND METHODS

Mature ripe breadnut fruits were harvested or collected at St. Aubin and Pamplemousses over a period of 4 years during the production season that ranged between October–February (summer) and April–June (early winter) in Mauritius. Immature fruits were harvested for evaluation as a vegetable.

Fully mature firm fruits that contained mature seeds and ripe fruits were harvested or collected for evaluation of the mature breadnut seeds as a staple, vegetable, or healthy snack and for processing.

The breadnut seeds were extracted from the ripe pulp, the seeds were thoroughly washed, and the surface microbial load was reduced by soaking the seeds for 5 minutes in chlorinated water (25 mL of 3.25% active chlorine/5 liters of water); subsequently, they were (i) characterized, (ii) evaluated as fresh produce or processed immediately, or (iii) stored for 1 day at ambient prior to processing into selective products. The fruits were characterized after harvest for weight, circumference, length, number of seeds, total seed weight; the seeds were also characterized for the same parameters.

5.2.1   EVALUATION OF BREADNUT AS A VEGETABLE, STAPLE, AND NUT

For the evaluation as a vegetable, both immature and mature green fruits were harvested. The immature fruits were thoroughly washed, peeled and pulped, and seeds cut into small pieces prior to cooking. Seeds were extracted from ripe fruits, washed and cooked after removal of the endocarp and testa (peeling) or boiled, peeled, and cooked.

5.2.2   EVALUATION OF BREADNUT FOR SELECTED PROCESSED PRODUCTS

For processing, different treatments and methods were evaluated, and the processing methods were reviewed based on outcomes of each trial until

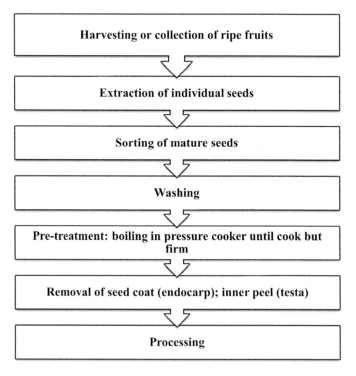

FIGURE 5.1   Preparation of breadnut seeds for cooking or processing.

the optimum protocol was developed for each processed product. Sensory evaluation was carried out, and shelf-life determined for each selected product. The breadnut seeds were prepared and pretreated prior to processing as shown in Figure 5.1. For use as a vegetable, snack, or frozen nuts, the breadnut seeds were precooked in a pressure cooker until fully cooked but firm (15 minutes), while for processing, the seeds were pressure cooked (10 minutes) until 80% cooked to maintain firmness. Boiling the breadnut seeds and removal of the testa prior to processing was found to be the most effective processing steps as it eliminated the "raw" and unappealing flavor associated with raw breadnut seeds.

5.2.2.1 Breadnut Seeds in Brine

The breadnut was boiled in pressure cooker in salted water (1.5% w/v) until cooked but firm; two treatments were evaluated: fully peeled boiled breadnut seeds and boiled breadnut seeds with testa. The breadnut seeds were canned in hot 3.0% brine (salinity of 3.1–3.2) to which citric acid (0.075% w/v, pH 4.5) had been added, pasteurized for 10 minutes at 85°C, cooled, and stored at ambient.

5.2.2.2 Frozen Breadnut Seeds

Peeled raw and pre-boiled (0.5% salt water) peeled breadnut were subjected to individual quick-freezing technique, packed in polyethene vacuum bags, and stored at −15°C up to 15 months; they were evaluated at every 3 months interval to check product quality.

5.2.2.3 Breadnut Seeds Flour

Different pretreatments (eight methods) were investigated for flour production with a view to reduce the labor-intensive procedure of cleaning the nuts and to yield dried breadnut/flour of good acceptability and quality.

Mature breadnut seeds were peeled (removal of endocarp, with and without testa), sliced (1 mm) or shredded (2 × 2 mm) and dried or blanched, prior to drying or boiled, before drying at 50 and 55°C. The pre-treated breadnut seeds were dried until brittle. The dried breadnut shreds generated from the different combination treatments were grinded into flour (0.5–0.8 µm) or granules (0.75 mm, 1.0–1.5 mm diameter; similar to semolina and broken wheat) using a grinding mill and stored in high density polyethylene (HDPE) packs.

The production method that yielded promising results in terms of product's taste, flavor, color, and texture and that was less labor intensive was selected. Based on the recommended method, breadnut seed flour, and breadnut seed granules were produced for further evaluation. The breadnut flour of different grades (coarse granules, small granules, and flour) was assessed into a range of products (sweet and savory). The nutritional value and shelf life of the flour was determined.

5.3 RESULTS AND DISCUSSION

5.3.1 FRUIT AND SEED CHARACTERISTICS

Fruits are oval to round in shape and texture varied according to maturity of fruits.

Immature fruits were green with sharp pointed firm spines and hold immature seeds; these seeds are susceptible to oxidative browning after extraction and when cut exudates a latex. The immature seeds have a very low acceptability. Mature and ripe fruits were light green to greenish brown in color with soft spines and pulp and contain the mature seeds. The seeds are embedded within the soft pulp; the edible part is enclosed within a brown thick pattern veined endocarp and wrapped in a thin layer membrane (testa).

TABLE 5.1   Breadnut Fruit Characteristics

Fruit circumference	32–43 cm
Fruit length	13–22 cm
Fruit weight	500–1170 g
Fibrous core	59–68 g
Number of seeds/fruit	13–61
Seed weight	160–680 g

Fruits' characteristics are shown in Table 5.1. Characterization indicated that a fruit weighed between 500 and 1170 g with around 13–61 seeds. The seed recovery rate was between 30% and 45%.

Each seed weighed between 5.25 and 10.50 g with a diameter of 2.20–3.73 cm and length of 2.7–3.5 cm; the seeds formed 22–45% of the total fruit weight. The endocarp constituted an average of 21% of fruit weight, while the testa weight can vary between 10 and 14% of fruit weight. The physical characteristics are within the range reported by Ragone (2006) and Roberts-Nkrumah (2005).

The peeled breadnut seeds that are of main commercial interest are off white in color, firm with light trace of latex. The recovery rate of peeled seeds varied between 25% and 30% of fruit weight.

5.3.2 USE OF BREADNUT

5.3.2.1 Immature Fruits as a Vegetable

The cooked dish was slightly bitter and was rated as moderate by consumers, thereby indicating its limited use locally.

5.3.2.2 Mature Seeds as Vegetable, Staple, and Snack

Mature breadnut seeds were cooked into dish similar to local potato-based dishes. Raw breadnut seeds and boiled breadnut seeds were evaluated. The curry from raw breadnut had a longer cooking time, required more water, and maintained a very firm and rigid texture. Boiling of the seeds with salted water (1.5–2.0%) in a pressure cooker and removal of the seed peel prior to cooking/preparation of different dishes improved the taste with the product having a sweet cooked flavor similar to chestnut and edoes, and soft and creamy texture. These observations are in line with that of Mathews et al. (2001) wherein cooked breadnut steamed at atmospheric pressure and above atmospheric pressure for 10–15 minutes were most preferred. Boiled seeds consumed as a snack was rated with high acceptability as indicated by a consumer perception study. The findings are in line with those reported by Roberts-Nkrumah (2015), Ragone (2006), and Williams and Badrie (2005).

The seeds may also be roasted in pan or oven at 200°C for 15–20 minutes or steam cooked yielding cooked nuts of very good acceptability.

Seeds from over ripe fruits that were collected (brown pulp and strong fermentation flavor and seeds were enclosed in a dark brown endocarp) had a slight bitter taste and astringent off taste after boiling, indicating declining quality of the seeds due to physiological changes.

5.3.3   NUTRITIONAL VALUE OF BOILED BREADNUT SEEDS

Table 5.2 shows the proximate composition of boiled breadnut seeds and compares it with the nutritional value of that reported by Williams and Badrie (2005) in West Indies.

TABLE 5.2   Proximate Composition of Boiled Breadnut Seeds for Mauritius and Comparison with Values for West Indies

Composition	Breadnuts seeds (Mauritius)	Breadnuts seeds (West Indies)
Moisture (g)	62.46	61.59
Protein (g)	4.08	6.89
Fat (g)	1.87	4.20
Carbohydrate (g)	28.36	23.90
Ash (g)	1.14	3.42
Crude fiber (g)	2.09	N/A
Gluten (g)	NIL	N/A
Reducing sugar (g)	1.53	N/A
Calcium (mg)	47	9.62
Magnesium (mg)	36	44.78
Phosphorus (mg)	120	4.28
Sodium (mg)	43	204.61
Potassium (mg)	420	734.62
Iron (mg)	1.25	1.18
Zinc (mg)	2.14	0.69
Copper (mg)	0.29	0.34
Manganese (mg)	0.14	0.36
Energy Kcal	150	-

Source: FAREI (2014); Williams & Badrie (2005).

The boiled breadnut is rich in carbohydrates and a good source of minerals like calcium, phosphorous, and potassium.

The fat content of the breadnut seeds collected in Mauritius is lower than that reported for West Indies, and calcium, phosphorous and iron concentration is slightly higher for nuts under the current study.

The high carbohydrate and protein content and presence of minerals confirms that breadnut seeds has a good nutritional profile and can be considered and included as a food and nutrition security crop in the Mauritian diet as a vegetable, staple, and healthy snack.

5.3.4 BREADNUT IN BRINE

An average product yield of 65% was recorded (seed weight basis). Sensory evaluation after 1 week yielded a product of moderate acceptability. Breadnut preserved without the testa appeared to disintegrate slightly after pasteurization and during storage as observed by the slightly turbid brine and tiny bits of cooked seeds. The brine solution for breadnut seeds with the testa or thin peel was clear, and these breadnuts had a better sensory value; it is therefore recommended to remove only the endocarp. These observations are in line with that of Mathews et al. (2001) whereby precooked, thin peel covered breadnuts canned in 2.5% brine were most preferred. Salinity and pH were stable, brine was clear, and a shelf life of 12 months is recommended for the breadnut seeds with the thin membrane or testa that should be removed prior to consumption.

5.3.5 FROZEN BREADNUT

Raw frozen seeds suffered from chilling injury with water accumulation, the seeds became spongy, and when cooked (unthawed and thawed) yielded a product with low acceptance. However, precooked or preboiled peeled breadnut was stable when frozen to −15°C and keep up to 15 months.

The boiled frozen breadnut is not affected by chilling injury and after thawing; the breadnut can be used for cooking, as a snack, or for processing into flour with very good results. Assessment of the product indicated that it was suitable for preparation of curries with its taste close to fresh ones. A yield of 62–72% (seed weight basis) or 30% (fruit weight basis) was recorded.

5.3.6   BREADNUT FLOUR

5.3.6.1   Flour Production

Flour from raw breadnut seeds had a slight raw, astringent, and off flavor, and the same was detected in the prepared products. However, flour and prepared dishes from boiled breadnut seeds had a high sensory value indicating that boiling improves the flour quality by inactivation some anti-nutritional factors. This is in line with the findings of Fagbemi et al. (2005) who showed that breadnut seeds contain some level of phytic acid, trypsin inhibitors, and tannins; boiling was the most effective pretreatment technique to reduce the tannin and trypsin inhibitor activity by 37% and 20%, respectively, thereby improving the digestibility of the breadnut seed flour by at least 71% when compared to raw dried flour. Drying the seeds with the testa yielded flour that was lightly brown in color and slightly bitter while drying fully peeled breadnut shreds resulted into off white colored flour with a higher sensory appeal and flavor although manual peeling is highly intensive. The following protocol is recommended for flour production:

The mature seeds are cooked in a pressure cooker (up to 3–4 whistles are recommended). The boiled seeds are dried for 30 minutes at 50°C to facilitate removal of the endocarp (brittle and easy to crack); the seeds are peeled to remove the testa, reduced into shreds of 2 × 2 mm, dried at 50°C until brittle, and grinded coarsely into granules or finely into flour.

Yield of dried breadnut seeds was estimated at 23–29% (seed weight basis) at a moisture content of 8–10%. It was found that the dried breadnut when grinded could yield three categories of flour: granules similar to broken wheat, coarse powder like semolina, and flour. The three types of products with moisture content of 10–12% and packed in moisture proof packages can keep above 12 months.

5.3.6.2   Uses of Flour

The breadnut flour of different grades (coarse granules, similar to broken wheat; small granules; and fine particles) were assessed into a range of products (sweet and savory). For best results, it was found that rehydrating the coarse and small granules in hot water for 1 h prior to cooking reduced cooking time and improved taste. The recovery rate of the soaked granules was three times more than the dried granules. The soaked granules and the breadnut coarse granules were found to be highly suitable for the preparation of porridge (sweet or savory), and the dish was similar to oatmeal or rice meal or "couscous." The coarse breadnut granule was found to be highly suitable for thickening soups and stews and preparation of *dal* (Indian pulse-based dish) with high sensory value.

The coarse breadnut was also added to breadfruit-based burgers with a view to improve as coating in the form of roasted granules.

With a view to improve protein content of breadfruit burger, breadfruit burgers were coated with the (roasted) breadnut granules and breadfruit mash was mixed with pre-soaked breadnut granules at 5%.

The smaller granules were found to be suitable for the preparation of semolina-based dish and dessert cream. Both breadnut smaller granules and flour after presoaking in hot water and blending were found to be highly suitable for the preparation of a delicious milk-based beverage and therefore constitute a good nutritional supplement.

The breadnut seeds flour can be used up to 50% as a supplement with wheat flour for a wide range of prepared dishes like chapatti, tortillas, pizza, cookies, and pastries with high acceptability.

5.3.6.3   Nutritional Value of Breadnut Flour

The proximate composition of breadnut flour is shown in Table 5.3 and indicates that dried breadnut seeds are highly nutritious. These results are comparable with those of previous studies (Ragone, 2003; Malamo et al., 2011). The high carbohydrate (74.07 g) and energy value (350 kcal) confirm breadnut as a high-energy food. The protein content of

TABLE 5.3   Proximate Composition of Dried Breadnut Seeds (100 g)

Composition	Breadnut seeds flour
Moisture (g)	11.25
Protein (g)	9.74
Fat (g)	1.65
Saturated fat (g)	38
Carbohydrate (g)	74.07
Ash (g)	2.27
Crude fiber (g)	1.02
Gluten (g)	NIL
Reducing sugar (g)	1.90
Calcium (mg)	110
Magnesium (mg)	70
Phosphorus (mg)	260
Sodium (mg)	10
Potassium (mg)	390
Iron (mg)	1.42
Zinc (mg)	1.54
Copper (mg)	0.48
Manganese (mg)	0.28
Energy Kcal	350

Sample is analyzed in 2014.

breadnut seeds flour for Mauritius is much higher (9.74 g) than the one reported by Malamo et al. (2011), which could be due to differences in the production method and agro-climatic factors. Breadnut flour contains a significant amount of minerals, particularly potassium (390 mg), phosphorous (260 mg), calcium (110 mg), and magnesium (70 mg), and low fat (1.65 g/100 g). Breadnut being gluten free can be a good healthy and nutritious source of food for celiac patients. Breadnut flour can be used as a nutritional supplement or for the preparation of food products and as a supplemental flour with wheat and other flour for a wide number of uses.

5.4 CONCLUSION

This study showed that breadnut seeds can be used as a vegetable, staple, and healthy snack in the fresh form and has wide application in the processed form either as processed products or as nutritional supplements. Being gluten free, it is also highly suitable for use by celiac patients. Its commercial cultivation in agro-forestry systems and exploitation will be highly beneficial in terms of sustainable development and food nutrition security.

KEYWORDS

- **breadnut**
- **nutritional value**
- **processing**
- **uses**

REFERENCES

Fagbemi, T. N., Oshodi, A. A., & Ipimmoroti, K. O., (2005). Processing effects on some anti-nutritional factors and *In vitro* multienzyme protein digestibility (IVPD) of three tropical seeds: Breadnut (*Artocarpus altilis*, cashewnut (*Anacardium occidentale*) and Fluted Pumpkin (*Telfairia occidentalis*). *Pakistan Journal of Nutrition, 4*(4), 250–256.

Malomo, S. A., Eleyinmi, A. F., & Fashakin, J. B., (2011). Chemical composition, rheological properties and bread making potential of composite flours from breadfruit, breadnut and wheat. *African Journal of Food Science, 5*(7), 400–410.

Mathews, R., Commissiong, E., Baccus-Taylor, G., & Badrie, N., (2001). Effect of peeling methods on breadnut (*Artocarpus altilis*) seeds and acceptability of canned seeds in brine. *Journal of Food Science and Technology, 38*(4), 402–404.

Negron de Bravo, E., Graham, H. D., & Padovani, M., (1983). Composition of breadnut (seeded breadfruit), *Caribb. J., Sci., 19*(3–4), 27–32.

Quijano, J., & Arango, G. J., (1979). The breadfruit from Colombia - A detailed chemical analysis, *Econ. Bot., 33*(2), 199–202.

Ragone, D., (2006). *Species Profiles for Pacific Island Agroforestry.* www. traditional tree. org (accessed 15 March 2012).

Roberts-Nkrumah, L. B., (2005). Fruit and seed yields in chataigne (*Artocarpus camansi* Blanco) in Trinidad and Tobago. *Fruits, 60*(6), 387–393.

Roberts-Nkrumah, L., (2015). *Breadfruit and Breadnut Orchard Establishment and Management* (A manual of commercial production). FAO.

William, K., & Badrie, N., (2005). Nutritional composition and sensory acceptance of boiled breadnut (*Artocarpus camansi* Blanco) seeds. *Journal of Food Technology, 3*(4), 546–557.

Woodroof, J. G., (1979). *Tree Nuts,* 2nd edn., AVI Publ. Co Inc.: Westport, Connecticut, USA.

CHAPTER 6

STORAGE STUDY OF JAMUN-AONLA BLENDED READY-TO-SERVE BEVERAGES

PREETI SINGH, NEELIMA GARG, and SANJAY KUMAR

Central Institute for Subtropical Horticulture, Rehmankhera P.O., Kakori, Lucknow–226101, India, E-mail: neelimagargg@rediffmail.com

CONTENTS

ABSTRACT

Jamun (*Syzygium cumini*) and *aonla* (*Emblica officinalis*) are two important medicinal plants of India, and their respective juices are used for curing number of health ailments. Juice blending is one of the methods to improve the nutritional quality of the product. Ready-to-serve (RTS) beverages were prepared by blending juices of jamun and aonla in three ratios (9:1, 8:2, and 7:3). Pure aonla and jamun juice were kept as control. Microbiological, biochemical, and sensory evaluations of RTS beverages were carried out at 0 day and after 3 and 6 months of storage at $12 \pm 2°C$. No microbial growth could be detected in any of the samples during the storage period. Among the treatments, RTS beverage prepared from jamun and aonla juices in 9:1 ratio was found to be the best on sensory score (8.72 out of 9), followed by blend in the ratio of 8:2 (8.3 out of 9). Increasing the aonla juice further reduced the acceptability of the RTS beverage. After 6 months of storage, the RTS beverage blend of 9:1 ratio had 15.8 °B total soluble solids (TSS), 0.27% titratable acidity, 9.18 mg/100 mL ascorbic acid, 6.17% reducing sugar, 23.15 mM/mL antioxidants (fluorescence recovery after photobleaching (FRAP) value), and 1.96 mg/100 mL anthocyanin content compared to pure jamun juice (15.8 °B TSS, 0.32% titratable acidity, 11.9 mg/100 mL ascorbic acid, 5.73% reducing sugar, 19.39 mM/mL antioxidants, and 2.29 mg/100 mL anthocyanins. The study reflected that jamun–aonla blend in the ratio of 9:1 could be stored up to 6 months with higher antioxidant level and better sensory acceptability.

6.1 INTRODUCTION

Jamun *(Syzygium cumini)* is a minor fruit crop belonging to family Myrtaceae. It is considered to be indigenous to India and West Indies, being cultivated in Philippines, West Indies, and Africa (Shrivastava and Kumar, 2009). It is gaining popularity among the rural as well as urban masses due to its high nutraceutical values. The ripe berries are rich sources of iron and pectin with a fair amount of ascorbic acid. It is used as an effective therapeutic medicine against diabetes, heart, and liver trouble (Garande and Joshi, 1995). However, jamun fruit is highly perishable; the short shelf-life of fruit makes it available only for a short period, which makes its popularity unrealized. Jamun is very popular as a dessert fruit, because of its slight astringent but sweet-sour taste and excellent color. In jamun growing regions, a glut of surplus fruit is available that needs to be processed and converted into value-added products. As the fruit is a rich source of anthocyanin, it imparts antioxidant properties too. Aonla (*Emblica officinalis* Gaertn.), also known as Indian gooseberry, is famous for medicinal values including antisorbutic, diuretic, laxative, and antibiotic properties. The fruits also possess pronounced expectorant, antiviral, cardiotonic, and hypoglycemic activity (Mehta and Tomar, 1979). Fresh aonla is too sour to consume and hence is preferred in the form of preserves, dried aonla, *trifala*, jam, juice, pickle and *chavyanprash*, toffees, and fruit bar (Singh and Kumar, 1995). Although aonla juice and beverages have poor consumer acceptance, it could be utilized for vitamin C enrichment of other fruit juice-based beverages. These vitamin C and antioxidant-rich natural drinks if given due publicity can replace synthetic drinks and overcome the vitamin and antioxidant deficiency in people as well.

6.2 MATERIALS AND METHODS

The mature fully ripened jamun and aonla fruits procured from the experimental farm of Central Institute for Subtropical Horticulture, Lucknow, were collected in clean polythene bags and brought to laboratory. The fruits were washed separately under running tap water, crushed in fruit mill, and the juice was extracted by applying pressure of 1500–2000 kg/ square inch with a hydraulic press. Ready to serve beverages were prepared by blending juices of jamun and aonla in ratios, viz., control jamun, 9:1 (T_1), 8:2 (T_2), 7:3 (T_3), and control aonla. The treatments as well as controls were adjusted with requisite proportion of water, sugar, and citric acid in order to maintain 10% juice, 14 °B total soluble solids (TSS), and 0.24% acidity and pasteurized at 90°C for 1 min before packing in sterilized glass bottles. The beverages were stored for 6 months at 12 ± 2°C. Biochemical analysis of jamun–aonla RTS beverage was carried out for TSS by using hand refractometer (Erma, Japan), acidity, and ascorbic acid, according to the methods of Ranganna (2000). The antioxidant property of juice in terms of FRAP values was determined according to Benzie and Strain (1999). The amount of reducing sugars was determined by the spectrophotometric method according to Folin and Wu (1920). Microbiological quality of jamun–aonla RTS beverage was carried out as described by Speck (1985). For judging the sensory attributes of the RTS drink, sensory evaluation was conducted by a panel of seven semi-skilled judges. The attributes considered in the scoring were color, clarity, aroma, taste, tannin, astringency, freedom from acetic acid, sugar, and impression evaluated (Amerine et al., 1965). The overall final rating was obtained by calculating the average of scores.

6.3 RESULTS AND DISCUSSION

The microbial quality assessment of the jamun-aonla RTS beverage revealed that all the treatments as well as controls were microbiologically safe during the storage period. No significant increase in TSS was observed during storage (Table 6.1).

Slight increase in total acidity was observed at the end of 6 months of storage (Table 6.1). Priyadevi et al. (2002) reported that pectic substances are responsible for increasing the acidity of fruits. Hence, in the present study, degradation of pectic substances into soluble solids might have contributed toward increase in the acidity of jamun beverages. Ascorbic acid content of all the treatments decreased continuously during the entire period of storage (Table 6.1). Highest ascorbic acid (19.74 mg/100 mL) was observed in pure aonla juice and lowest (4.23 mg/100 mL) in T_3 (Table 6.1) This reduction might be due to oxidation of ascorbic acid into dehydroascorbic acid by oxygen (Sethi et al., 1980). The reducing sugar content slightly increased during storage of jamun–aonla RTS beverages (Table 6.1). The increase is attributable to the hydrolysis of sucrose in glucose and fructose by the acid present in the beverages (Lotha, 1992) or gradual inversion of nonreducing sugars into reducing sugars in the acidic medium (Malav et al., 2014). Anthocyanins and antioxidants decreased in all treatments during storage. Among the beverages containing jamun, the lowest anthocyanin content (1.54 mg/100 mL) was observed in T_3 treatment (Figure 6.1), while the highest content (2.16 mg/100 mL) was observed in pure jamun juice followed by T_1 (1.90 mg/100 mL) after 6 months of storage. Total antioxidants content in the samples ranged from 19.39 to 55.49 mM/mL of jamun-aonla RTS at 0 day, being highest in pure aonla juice RTS and lowest in pure jamun juice RTS (Figure 6.2).

Antioxidants gradually decreased after 6 months of storage. Raj et al. (2011) have reported gradual decline in polyphenol contents from 332 to 305 mg % in sand pear and pear-apple juice beverage during storage period of 6 months.

TABLE 6.1 Biochemical Analysis of Jamun–Aonla Blended Ready-to-Serve Beverage after Six Months of Storage

Parameters	Storage period (Months)	Treatments				
		Control jamun	T_1	T_2	T_3	Control aonla
T.S.S (°B)	0	15.8	15.8	15.8	15.8	15.6
	2	15.8	15.8	15.8	15.4	15.6
	4	15.8	15.8	15.8	15.4	15.8
	6	16.0	16.0	15.8	15.6	15.8
Acidity (%)	0	0.24	0.24	0.24	0.24	0.23
	2	0.24	0.29	0.31	0.27	0.27
	4	0.26	0.30	0.32	0.32	0.30
	6	0.29	0.30	0.33	0.38	0.34
Ascorbic acid (mg/100ml)	0	11.9	9.52	4.76	4.76	21.4
	2	11.22	9.18	4.59	4.59	20.4
	4	11.01	8.89	4.44	4.37	19.88
	6	10.34	8.46	4.37	4.23	19.74
Reducing sugar (%)	0	5.73	6.01	5.61	5.73	5.53
	2	5.77	6.17	5.67	5.80	5.71
	4	5.85	6.21	7.71	5.83	5.76
	6	5.96	6.23	5.77	5.88	5.81

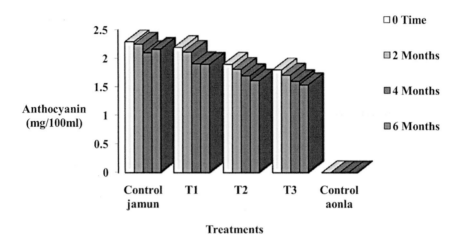

FIGURE 6.1 **(See color insert.)** Changes in anthocyanin content during storage of blended jamun-aonla ready-to-serve beverages.

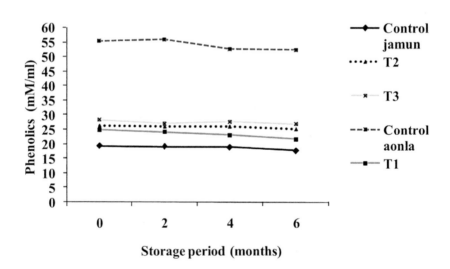

FIGURE 6.2 **(See color insert.)** Changes in phenolics in blended jamun–aonla ready to-serve beverages during storage.

Organoleptic analysis indicated treatment T_1, RTS prepared from jamun and aonla juices in 9:1 ratio, scored as best (8.72 out of 9) during storage (Figure 6.3 and Plate 1), followed by blend in the ratio of 8:2 (8.3 out of 9).

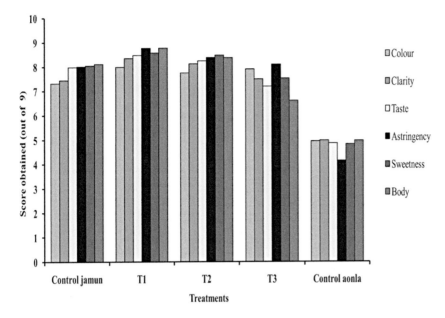

FIGURE 6.3   (See color insert.) Sensory evaluation of blended jamun–aonla ready-to-serve beverages during storage.

PLATE 1   Blended jamun–aonla ready-to-serve beverages.

Increasing the aonla juice further reduced the acceptability of the RTS beverage.

6.4 CONCLUSION

The study reflected that jamun–aonla blend in the ratio of 9:1 could be stored up to 6 months at room temperature with higher antioxidant level and better sensory acceptability.

ACKNOWLEDGMENT

The authors are thankful to Director, Central Institute for Subtropical Horticulture, Lucknow, for his keen interest in the work and constant support. The research was conducted under the Women Scientist program funded by Department of Science and Technology, New Delhi.

KEYWORDS

- **anthocyanins**
- **antioxidants**
- **beverage**
- **jamun**
- **juice blends**
- **ready-to-serve**
- *syzygium cumini*

REFERENCES

Amerine, M. A., Pangborn, R. M., & Roessler, E. B., (1965). *Principles of Sensory Evaluation of Food*. New York, Academic Press.

Benzie, F. F., & Strain, J. J., (1999). Ferric reducing/antioxidant power assay: direct measure of total antioxidant activity of biological fluids and modified version for simultaneous measurement of total antioxidant power and ascorbic acid concentration. *Methods in Enzymology, 299,* 15.

Folin, O., & Wu, H., (1920). Estimation of blood sugar by alkaline copper reducing method. *J., Biol. Chem., 41,* 367.

Garande, V. K., & Joshi, G. D., (1995). Storage of *jamun* fruit products. *Asian Food J., 10,* 54–56.

Lotha, R. E., (1992). Studies on processing and storage of kinnow mandarin juice. *PhD, Thesis,* IARI, New Delhi.

Malav, M., Gupta, R., & Nagar, T., (2014). Studies on bio-chemical composition of orange based blended ready-to-serve (RTS) beverages. *Biosci. Biotech. Res. Comm., 7*(1), 78–83.

Mehta, G. L., & Tomar, M. C., (1979). Studies on simplification of preserve making II. Amla (*Phyllanthus emblica* L.). *Indian Food Packer, 33* (5), 27–30.

Priyadevi, S., Thangam, M., & Desai, A. R., (2002). Studies on variability in physico-chemical characters of different jamun (*Syzygium cuminii*). *Indian J., Hort., 59,* 153–156.

Raj, D., Sharma, P. C., & Vaidya, D., (2011). Effect of blending and storage on quality characteristics of blended sand pear-apple juice beverage. *J., Food Sci. Technol., 48*(1), 102–105.

Ranganna, S., (2000). *Handbook of Analysis and Quality Control for Fruit and Vegetable Products.* IInd edn. Tata Mc Graw Hill Publication Co Ltd, New Delhi, pp. 1112.

Sethi, V., Anand, J. C., & Saxena, S. K., (1980). Kinnow orange in juice and beverage making. *Indian Hort., 25,* 13.

Shrivastava, R. P., & Kumar, S., (2009). *Fruit and Vegetable Preservation.* Principles and practices, IBDC, New Delhi.

Singh, I. S., & Kumar, S., (1995). Studies on processing of *aonla* fruits II. Aonla products. *Prog Horticult., 27*(1/2), 39–47.

Speck, M., (1985). Compendium of methods for the microbiological examination of foods. Second edition, *American Public Health Association Inc.,* pp. 644–649.

CHAPTER 7

DEVELOPMENT OF BLENDED AONLA SQUASH

NEELIMA GARG, SANJAY KUMAR, PREETI SINGH,
and KAUSHLESH K. YADAV

*Central Institute for Subtropical Horticulture Rehmankhera,
P.O. Kakori, Lucknow–227107, India,
E-mail: neelimagargg@rediffmail.com*

CONTENTS

ABSTRACT

Aonla (*Emblica officinalis*), despite being a highly nutritious and thera-peutically important fruit, is highly acidic. Beverages prepared from aonla have low acceptability owing to their high acidity. Blending appears to be one of the tools for the preparation of acceptable beverage from aonla. Litchi possesses pleasant aroma, while grape contains good flavor in addi-tion to attractive color in purple variety. In the current study, blended squash from aonla was prepared with litchi and grape (purple variety) separately in 3:1 and 1:1 ratio each. The ascorbic acid and total phenolic content was highest in aonla–litchi squash with 3:1 ratio (152 mg/100 g and 652 mg/100 g, respectively). In general, the ascorbic acid content decreased while phenolics increased after 6 months of storage at $12 \pm 2°C$. The anthocyanin content of aonla–grape squash was 3.7 and 7.7 mg/100g in 3:1 and 1:1 blends, respectively, which declined to 1.2 and 2.9 mg/100 g, respectively, after 6 months of storage. Initial sensory evaluation of the product revealed that aonla–litchi (1:1) blend was liked most, scoring 8.4 out of 9, followed by aonla–grape (1:1) with 8.1 score. Sensory scoring of squashes after 3 and 6 months of storage also indicated higher preferences for 1:1 blend than for 3:1 blend in both aonla–litchi and aonla–grape squashes. It may be inferred that 50% litchi or grape juice could be mixed with aonla juice to obtain acceptable quality of squash.

7.1 INTRODUCTION

Aonla *(Emblica Officinalis,* Gaertn), commonly known as Indian gooseberry, belongs to family Euphorbiaceae. India produces a huge quantum of aonla as the tree is well adapted to various kinds of soils and climatic conditions. The glistening, translucent, pale green fruits are highly fibrous in nature. The astringent flesh contains fairly rich amount of vitamin C and polyphenols and has high antioxidant property. The ascorbic acid in aonla is considered highly stable, apparently protected by tannins, which retards oxidation (Morton, 1987). The juice extracted from aonla fruit, despite having unparallel nutritional and medicinal qualities, is not suitable for fresh consumption due to lack of any flavor, high acidity, and astringent taste. Even beverages prepared from aonla juice are not very appealing due to poor sensory properties. Acceptability of aonla drink, however, could be drastically increased by addition of juices of some other fruits possessing attractive color, pleasant aroma, and taste. Fruits like litchi (*Litchi chinensis*) and grape (*Vitis vinifera*) may find good suitability in this context. Both the fruits are easily available in the market. Litchi bears a good blend of taste and aroma. Similarly, purple varieties of grape possess attractive color due to the presence of anthocyanin pigments. It also has soothing taste and flavor. The beverages from these two fruits are already well established in the market and have good public demand.

In the present investigation, an attempt has been made to enhance the sensory qualities of aonla squash through blending of litchi or grape juice and evaluate the biochemical and sensory qualities of products during storage.

7.2 MATERIALS AND METHODS

Aonla, litchi, and grape fruits were brought from Central Institute for Subtropical Horticulture experimental farm located at Rehmankhera, Lucknow. Healthy fruits were selected and washed thoroughly with tap water. Whole aonla fruits were subjected to fruit mill, and coarse pulp containing ground seeds was obtained. The pulp was wrapped in thick cloth and pressed in a hydraulic press to extract juice. The filtered juice was pasteurized at 90°C and preserved with 500 ppm SO_2 in the form of potassium metabisulfite. Litchi fruits were peeled, stones separated manually, and the pulp was homogenized in a blender. It was then squeezed through a muslin cloth and the juice was collected. The juice was pasteurized and preserved like aonla juice. Healthy grape berries were wrapped in thick cloth and pressed in a hydraulic press to obtain juice. Grape juice was also pasteurized in the similar manner and preserved with 500 ppm sodium benzoate. Aonla juice was blended with litchi and grapes juice separately in 3:1 and 1:1 ratios, and a squash was prepared from each of the blends using 40% juice, 50% sugar, and 0.9% acidity. The final combinations were as follows:

1. Aonla-litchi (3:1): Aonla-Litchi I
2. Aonla-litchi (1:1): Aonla-Litchi II
3. Aonla-grape (3:1): Aonla-Grape I
4. Aonla-grape (1:1): Aonla-Grape II

The squashes were filled hot in glass bottles, sealed, and stored in dry and cool place. They were analyzed for various biochemical and sensory parameters at 0, 3, and 6 months of storage at 12 ± 2°C.

The total soluble solids (TSS) of squashes were recorded using a hand refractometer (Erma, Japan). Titratable acidity, ascorbic acid, and total phenolics were determined according to the methods described by Ranganna (2000). Ascorbic acid content of beverage was measured by titrating samples against dye (2, 6-dichloro phenol indophenol, sodium salt) solution, while total phenolics were estimated spectrophotometrically using Folin–Ciocalteu's reagent. The microbial examination of fermented beverage was carried out according to the method detailed by Speck (1985). The organoleptic evaluation of the beverage was carried out on the basis of color, flavor, and taste by a panel of semi-skilled judges by using a 9-point hedonic scale as reported by Amerine et al. (1965).

7.3   RESULTS AND DISCUSSION

The microbial examination carried out at 0, 3, and 6 months of storage revealed no microbial growth at any stage in any sample. The TSS of squash samples ranged from 50 to 55 °B, while titratable acidity ranged from 0.92% to 0.95% at 0 day. Minor changes in these values were observed during storage. The ascorbic acid content was found to be the highest in 3:1 aonla–litchi blend (152 mg/100 mL), while lowest was in 1:1 aonla-grape blend (92 mg/100 mL). The amount of ascorbic acid in the samples was directly proportional to the volume of aonla juice added, as it was the main source of ascorbic acid. Decrease in the ascorbic acid content to an extent 33–35% and 24–29% was observed in aonla–litchi and aonla–grape blends, respectively, after 6 months of storage at 12 ± 2°C (Figure 7.1).

Decrease in ascorbic acid content of aonla juice was also reported by Bhattacherjee et al. (2013). Loss in ascorbic acid is attributed to oxidation of ascorbic acid molecules during storage. The total phenolic content in the samples ranged from 443 to 652 mg per 100 mL of squash at 0 day, being highest in 3:1 aonla–litchi blend and lowest in 1:1 aonla–litchi blend. It increased to a range of 613 to 875 mg per 100 mL after 6 months of storage (Figure 7.2).

The increase in phenolic content might be due to gradual solubilization of phenolics or transformation of molecules into some other forms

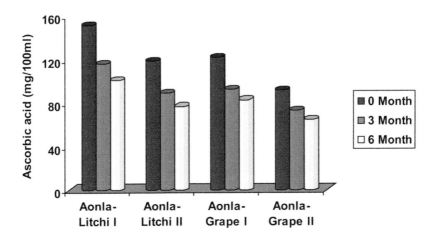

FIGURE 7.1   (See color insert.) Changes in ascorbic acid content of blended aonla squash during storage.

during storage. The total anthocyanin content of aonla–grape squash was 3.7 and 7.7 mg/100g in 3:1 and 1:1 blends, respectively, which declined to 1.2 and 2.9 mg/100 g, respectively, after 6 months of storage (Figure 7.3).

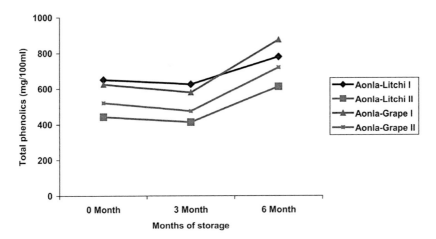

FIGURE 7.2   (See color insert.) Changes in total phenolic content of blended aonla squash during storage.

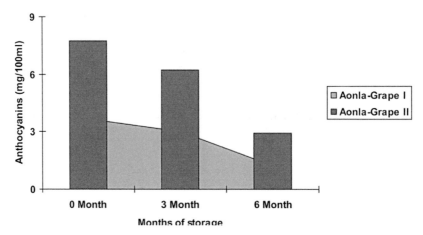

FIGURE 7.3   (See color insert.) Changes in anthocyanin content of aonla–grape squash during storage.

Decrease in anthocyanins was also observed by Yadav et al. (2014) in mulberry juice during storage. The sensory evaluation of squash, carried out at 0 day by semi-skilled judges on the basis of color, flavor, and taste revealed that aonla–litchi (1:1) blend was liked most, scoring 8.4 out of 9, followed by aonla–grape (1:1) with 8.1 score (Figure 7.4).

Similar trends were observed during sensory scoring of squashes after 3 and 6 months of storage, indicating higher preferences for 1:1 blend than for 3:1 blend in both aonla–litchi and aonla–grape squashes. This indicated that the addition of higher quantity of litchi or grape juice resulted in better quality of product due to enhanced color, flavor, and taste. It may therefore be concluded from the study that 50% litchi or grape juice could be mixed with aonla juice to obtain an acceptable quality of squash with improved color, flavor, and taste.

ACKNOWLEDGMENT

The authors are thankful to Director, Central Institute for Subtropical Horticulture, Lucknow, for providing infrastructure facility for carrying out the research work.

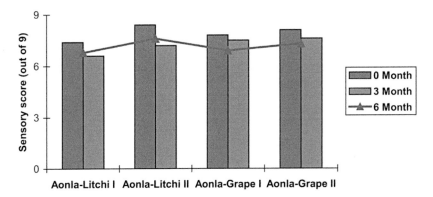

FIGURE 7.4 **(See color insert.)** Changes in sensory scores of blended aonla squash during storage.

KEYWORDS

- aonla
- beverages
- grape
- litchi
- sensory qualities

REFERENCES

Amerine, M. A., Pangborn, R. M., & Roessler, E. B., (1965). *Principles of Sensory Evaluation of Food.* Academic Press. New York and London.

Bhattacherjee, A. K., Dikshit, A., Kumar, S., Shukla, D. K., & Tandon, D. K., (2013). Quality evaluation in storage of aonla (*Emblica officinalis* Gaertn.) juice extracted from fruits preserved by steeping in water. *Internatl. Food Res. J.,* 20(4), 1861–1865.

Morton, J., (1987). *Emblic.,* In: '*Fruits of Warm Climate*', by Julia, F., Morton, Miami, FL., pp. 213.

Ranganna, S., (2000). *Handbook of Analysis and Quality Control for Fruit and Vegetable Products.* IInd edn. Tata Mc Graw Hill Publication Co Ltd, New Delhi, pp. 1112.

Speck, M. L., (1984). *Compendium of Methods for the Microbiological Examination of Foods.* American Public Health Association, Washington DC.

Yadav, P., Garg, N., & Kumar, S., (2014). Improved shelf stability of mulberry juice by combination of preservatives. *Indian J., Natural Products Resources,* 5(1), 62–66.

CHAPTER 8

SACCHAROMYCES CEREVISIAE POSTHARVEST DIP TREATMENT FOR IMPROVING QUALITY AND STORABILITY OF MANGO cv. DASHEHARI

BHARATI KILLADI, NEELIMA GARG, REKHA CHAURASIA, KAUSHLESH K. YADAV, and D. K. SHUKLA

Division of Post Harvest Management, ICAR–CISH Lucknow 226106, India, E-mail: bharatipal.pal@gmail.com

CONTENTS

ABSTRACT

Bio-agents are natural antagonists capable of inhibiting and keeping the target pathogens at low level, besides being nontoxic and environmentally safe. Anthracnose is one of the major postharvest diseases of mango.

The use of chemical pesticide to treat this disease is avoided from food safety point of view. The objective of the present study is to understand the use of *Saccharomyces cerevisiae* as a bio-agent for controlling spoilage and improving the quality of mangoes. Mature green fruits of Dashehari were treated with three strains of *S. cerevisiae* as (T_1), (T_2), and (T_3) @ 10^8 cells/mL for 10 minutes and control (T_4) dip treated with water and stored under ambient conditions (34 ± 2°C and 85% to 90% RH). Fruits were assessed for physico-chemical parameters at regular intervals of 0, 4, 6, 8, and 10 days. The cumulative physiological loss in weight (CPLW) was highest in T_4 (12.87%) followed by T_2 (12.44%), T_1 (12.41%), and T_3 (11.57%) on the 10th day of storage. The total soluble solids (TSS) increased and titratable acidity (TA) and firmness of the fruits decreased during the course of fruit storage. The total carotenoid content was highest in T_2 (7.22 mg/100 g), followed by T_3 6.88 (mg/100 g), T_1 (5.43 mg/100 g) and T_4 (4.88 mg/100 g) on the 10th day of storage. The antioxidant content estimated by fluorescence recovery after photobleaching (FRAP) was maximum in T_1 (4602.22 μmolar trolox equivalent/g), followed by T_2 (3353.02 μmolar trolox equivalent/g) and T_3 (2975. 24 μmolar trolox equivalent/g) on the 10th day of storage. Percent inhibition of antioxidants estimated by 2,2-diphenyl-1-picrylhydrazyl (DPPH) was maximum in T_3 (76.47%), followed by T_1 (61.52%) and T_2 (58.36%) on the 10th day of storage. The spoilage of the fruit was minimum in T_3 (7.33%), followed by T_2 (11.33%), T_1 (11.66%), and T_4 (15.33%) on the 10th day of storage. In vitro studies on the efficacy of *S. cerevisiae* for controlling *Colletotrichum gloeosporioides* (pathogen for anthracnose) confirmed the efficacy of the bio-agent. Thus, *S. cerevisiae* may be used as a potential bio-agent for improving the postharvest quality of mango.

8.1   INTRODUCTION

Mango (*Mangifera indica* L.) is one of the major and important fruits of the tropics and subtropical areas. India is the largest producer of mango, with the production of about 12 million tons of fruits. About 21 million tons of vegetables are spoiled each year, with a total estimated value of 240 billion rupees. India loses about 35–40% of the produce estimated at Rs. 40,000 crores per year due to improper postharvest management. Postharvest pathogen developed may be considered as a minor problem for local markets with short time period of harvest and selling; however, when the fruit is exported to foreign countries, prolonged storage requires control of the postharvest pathogens. Mango cv. Dashehari is harvested in the month of June, with the onset of monsoon accompanied with high temperature and relative humidity.

Postharvest pathogens are the causes of loss during storage of fruits. The most important and major postharvest pathogen of mango is anthracnose. The primary means to control postharvest disease of fruits is the use of synthetic fungicides. Yeasts have been explored as one of the promising alternatives for the control of postharvest pathogen (Rosa et al., 2010; Liu et al., 2013). Biocontrol of postharvest spoilage in mango by antagonistic microorganisms seems promising in reducing the use of synthetic fungicides (Lima et al., 1999; Janisiewicz and Korsten, 2002). Incidence and severity of postharvest disease significantly reduced in mango (Kefalew and Ayalew, 2008). Several yeasts have been reported as bio-control agents in sweet cherry (Wang and Tian, 2008); apple (Yu et al., 2008); jujube fruit (Cao et al., 2012), pear fruit (Yu et al., 2012); mandarins (Guo et al., 2014), and mango (Bautista-Rosales et al., 2014). *S. cerevisiae* has been used for postharvest management of postharvest diseases of grape (Suzzi et al., 1995) and apple (Scherma et al., 2003). Antifungal activity of *S. cerevisiae* and *S. pombe* against *Botrytis cineraria* on grape was reported for the first time (Nally et al., 2012). In the present investigation, *S. cerevisiae* was used as pretreatment for improving the quality and storability of mango cv. Dashehari.

8.2   MATERIALS AND METHODS

S. cerevisiae strains were obtained from the culture collection of microbiology laboratory in the Division of Post-Harvest Management, CISH (Central Institute for Subtropical Horticulture). The yeast culture was maintained on yeast potato dextrose agar (YPDA). Three yeast strains, viz., bakers' yeast, industrial yeast, and *S. cerevisiae* were tested for anti-pathogenicity for *Colletotrichum gloeosporioides* according to the standard protocol by dual culture technique, and growth inhibition was measured.

Yeast strains were multiplied on YPDA medium, and the plates were incubated at $30 \pm 2°C$ for 3 days. The cells were collected using a cell scrapper, and the final cell number was maintained @10^8 cells/mL in distilled water.

Green hard fruits of mangoes cv. Dashehari were harvested with stalks of 8–10 mm; they were washed with water and dip treated for 10 minutes with three yeast strains of *S. cerevisiae*, viz., T_1, T_2, T_3, and T_4 (control). The fruits were surface dried, packed, and stored under ambient conditions ($34 \pm 2°C$ and 85% to 90% RH). The observations of physiochemical parameters were recorded at an interval of 0, 4, 6, 8, and 10 days of storage. The fruit weight was recorded at the time of packaging and subsequently at each withdrawal. The difference in weight was expressed as percent weight loss. After withdrawal of fruits from each treatments, cell cfu/mL microbial load were counted by serial dilution method with observed effect on spoilage fruit. Firmness of the fruit was measured using a penetrometer (8 mm probe, USA) and expressed as kg/cm^2. Total soluble solids (TSS) was measured using a digital refractometer, model PAL-1 (Atago, Tokyo, Japan). Titratable acidity (TA) was estimated by the methodology propounded by Rangana (2000). Five grams of sample was diluted with 50 mL of distilled water and titrated with 0.1 mol/L NaOH solution, and the results were expressed as percent citric acid.

Fruit pulp was macerated for the estimation of total carotenoids according to the methods of Rangana (2000). Samples of 2 g each in triplicates were extracted in 15 mL acetone thrice and filtered through cotton wool in a conical flask. The extraction was done till colorless. Petroleum ether (15 mL) was added to the extract and diluted with 2% (15 mL) sodium chloride solution. All the extracts were transferred in a separating funnel and washed with 10 mL of 2% sodium chloride. The nonaqueous layer was extracted and collected in a 50-mL volumetric flask, and the volume was made up with 3% acetone in petroleum ether; the observations were recorded at 452 nm and expressed as mg/100 g.

The FRAP assay was done according to the methodology of Benzie and Strain (1996). The reduction of a ferric–tripyridyltriazine complex to its ferrous, colored form in the presence of antioxidants is the principle of the assay. The FRAP agent contained 2.5 mL of a 10 mmol/L TPTZ (2,4,6-tripyridy-s-triazine, Sigma) solution in 40 mmol/L HCL plus 2.5 mL of 20 mmol/L FeCl$_3$ and 25 mL of 0.3 mol/L acetate buffer, pH 3.6 and was prepared freshly and warmed at 37°C. Aliquots of 40 μL sample supernatant were mixed with 0.2 mL distilled water and 1.8 mL FRAP reagent, and the reaction mixture was incubated at 37°C for 10 min. Absorbance measured spectrophotometrically at 593 nm. The standard solution used was 1 mmol/L Trolox, and the final result was expressed as the concentration of antioxidants μmol trolox equivalent/g. If the FRAP value measured was beyond the linear range of standard curve, then adequate dilutions were made.

The 2,2-diphenyl-1-picrylhydrazyl (DPPH) estimation was done according to the method of (Brand-Williams et al., 1995). DPPH was weighed (24 mg) and dissolved in 100 mL methanol; this served as stock solution and was stored at −20°C until needed. The working solution was obtained by mixing 10 mL of stock solution with 45 mL methanol to get an absorbance of 1.1 ± 0.02 units at 515 nm using the spectrophotometer. Fruit extracts of 150 μL were allowed to react with 2850 mL of DPPH solution for 24 h in the dark. Then, the absorbance was read at 515 nm. The standard was a linear curve between 25 and 800 μM Trolox. Additional dilutions were made if the DPPH value measured was over the linear range of the standard curve.

Fruits were evaluated for spoilage according to McDonald et al. (1998) by measuring the rotten area in relation to the total surface area and expressed as percentage.

Surface microbial counts were monitored at 2 days intervals according to the method of Collins and Lyne (1984).

All the analysis was carried out in triplicates, and the data recorded during the course of investigation were subjected to statistical analysis by SAS 9.3 and CD at the significance level of 0.05.

8.3   RESULTS AND DISCUSSION

Results from the dual culture assay showed that all antagonistic microorganisms inhibited the mycelial growth of *C. gloeosporioides* with varying efficiencies. Yeast antagonists presented a moderate inhibition against the majority of *Colletotrichum* isolates. Treatments of yeast significantly exhibited strong antagonism against the isolates of *C. gloeosporioides* and were found maximum growth inhibition by effect of baker yeast (12 mm), industrial yeast (7 mm), *S. cerevisiae* (9 mm) compared with T_4 as control (18 mm), respectively.

The cumulative physiological loss in weight percent (CPLW %) differed significantly ($p \leq 0.5$) among the treatments and the storage period (Figure 8.1).

The CPLW percent increased with an increase in storage period. The CPLW was maximum (13.08%) in control followed by T_1 (12.87%) and T_2 (12.44%) and minimum in T_3 (11.27%) on the 10th day of storage. The increase in CPLW may be due to evapo-transpiration from the surface of fruits. Fruit transpiration vary significantly according to climactic conditions and characteristic of fruit that affects water balance; these findings are in concomitance with the findings of Leonadi et al. (1999) in tomato.

There was a significant difference ($p \leq 0.5$) in the firmness of fruits among the treatments and the storage period (Figure 8.2).

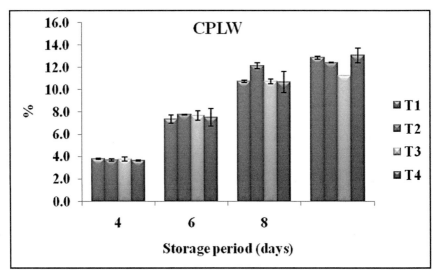

FIGURE 8.1   (See color insert.) Effect of yeasts on the CPLW percent of mango cv. Dashehari during storage. T_1 (baker's yeast), T_2 (industrial yeast), T_3 (*Saccharomyces cerevisiae*), and T_4 (control).

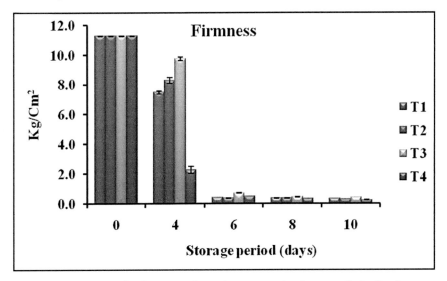

FIGURE 8.2 **(See color insert.)** Effect of yeasts on the firmness (kg/cm^2) of mango cv. Dashehari during storage. T_1 (baker's yeast), T_2 (industrial yeast), T_3 (*Saccharomyces cerevisiae*), and T_4 (control).

With the increase in storage period, the firmness of the fruits decreased. The firmness was maximum in T_3 (0.45 kg/cm^2), followed by T_1 and T_2, while minimum firmness was noted in T_4 (0.32 kg/cm^2) on the 10th day of storage. The decrease in firmness may be due to rapid loss of cell structure because of evapo-transpiration and water loss, causing ripening of fruits. As the ripening stages advanced, fruit softening increased as reported by Razzaq et al. (2013) in "Samar Bahisht Chaunsa" mangoes. Breakdown of cell wall polymers occurs during fruit ripening, resulting in softening of mango fruits (Zaharah and Singh, 2011). Firmness is affected by cell wall modification, and polygalacturonase and pectin methyl esterase degradation activity increased during ripening (Gonzalez-Aguilar et al., 2008). Similar reports on firmness of "Ataulfo" mango fruits (Robies-Sanchez et al., 2009b).

The growth and development of microbes increased with an increase in storage period (Table 8.1).

The spoilage of fruits increased with an increase in storage period and varied significantly ($p \leq 0.5$) among the treatments (Figure 8.3).

Maximum spoilage (15.33%) was observed in T_4, followed by T_2 (12.03%), T_1 (11.33%), and minimum in T_3 (7.99%) on the 10th day of stor-

TABLE 8.1   Effects of Yeasts on Microbial Growth during Storage and Ripening of Fruits

Treatments	Bacterial, cfu/gm	Yeast, cfu/gm	Fungus
0th day			
Baker Yeast	45	8.1×10^6	0
Industrial Yeast	33	8.23×10^6	0
S. cerevisiae	20	9.16×10^6	0
Control	1.10×10^2	0	0
4th day			
Baker Yeast	2.85×10^2	5.31×10^4	0
Industrial Yeast	36	9.52×10^4	0
S. cerevisiae	1.21×10^2	3.10×10^5	0
Control	8.51×10^2	12	0
6th day			
Baker Yeast	5.62×10^5	6.25×10^3	0
Industrial Yeast	6.23×10^2	5.51×10^4	0
S. cerevisiae	2.13×10^4	7.36×10^3	0
Control	7.71×10^7	26	1
8th day			
Baker Yeast	3.46×10^6	5.37×10^2	0
Industrial Yeast	2.41×10^3	4.35×10^3	0
S. cerevisiae	6.35×10^5	2.25×10^2	0
Control	8.31×10^8	65	4
10th day			
Baker Yeast	9.74×10^5	8.35×10^2	5 – Ripening
Industrial Yeast	4.93×10^5	5.68×10^2	0 – Ripening
S. cerevisiae	2.57×10^6	3.63×10^2	1 – Ripening
Control	7.43×10^9	0	10 & spoilt

age. In control fruits, spoilage was noticed from the 6th day of storage, while in T_1 and T_2, the spoilage of fruits was noticed on the 8th day of storage. The spoilage was mainly due to anthracnose and stem end rot. S. cerevisiae has been recognized as GRAS (generally recognized as safe) in Bio-safety level 1 in Europe (Murphy and Kavaragh, 1999). The main mode of action of this yeast is competition for space and nutrients. Competition between yeast and fungi was previously reported for grapes (Mc Laughlin et al., 1992),

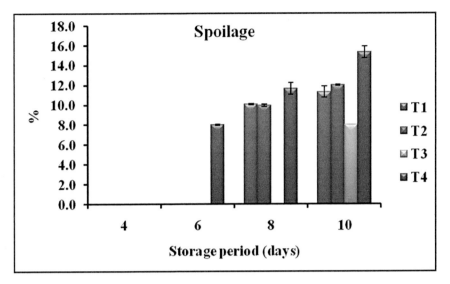

FIGURE 8.3 **(See color insert.)** Effect of yeasts on the spoilage (%) of mango cv. Dashehari during storage. T_1 (baker's yeast), T_2 (industrial yeast), T_3 (*Saccharomyces cerevisiae*), and T_4 (control).

apple (Filonow et al., 1996; Ippolito et al., 2000) and tomato (Kalogiannis et al., 2006). The mechanism of action *S. cerevisiae* is by production of more pseudohyphae and invasive growth than the food and industrial strains (de Lianos et al., 2006). Mode of action of antagonistic yeasts are utilized for the management of postharvest fungal diseases of fruits as reviewed by Liu et al. (2013). Siderophore production was the antifungal pattern observed for the consistent control of gray and sour rot with *Saccharomyces* as reported by Nally et al. (2015). The production of volatile organic compounds (VOCs) with in vitro and in vivo inhibitory effect on pathogen growth was observed for *S. cerevisiae* as studied by Parafita et al. (2015).

The TSS and TA of the fruits significantly varied ($p \leq 0.5$) among the treatments (Figures 8.4 and 8.5).

With the increase in storage period, the TSS increased, while the TA decreased among all the treatments. The TA was highest in T_3 (0.10%), while it was at par in T_1, T_2, and T_4. TSS was highest in T_1 (24.07 °B), while it was at par in T_2 and T_4 at the end of the storage. TSS was lowest in T_3 (21.58 °B) on the 10th day of storage. Increase in sugar content and decrease in TA were observed during ripening of mango cv. "Cogshall"

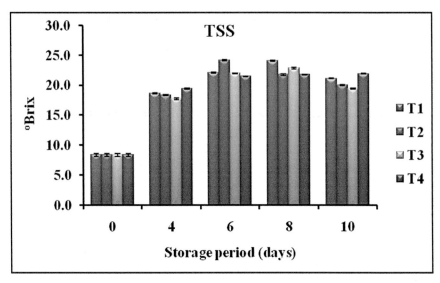

FIGURE 8.4 **(See color insert.)** Effect of yeasts on the TSS (°Brix) of mango cv. Dashehari during storage. T_1 (baker's yeast), T_2 (industrial yeast), T_3 (*Saccharomyces cerevisiae*), and T_4 (control).

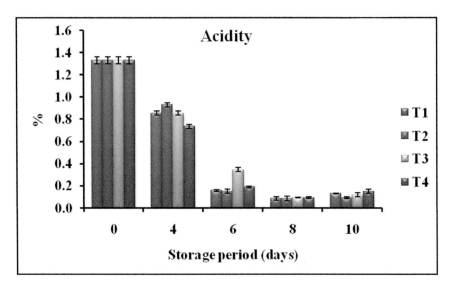

FIGURE 8.5 **(See color insert.)** Effect of yeasts on the titratable acidity (%) of mango cv. Dashehari during storage. T_1 (baker's yeast), T_2 (industrial yeast), T_3 (*Saccharomyces cerevisiae*), and T_4 (control).

(Jacques et al., 2009). During ripening, the increase in TSS is attributed to free sugar accumulation from the hydrolysis of starch (White, 2002).

There was a significant difference ($p \leq 0.5$) in the total carotenoid content among the treatments and the storage period (Figure 8.6).

The total carotenoid content was maximum in T_2 (7.22 mg/100 g), followed by T_3 (6.88 mg/100 g) and T_1 (5.43 mg/100 g) and minimum in T_4 (4.88 mg/100 g) on the 10th day of storage. Increase in carotenoid biosynthesis in mango varieties is associated with the increase in respiration, which is ethylene dependent (Saltveit, 1999).

Antioxidants varied significantly ($p \leq 0.5$) among the treatments and the period of storage. The antioxidants estimated by the FRAP method (Figure 8.7) did not follow a particular trend as indicated by the carotenoid content of the fruits. It was maximum in T_2 (7197.46 μmolar trolox equivalent/g), followed by T_3 (5011.75 μmolar trolox equivalent/g) on the 8th day of storage; thereafter, it decreased. Antioxidants estimated by DPPH (Figure 8.8) and expressed as percent inhibition was highest in T_2 (77.44%), followed by T_1 (72.62%) on the 8th day of storage; thereafter, it decreased. Percent inhibition was highest in T_3 (76.48 %) on the 10th day

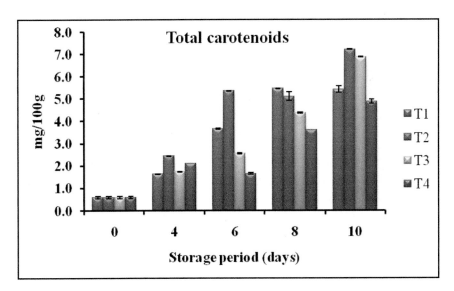

FIGURE 8.6 (See color insert.) Effect of yeasts on the total carotenoids (mg/100 g) of mango cv Dashehari during storage. T_1 (baker's yeast), T_2 (industrial yeast), T_3 (*Saccharomyces cerevisiae*), and T_4 (control).

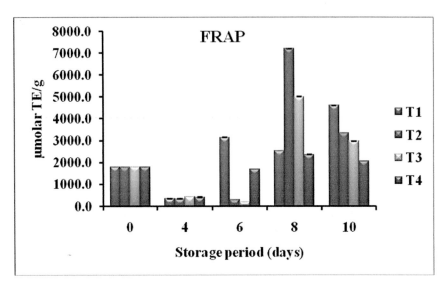

FIGURE 8.7 (See color insert.) Effect of yeasts on the antioxidant FRAP (μmolar TE/g) of mango cv. Dashehari during storage. T_1 (baker's yeast), T_2 (industrial yeast), T_3 (*Saccharomyces cerevisiae*), and T_4 (control).

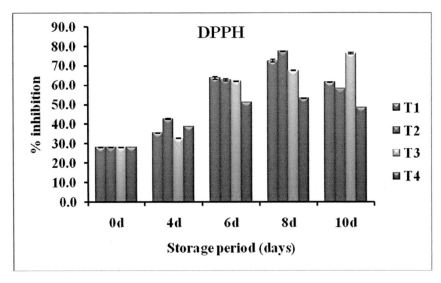

FIGURE 8.8 (See color insert.) Effect of yeasts on the antioxidant DPPH (% inhibition) of mango cv. Dashehari during storage. T_1 (baker's yeast), T_2 (industrial yeast), T_3 (*Saccharomyces cerevisiae*), and T_4 (control).

of storage. The antioxidant potential of mango varieties could be due to complicated bioactive compounds and its synergistic actions contributing to the antioxidant activity in mango (Liu et al., 2013). Total antioxidant scavenging activities linearly increased up to 7 days and then decreased in Chaunsa mango (Razzaq et al., 2013).

8.4   CONCLUSION

The CPLW was highest in untreated control and least in *S. cerevisiae*-treated fruits on the 10th day of storage. The TSS increased and TA and firmness of the fruits decreased during the course of fruit storage. The total carotenoid contents were highest in yeast-treated fruits and minimum in control fruits on the 10th day of storage. The antioxidant content estimated was maximum in yeast-treated fruits on the 10th day of storage. The spoilage of the fruit was minimum in *S. cerevisiae*-treated fruits compared to control fruits on the 10th day of storage. In vitro studies on the efficacy of *S. cerevisiae* for controlling *C. gloeosporioides* (pathogen for anthracnose) confirmed the efficacy of the bio-agent. Thus, *S. cerevisiae* may be used as a potential bio-agent for improving the postharvest quality of mango.

KEYWORDS

- **antioxidants**
- **carotenoids**
- **post-harvest**
- **quality**
- ***Saccharomyces cerevisiae***
- **storability**

REFERENCES

Bautista-Rosales, P. U., Calderon-Santoyo, Servin-Villegas, R., Ochoa-Alvarez, N., Angelica, Vazquez-Juarez, E., Ricardo, & Ragazzo-Sanchez, J. A., (2014). Biocontrol action mechanisms of *Cryptococcus laurentii* on *Colletotrichum gloeosporioides* of mango. *Crop Protection, 65,* 194–201.

Brand-Williams, W., Cuvelier, M. E., & Berset, C., (1995). Use of free radical method to evaluate antioxidant activity. *Lebensmittel Wissenscaft and Technologies, 28,* 25–30.

Cao, B., Li, H., Tian, S., & Qin, G., (2012). Boron improves the biocontrol activity of *Cryptococcus laurentii* against *Penicillium expansum* in jujube fruit. *Postharvest Biol. Technol., 68,* 16–21.

Collins, C. H., & Lyne, P. M., (1984). *Microbiological Methods*.5ᵗʰ edn. Butterworth and Co. Publisher Ltd., London, 331–345.

De Lianos, R., Fernadez-Espinar, M. T., & Querol, A., (2006). A comparison of clinical and food *Saccharomyces cerevisiae* isolates on the basis of potential virulence factors. *Antonie Leeuwenhoek, 90,* 221–231.

Filonow, A. B., Vishiac, H. S., Anderson, J. A., & Janisiewiez, W. J., (1996). Biological control of *Botrytis cinerea* in apple by yeasts from various habitats and their putative mechanisms of antagonism. *Boil. Control, 7,* 212–220.

Gonzalez-Aguilar, G. A., Celis, J., Stotelo-Mundo, R. R., De La Rosa, Rodrigo-Garcia, J., & Alvarez-Parrilla, E., (2008). Physiological and biochemical changes of different fresh cut mango cultivars stored at 5°C. Int. *J., Food Sci. Technol., 43,* 91–101.

Guo, J., Fang, W., Lu, H., Zhu, R., Lu, L., Zheng, X., & Yu, T., (2014). Inhibition of green mold disease in mandarins by preventive applications of methyl jasmonates and antagonistic yeast Cryptococcus laurentii. *Postharvest Biol. Technol., 88,* 72–78.

Ippolito, A. E. L., Ghaouth, A., Wilson, C. L., & Wisniewski, M., (2000). Control of postharvest decay of apple fruit by *Aureobasidium pullulans* and induction of defence responses. *Postharvest Biol. Technol., 19,* 265–272.

Jacques, J., Caro, Y., & Leehandel, M., (2009). Comparison of postharvest changes in mango (cv. Cogshall) using a ripening class index (Rci) for different carbon supplies and harvest dates. *Postharvest Biology and Technology, 54,* 25–31.

Janisiewicz, W. J., & Korsten, L., (2002). Biological control of post harvest diseases of fruits. *Annu. Rev. Phytopathol., 40,* 411–441.

Kalogiannis, S., Tjamos, S. E., Stergiou, A., Antoniou, P. P., Ziogas, B. N., & Tjamos, E. C., (2006). Selection and evaluation of phyllosphere yeasts as biocontrol agent against grey mould of tomato. *Eur. J., Plant Pathol., 116,* 69–76.

Leonadi, C., Bailie, A., & Guichard, S., (1999). Effects of fruit characteristic and climactic conditions on tomato transpiration in greenhouse. *J., Hortic. Sci. Biotech., 74,* 748–756.

Lima, G., Arru, S., De Curtes, F., & Arras, G., (1999). Influence of antagonist host fruit and pathogen on the biological control of postharvest fungal diseases by yeast. *J., Int. Microbiol. Biotechnol., 23,* 223–229.

Liu Feng-Xia, Shu Fang Fu, Xiu-Fang Bi, Fang Chen, Xiao-Jun Liao, Xiao-Song Hu, & Ji-Hong Wu, (2013). Physico-chemical and antioxidant properties of four mango (*Mangifera indica* L.) cultivars in China. *Food Chemistry, 138,* 396–405.

Mc Laughlin, R. J., Wilson, C. L., Derby, S., Ben Arie, R., & Chalutz, E., (1992). Biological control of postharvest diseases of grapes, peach and apple with yeast *Kloeckera apiculata* and *Candida guilliermondii*. *Plant Dis., 76*, 470–473.

McDonald, R. E., McCollum, T. G., & Baldwin, E. A., (1998). Heat treatment of mature-green tomatoes: differential effects of ethylene and partial ripening. *J., Amer. Soc. Hort. Sci., 123*, 457–462.

Murphy, A., & Kavanagh, K., (1999). Emergence of *Saccharomyces cerevisiae* as a human pathogen. Implication for biotechnology. *Enzyme Microb. Technol., 25*, 551–557.

Nally, M. C., Pesce, V. M., Maturano, Y. P., Assaf Rodriguez, L. A., Figueroa de Castellanos, L. I., & Vazquez, F., (2015). Antifungal mode of action of *Saccharomyces* and other biocontrol yeasts against fungi isolated from sour and grey rots. *International Journal of Food Microbiology, 204*, 91–100.

Nally, M. C., Pesce, V. M., Maturano, Y. P., Munoz, C. J., Combina, M., Toro, M. E., Castellanos de Figueroa, L. I., & Vazquez, F., (2012). Biocontrol of *Botrytis cinerea* in table grapes by non-pathogenic indigenous *Saccharomyces cerevisiae* yeasts isolated from viticultural environments in Argentina. *Postharvest Biol. Technol., 64*, 40–48.

Parafita, L., Vitale, A., Restuccia, C., & Cirvilleri, G., (2015). Biocontrol ability and action mechanism of food–isolated yeast strains against *Botrytis cinerea* causing postharvest bunch rot of table grape. *Food Microbiology, 47*, 85–92.

Razzaq, K., Ahmad, S. K., Aman, U. M., & Muhammad, S., (2013). Ripening period influences fruit softening and antioxidative system of 'Samar Bahisht Chaunsa' mango. *Scientia Horticulturae, 160*, 108–114.

Robies-Sanchez, R. M., Islas-Osuna, M. A., Astiazaran-Garria, H., Vazquez-Ortiz, F. A., Martin-Belloso, O., Gorinstein, S., & Gonzalez-Augilar, G. A., (2009b). Quality index consumer acceptability, bioactive compounds and antioxidant activity of fresh cut 'Ataulfo' mangoes (*Mangifera indica* L) as affected by low temperature storage. *J., Food Sci., 74*, 126–134.

Saltveit, M. E., (1999). Effect of ethylene on quality of fresh fruits and vegetables. *Postharv. Biol. Technol., 15*, 279–292.

Scherm, B., Ortuzzu, A., Budroni, M., Arras, C., & Migheli, Q., (2003). Bio-control activity of antagonistic yeasts against *Penecillium expansum* on apple. *J., Plant Pathol., 85*, 1–9.

Suzzi, G., Romano, P., Ponti, I., & Montuschi, C., (1995). Natural wine yeasts as biocontrol agents. *J., Appl. Bacteriol., 78*, 304–308.

White, P. J., (2002). Recent advances in fruit development and ripening: an overview. *J., Exp. Bot., 53*, 1995.

Yu, T., Zhang, H., Lu, X., & Zheng, X., (2008). Biocontrol of *Botrytis cinerea* in apple fruit by *Cryptococcus laurentii* and indole-3-acetic acid. *Biol. Control, 46*, 171–177.

Zaharah, S. S., & Singh, Z., (2011). Mode of action of nitric oxide in inhibiting ethylene biosynthesis and fruit softening during ripening and cold storage of 'Kensington Pride' mango. *Postharvest Biol. Technol., 62*, 258–266.

CHAPTER 9

POSTHARVEST QUALITY EVALUATION OF WINTER ANNUALS

SELLAM PERINBAN,[1] BABITA SINGH,[1] JAYOTI MAJUMDER,[2] and PUJA RAI[1]

[1]*Directorate of Floricultural Research, Indian Agricultural Research Institute, Pusa Campus, New Delhi–110012, India, E-mail: chella.perinban@gmail.com*

[2]*Bidhan Chandra Krishi Viswavidyalaya (BCKV), Kalyani, West Bengal, India*

CONTENTS

ABSTRACT

Most of the winter annual flowers have very attractive flowers, and these winter annuals are generally grown as land cover or border plants in gardens. Due to their unique flower qualities, striking colors, and seasonality, many of these flowers are sold in the retail markets as cut flowers. Hence, this work was carried out at the Directorate of Flori-cultural Research, IARI, New Delhi, during 2013–2014 to evaluate the postharvest keeping quality of winter annuals. Five winter annual flow-ers, namely antirrhinum, dimorphotheca, lupin, larkspur, and Sweet William, were evaluated for vase life under five different preserva-tives like sucrose, 8-hydroxyquinoline citrate (8-HQC), aminooxy ace-tic acid (AOA), benzyl adenine (BA), and aluminum sulfate (Al_2SO_4) in different combinations. From the results, it was observed that in all flowers, the treatments with plant bioregulators (AOA & BA) and aluminum sulfate significantly improved the vase life of flowers than other treatments. Vase life of flowers treated with BA was maximum in antirrhinum (9.3 days), dimorphotheca (8 days), and larkspur (8 days); lupin treatment with 8-HQC improved the vase life up to 6 days, and in Sweet William, the vase life of 9.67 days was observed after treat-ment with AOA. Maximum stem elongation was observed in lupin (19 cm) after treatment with Al_2SO_4, whereas maximum flower diameter of dimorphotheca was recorded in treatment with AOA. Other quality parameters such as flower weight, percentage flower opening, and vase solution uptake rate (VSUR) were significantly improved in treatments with AOA, BA, and Al_2SO_4. Based on these results, it was evident that among the annual flowers evaluated, antirrhinum and Sweet William were found to be most suitable for commercial use as cut flower on the basis of their vase life.

9.1   INTRODUCTION

Annual flowers in general are a group of herbaceous plants that grow from seeds and complete their life cycle within 1 year or one season. In particular, winter annuals provide a beautiful display of colors in the gar-den during winter season in India. They enhance the decorative value of a garden within a short span of time. But most of the annual flowers are generally grown only for garden display purpose in various ways. These flowers can be used as specialty cut flowers for their unique flowers and distinct and attractive colors. Further, adding new specialty cut flowers will improve the floriculture trade.

Vase life is an important criterion that determines the suitability of the flower as a specialty cut flower. Earlier studies show that adding chemicals to the vase solution will increase the vase life of cut flowers. Any vase solution should contain an energy source and an antimicrobial component. Sugar is the most commonly used energy source in vase solutions. After harvest, the availability of sugar is limited in flowers. Adding sugars mostly in the form of sucrose will delay the senescence and improve the vase life of the flowers. But adding sucrose in the vase solution favors the growth of microorganisms, which block the xylem vessels and reduce the water uptake, thereby causing stem bending. Hence, biocides like 8-hydroxyquinoline citrate (8-HQC) (Ali and Hassan, 2014) and aluminum sulfate (Al_2SO_4) (Jowkar et al., 2012) are very much important to reduce the microbial growth in the vase solution. Apart from these two basic elements, ethylene produced during senescence of flowers accelerates the senescence process that results in petal wilting, permeability of petal cells, and degradation of membrane lipids. The senescence effects can be reduced by inhibitors of ethylene biosynthesis like aminooxy acetic acid (AOA) (Chaturaphat et al., 2003; Zuliana et al., 2008) and benzyl adenine (BA) (Singh et al., 2008; Danaee et al., 2011).

This work was carried out with an objective to evaluate the postharvest vase life of winter annual flowers to find their suitability as cut flowers. For this study, five winter annuals, namely antirrhinum, dimorphotheca, larkspur, lupin, and Sweet William, were selected as these flowers have sturdy stem of more than 30 cm in length.

9.2 MATERIALS AND METHODS

The selected winter annuals were cultivated in the research field of the Directorate of Floricultural Research, IARI, New Delhi, during 2013–2014. The details of harvest stage for each flower are given in Table 9.1. After harvest, the flower stems were trimmed to 40 cm and dipped in water till they were placed inside the respective treatments.

Flower stems were held in centrifuge tubes filled with 50 mL distilled water (control) or other treatment vase solutions. All the chemicals, viz., sucrose, 8-HQC, BA, and AOA in distilled water were used

TABLE 9.1　Details of Harvest Stage for Each Flower

Winter annual	Stage of harvest	Defoliation
Antirrhinum	3–4 flower open in the spike	Defoliated
Dimorphotheca	9–10 mm	Defoliated
Larkspur	3–4 flower open in the spike	Not defoliated
Lupin	7–10 flowers open in a spike	Defoliated
Sweet william	7–10 flowers open in a spike	Non defoliated in the top near the flower head.

as vase solution in different combinations [T_0: control (distilled water), T_1: sucrose 2%, T_2: sucrose (2%) + 8-HQC (200 ppm), T_3: sucrose 2% + AOA (0.5 mM), T_4: sucrose 2% + BA (0.22 mM), T_5: sucrose 2% + Al_2SO_4 (200 ppm)]. The tubes were covered with a cotton plug and an aluminum foil to avoid evaporative losses and kept in a room with natural light (12 h with 1600 lux light intensity). The average room temperature and humidity during the study were recorded as 25 ± 2°C and 80% ± 5%, respectively. Each experiment was replicated three times with three spikes per replication.

Weight of each spike was measured at 2 days interval. The relative fresh weight of spikes was calculated as:

$$Relative\ fresh\ weight\ (RFW)\ (\%) = \frac{weight\ of\ flower\ spike\ at\ day\ t}{initial\ flower\ spike\ weight} \times 100$$

where t = 0, 2, 4, 6, 8, 10 (He et al., 2006).

Difference in the stem/spike height was recorded at 2 days interval in antirrhinum, larkspur, and lupin. Change in the volume of solution and weight of tubes without spikes was recorded at 2 days interval. The following formulae were used to find the water relations and percent flower opening:

Vase solution uptake rate (VSUR) (mL g^{-1} initial fresh weight (IFW) day^{-1}) is given by,

$$VSUR = \frac{(S_{t-2}) - (S_t)}{(IFW \times 2)}$$

where t = 0, 2, 4, 6, 8, 10 (He et al., 2006).

$$Percentage\ flower\ opening = \frac{\left(No.\ of\ flowers\ at\ day\ t - Initial\ number\ of\ flowers\right)}{Initial\ number\ of\ buds}$$

Percentage flower opening was recorded in antirrhinum, larkspur, lupin, and Sweet William flowers. The flower diameter of dimorphotheca was recorded at 2 days interval using a Vernier caliper.

Vase life of the flowers was characterized based on petal wilting, flower drop, bud drop, stem bending and stem rotting. The average vase life of flowers was assessed as terminated when 50% of the flowers were senesced /dropped. The experiment was carried out in completely randomized block design (CRBD) with five treatments and three replications.

9.3   RESULTS AND DISCUSSION

9.3.1   CHANGE IN RELATIVE FRESH WEIGHT OF FLOWERS

From the results, it is evident that maximum increase in fresh weight was observed in lupin (150.84%), and with respect to the treatments, maximum increase in weight during vase life was observed in treatment with BA (T_4) (Table 9.2).

In treatment with sucrose 2% (T_2), there was an increase in RFW of all flowers during initial 2 days of study. However, after that, the rate of increase in fresh weight was reduced in all flowers, which could be due to the microbial growth in the vase solution that might have caused physical plugging in the stems and blockage of xylem vessels (Danaee et al., 2011). In lupin, the maximum RFW of 208.2% was recorded in treatment with 8-HQC (T_2) after 6 days of vase life. In treatments with HQC (T_2) and Al_2SO_4 (T_5), both the agents acted as bactericides that reduced the growth of stem plugging microorganisms and increased the water uptake for long time than in T_2 (Dole et al., 2009; Jowkar et al., 2012). In antirrhinum, maximum RFW (160.6%) was recorded on day 6 in treatment with BA (T_4), whereas in Sweet William, it was recorded as 146.1% after 8 days for the same treatment. Sakine et al. (2011) reported about delay in flower opening of roses due to treatment with BA as cytokinins reported for its negative effect on flower senescence.

TABLE 9.2 Changes in Relative Fresh Weight and Vase Solution Uptake Rate of Winter Annuals during Vase Life

Treatments	Relative Fresh Weight (%)						Mean
	Antirrhinum	Dimorphotheca	Larkspur	Lupin	Sweet willam		
T0 (Control)	79.30	95.51	89.41	96.47	99.58		92.05
T1 (Sucrose 2%)	118.78	94.45	141.59	118.65	111.77		117.05
T2 (Sucrose 2%+ 8-HQC 200 ppm)	78.78	150.58	141.92	208.22	125.00		140.90
T3 (Sucrose 2%+ AOA 0.5mM)	123.46	160.68	123.12	133.09	133.41		134.75
T4 (Sucrose 2%+ BA 0.22 mM)	157.05	158.95	141.44	168.82	142.82		153.81
T5 (Sucrose 2%+ Al$_2$SO$_4$ 200 ppm)	106.83	129.56	135.47	179.76	132.34		136.79
Mean	110.70	131.62	128.82	150.84	124.15		
Vase solution Uptake rate (ml/g/flower)							
T0 (Control)	0.10	0.28	0.82	0.85	0.41		0.49
T1 (Sucrose 2%)	0.25	0.23	1.06	1.09	0.36		0.59
T2 (Sucrose 2%+ 8-HQC 200 ppm)	0.21	0.33	1.04	1.12	0.32		0.60
T3 (Sucrose 2%+ AOA 0.5mM)	0.37	0.52	0.99	1.04	0.62		0.71
T4 (Sucrose 2%+ BA 0.22 mM)	0.59	0.39	1.01	1.06	0.81		0.77
T5 (Sucrose 2%+ Al$_2$SO$_4$ 200 ppm)	0.31	0.34	1.08	1.09	0.57		0.68
Mean	0.30	0.35	1.00	1.04	0.51		

9.3.2 CHANGES IN VASE SOLUTION UPTAKE RATE (VSUR)

Maximum VSUR was observed in lupin (1.04 mL/g/day) and larkspur (1 mL/g/day), and the minimum uptake was observed in antirrhinum (0.30 mL/g/day) and dimorphotheca (0.35 mL/g/day). With respect to treatment, maximum solution uptake (0.77 mL/g/day) was observed in treatment with BA (T_4).

9.3.3 CHANGES IN STEM ELONGATION, PERCENTAGE FLOWER OPENING, AND FLOWER DIAMETER

Maximum stem elongation was recorded in lupin in treatment with Al_2SO_4. Bending was not observed in lupin during vase life in all treatments except control. In antirrhinum and larkspur, maximum stem elongation was observed in treatment with BA (T_4) (Figure 9.1), and the finding is in accordance with Asil and Karimi (2010) who suggested that BA at a lower concentration is required to delay flower senescence. Maximum percentage flower opening in antirrhinum, larkspur and Sweet William was observed in treatment with BA (T_4), followed by treatment with AOA (T_3). This shows the ethylene sensitivity of flowers. The treatments with AOA and BA increased the RFW of flower stems due to delayed senescence, thus increasing the solution uptake and reduced tissue transpiration and respiration (Keramat et al., 2012). In lupin, the maximum percentage flower opening was observed in treatment with HQC (T_2) than in AOA and BA. This could be due the antimicrobial activity of 8-HQC that prevented stem blocking and increased water uptake, thereby reducing flower drop (Danaee et al., 2011). In dimorphotheca, maximum flower diameter was observed in treatments with AOA (T_3) (90.17 mm) followed by BA treatment (T_4) (87.73 mm) after 8 days of vase life. This clearly shows the ethylene sensitivity of the flower and the ethylene inhibition properties of AOA and BA, which improved the flower quality till 8 days in vase life when compared to the other treatments (Figure 9.2 a,b).

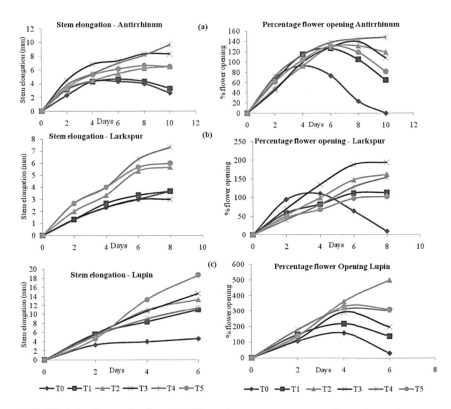

FIGURE 9.1 **(See color insert.)** Effect of treatments on stem elongation and percent flower opening of (a) antirrhinum, (b) dimorphotheca, (c) larkspur, (d) lupin, and (e) Sweet William.

9.3.4 CHANGES IN VASE LIFE

In antirrhinum, the maximum vase life of 9.33 days was recorded in treatment with BA (T_4), followed by 8.33 days in AOA treatment (T_3) (Figures 9.3 and 9.4).

In dimorphotheca, the maximum vase life of 8 days was recorded in treatment with BA (T_4), followed by AOA (T_3). In larkspur, the maximum vase life was observed in treatment with BA, whereas in lupin, it was observed that the treatment with 8-HQC (T_2) improved the vase life to 6 days as compared to other treatments. For Sweet William, the maximum vase life was recorded in treatment with AOA (T_3) (9.67 days), followed

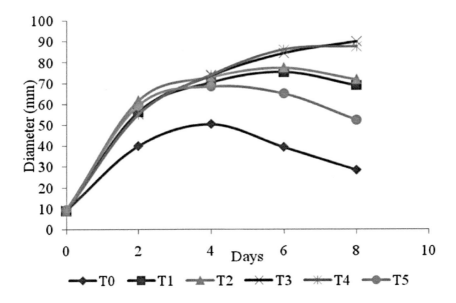

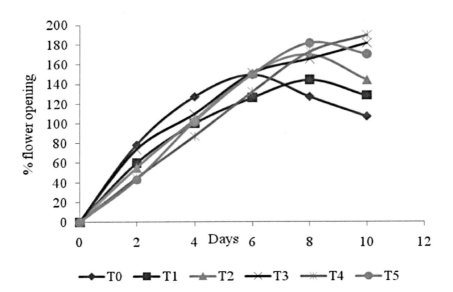

FIGURE 9.2 **(See color insert.)** (a) Effect of treatments on flower diameter (mm) of dimorphotheca and (b) effect of treatments on percent flower opening of Sweet William.

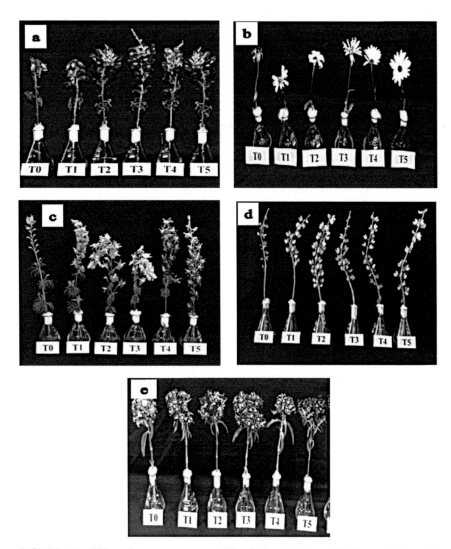

FIGURE 9.3 Effect of treatments on vase life of (a) antirrhinum, (b) dimorphotheca, (c) larkspur, (d) lupin and (e) Sweet William.

by BA (9.33 days). From the results, it is evident that the plant bioregulators AOA and BA significantly improved the vase life of antirrhinum, dimorphotheca, Sweet William, and antirrhinum flowers. A significant increase in vase life of flowers and the effect of AOA and BA on flower

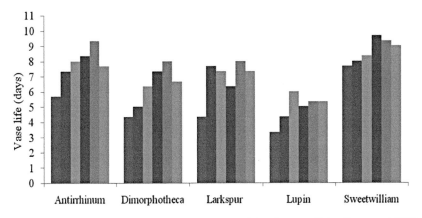

FIGURE 9.4 **(See color insert.)** Effect of treatments on vase life of winter annuals.

senescence were reported by Zuliana et al. (2008) and Sakine et al. (2011). Further, Sweet William and antirrhinum can be used as cut flower as they have maximum cumulative vase life of 8.67 days and 7.72 days, respectively, with respect to all treatments.

9.4 CONCLUSION

Winter annual flowers are not yet explored as cut flowers. The unique flower features and colors make these flowers very attractive. Among the five winter annual flowers evaluated in this study, antirrhinum and Sweet William had minimum vase life of ~6 days and ~8 days in control treatment itself. Adding sucrose to the vase solution improved the vase life of larkspur from 4.33 days to 7.67 days; dimorphotheca vase life also improved to 8 days by adding BA + sucrose in the vase solution. Based on the vase life of flowers, antirrhinum and Sweet William can be used as commercial cut flowers.

KEYWORDS

- **antirrhinum**
- **dimorphotheca**
- **larkspur**
- **lupin**
- **plant bioregulators**
- **Sweet William**
- **vase life**
- **winter annuals**

REFERENCES

Ali, E., & Hassan, F., (2014). Postharvest quality of *Strelitzia reginae* cut flowers in relation to 8-hydroxyquinoline sulphate and gibberellic acid treatments. *Sci. Agri., 5*(3), 97–102.

Asil, M. H., & Mahnaz, K., (2010). Efficiency of Benzyl adenine reduced ethylene production and extended vase life of cut *Eustoma* flowers. *Plant Omics Journal, 3*(6), 199–203.

Chaturaphat, R., Saichol, K., Wouter, G., & Van Doorn., (2003). Effect of aminooxyacetic acid and sugars on the vase life of Dendrobium flowers, *Postharvest Biology and Technology, 29*, 93–100.

Danaee, E., Ruhangiz, N., Sepideh, K., & Ali, R. L. M., (2013). Evaluation the effect salicylic acid and benzyl adenine on enzymatic activities and longevity of gerbera cut flowers, *Intl. Res. J., Appl. Basic. Sci., 7*(5), 304–308.

Fard, E. S., Khodayar, H., & Ahmad, K., (2010). Improving the keeping quality and vase life of cut alstroemeria flowers by pre and post-harvest salicylic acid treatments. *Not. Sci. Biol., 5*(3), 364–370.

He, S., Joyce, D. C., Irving, D. E., & Faragher, J. D., (2006). Stem end blockage in cut Grevillea 'Crimson Yul-lo' inflorescences. *Postharvest Biol. Technol., 41*, 78–84.

Jowkar, M. M., Mohsen, K. A. K., & Nader, H., (2012). Evaluation of aluminum sulfate as vase solution biocide on postharvest microbial and physiological properties of 'Cherry Brandy' rose, *Annals of Biological Research, 3*(2), 1132–1144.

Kazemi, M., Hadavi, E., & Hekmati, J., (2012). Effect of salicylic acid, malic acid, citric acid and sucrose on antioxidant activity, membrane stability and ACC-Oxidase activity in relation to vase life of carnation cut flowers. *Journal of Agricultural Technology, 8*(6), 2053–2063.

Sakine, F., Roohangiz, N., & Oru, D. V. I., (2011). Effects of post harvesting on biochemical changes in gladiolus cut flowers cultivars (white prosperity). *Middle East Journal of Scientific Research, 9*(5), 572–577.

Singh, A., Kumar, J., & Kumar, P., (2008). Effect of plant growth regulators and sucrose on post harvest physiology, membrane stability and vase life of cut spikes of Gladiolus. *J., Plant Growth Regulators, 55,* 221–229.

Sodi, A. M., & Antonio, F., (2005). Physiological changes during postharvest life of cut sunflowers. *Proc. VIIIth IS Postharvest Phys. Ornamentals, Acta. Hort., 669,* 219–224.

Zuliana, R., Boyce, A., Nair, H., & Chandran, S., (2008). Effects of aminooxyacetic acid and sugar on the longevity of pollinated *Dendrobium Pompadour. Asian Journal of Plant Sciences, 7*(7), 654–659.

CHAPTER 10

OPTIMIZING THE SHELF LIFE OF WHOLE AND FRESH CUT BREADFRUIT IN MAURITIUS

ROOP SOODHA MUNBODH

Food and Agricultural Research and Extension Institute (FAREI), Reduit, Mauritius, Tel.: 59799325, E-mail: soomunbodh@yahoo.com

CONTENTS

ABSTRACT

Breadfruit has a high potential for export as a fresh fruit, but the extreme perishability of this fruit hampers its export trade. High postharvest losses

(25–30%) are incurred during the export transit, whereby the fruits ripen on reaching export destination. Trials were carried out for the proper identification of optimum maturity stage for harvest of breadfruit and optimum storage conditions to extend shelf-life for export. Fruits were harvested from various sites, and fruit and peel physical characteristics were determined. For optimum shelf-life, the fruit should be pale green in color and firm in texture with latex stains on uniform protruding polygonal segments on the peel. The flesh of mature unripe breadfruit was observed to be ivory white in color and had a starchy taste when boiled. Mature ripe breadfruit had a yellowish green peel color, a cream flesh color with a soft texture, and was sweet when boiled. Handling practices were of utmost importance to minimize mechanical damage as bruised fruits will soften rapidly, causing reduction in shelf-life; thus, fruits were harvested early morning to avoid heat build-up by using either a fruit picker with mesh bag or a harvesting pole with a cutting scythe and a collecting bag to avoid fruit drop. The stems of the cut fruits of 3-4 cm length were pointed downward to allow latex flow away from fruit surface and thus prevent brown staining of the peel. Harvested fruits were placed in clean stackable plastic crates and transported in covered vehicles that provided adequate ventilation and protection (using a cover) to the harvested fruits. Storage trial showed that when whole fruit was washed in sterilized water (4 mL of javel/L of water), air-dried, cling filmed, and stored at 13°C with a relative humidity of 90–95%, breadfruit kept for 2 weeks, while noncling filmed fruits remained marketable for only 7 days. Cumulative % weight loss was only 2.23 % +/– S.E 1.33 and 0.93% +/– S.E 0.24 for Pamplemousses experimental station and Reduit crop research station, respectively, over 15 days of storage. At 13°C, cling filmed fruits retained their flavor and taste at 14 days of storage, while control fruits remained tasty for only 5–7 days. The color of fruit peel was maintained as bright pale green when cling filmed for 2 weeks as compared to noncling filmed fruits; the fruits assumed a dull green brownish tinge. Cling filmed fruits maintained their firmness at 8–9 kg/m². Postharvest disease incidence was observed after 2 weeks of storage and was caused by pathogens *Colletotrichum* and *Cladosporium* for cling filmed fruits at 5°C. Under ambient conditions, noncling filmed fruits ripened after 2 days, while cling filmed ones ripened after 4 days. Storage trial of minimally processed breadfruit showed that fruit cut into longitudinal slices of 1.5 cm thickness, soaked in sodium metabisulfite solution at 0.5 g/L for 5 minutes to prevent browning, air dried, and vacuum packed at 85% had a shelf-life of 12 days at 4 °C, wherein color, taste, firmness, and flavor were maintained. Under ambient conditions, the fresh cuts were observed to keep for only 2 days.

10.1 INTRODUCTION

Breadfruit is a climacteric fruit mostly grown in the tropics. The fruit is a starchy staple, and mature, unripe fruit is cooked and eaten in much the same way as tubers and root crops. Breadfruit has a high potential for export as a fresh fruit. During the last 5 years, export of fresh breadfruits from Mauritius to Europe has increased from 40 to196 tons, but the extreme perishability of this fruit hampers its export trade. High postharvest losses are incurred during the export transit. A large amount (25–30%) of exported fruits lose their firmness and become ripe and soft before reaching their retail outlet; as a result, there is an economic loss to the breadfruit exporters of Mauritius. Actually, in Mauritius, fresh breadfruit is currently being exported at an attractive remunerable price. The main goal is therefore to increase the amount of quality breadfruit for local and export markets and thus to provide growers and exporters with improved postharvest management practices to enable them to obtain quality breadfruits with longer marketable periods. There is therefore a need to devise strategies to minimize postharvest losses of breadfruit along the supply chain and to improve the actual packaging and storage conditions from the grower to the consumer for the local and export markets. Reduced postharvest losses along the export will definitely benefit the breadfruit growers and exporters economically. Improved innovative packaging and storage conditions that will lead to an increase in number of quality breadfruits from harvest to export for local and export markets and lengthening of shelf-life of the minimally processed breadfruit slices under different packing and storage conditions will also encourage supermarkets to display fresh-cut breadfruit chunks on the shelf, leading to local consumer acceptance of cut breadfruits. This will help them for the easy preparation of breadfruit curry and other dishes. Growers and exporters will also be able to tap new promising markets for export and local markets. There will be a larger area of land devoted to breadfruit plantation to satisfy the local demand, hence contributing to food security. A rise in number of breadfruit growers and exporters will also create more employment for the people in the business, including women entrepreneurs, and also provide income generation. It is expected to receive infrastructural government support as the breadfruit business expands. Extended areas of breadfruit plantation will also contribute to a greener Mauritius. Preliminary promising results will encourage the government to devote funds (food security fund of the Ministry of Agro Industry and Food Security) to develop the sector further for the benefit of vulnerable groups and for food security. Breadfruit therefore has an important role to play in food security, sustainable agriculture, and income generation and its profile need to be raised at the national level.

Hence, trials were carried out to (a) characterize of local accessions of breadfruit and to determine the optimum fruit maturity for storage and export of breadfruit from local germplasm in two agro-climatic zones of the island, (b) to evaluate the shelf-life of cling filmed and noncling filmed fresh breadfruit stored at 13°C, and (c) to evaluate the shelf-life of minimally processed cold stored breadfruits under different packaging conditions.

10.2   ACTIVITY 1: CHARACTERIZATION AND EVALUATION OF QUALITY ATTRIBUTES OF FRESH BREADFRUITS AT MATURE AND IMMATURE STAGES

Breadfruits were harvested early in the morning from a sub-humid agro-climatic zone at (a) Pamplemousses experimental station (ES) and from a humid agro-climatic site; (b) Reduit crop research station (CRS) at (i) mature green stage (young), characterized by light green skin and absence of external latex with the fruit outer segments closely packed and having a rough texture; and (ii) "fit" stage characterized by darker green skin with some browning, external dried latex, and fully flattened outer segments (Table 10.1). Fruits were harvested early morning to avoid heat build-up by using either a fruit picker with mesh bag or a harvesting pole with a cutting scythe and a collecting bag. Precautions were taken to avoid dropping harvested fruits to the ground as bruised fruits will soften rapidly, causing reduction in shelf-life. The stems of the cut fruits of 3–4 cm length

TABLE 10.1 Characterization of Breadfruits According to Site and Fruit Maturity

Site: Pamplemousses Experiment Station Sub-humid Agro climatic zone	Immature	Mature green	Site Reduit Crop research Station. Humid Agro climatic zone	Immature	Mature green
Fruit weight (g)	838	1287	Average Fruit size(g)	710	1360
Fruit color	Light olive green 144C	Pale green with brown streaks 152D	Fruit color	Olive green 144B	Pale green with brown streaks 144C
Type of fruit segment	Close knit segments, rough surface	Smooth surface with segments flattened out	Type of fruit segment	Close knit segments, rough surface	Smooth surface with segments flattened out
Fruit diameter (cm)	11.5	12.8	Fruit diameter	13.1	14.8
Fruit circumference (cm)	39.5	41	Fruit circumference (cm)	36.8	43.1
Fruit length (cm)	18.8	19.7	Fruit length	15	19.9
Fruit shape	round	oblong	Fruit shape	spherical	oblong
Firmness (kg/m²)	7.5	7	Firmness	9.0	8.0

TABLE 10.1 (Continued)

	Ivory white 151D	Cream, 155B
Pulp color	With slight browning on cutting	No browning on cutting
Core length	8.7 cm	8.66 cm
Flesh thickness (cm)	4.7	8.8
Sensory quality	Watery bland Taste, no aroma.	Full starchy taste

	Ivory white 152D	Cream,155D
Pulp color	With slight browning on cutting	No browning on cutting
Core length	7.15 cm	5.24
Flesh thickness (cm)	4.5	8.2
Sensory quality	Watery bland	starchy taste Taste, no aroma.

were pointed downward to allow latex flow away from fruit surface and thus prevent brown staining of skin. Harvested fruits were placed in clean stackable plastic crates to minimize handling damage. Fruits were loaded in a transport vehicle that provided adequate ventilation and protection (by using a cover) to the harvested fruits. Special care was taken that the fruits (free from pest and disease attack) were not bruised and subjected to mechanical injury by gently placing them in cushioned plastic crates and brought to the postharvest laboratory within 1 hour. The fruits were stored under ambient conditions in the postharvest laboratory and characterized for the physical and sensory quality attributes at both the immature and mature stages. A total of 25 fruits were characterized from each site representing the sub-humid and humid agro-climatic zones of the island.

Regarding harvest methods, Andrews and Mason (1992) reported that the "pick and catch" method is preferred in the eastern Caribbean islands from trees as much as 20 m tall.

The characterization process was carried out with the following instruments:

(a) Fruit weights were recorded using a "Sartorius" electronic balance.
(b) Fruit firmness was measured using an "Atago" penetrometer and the unit was kg/m^2.
(c) Firmness was measured as penetration force to depress 5 mm of fruit flesh.
(d) Fruit pulp and peel color were recorded according to the color chart of the Royal Horticultural Society.
(e) Flesh thickness and fruit diameter were recorded with a "caliper."
(f) Fruit circumference was measured using a standard tailor measuring tape.
(g) Sensory evaluation of boiled peeled chunks of mature and immature fruits was carried out using 10 untrained panelists on the station, and they were given random boiled samples and were asked to verbally describe the taste, texture, and overall appreciation of the sample. Ten tasting sessions were informally carried out for each maturity type. The descriptive terms used were:

(1) watery   –   starchy
(2) bland taste   –   strong breadfruit taste.

(3) No odor   —   strong aroma
(4) dislike   —   like

From the above results, it was observed that mature fruits are spherical to oblong shaped and have a higher mean weight, length, diameter, and circumference values than immature fruits. However, immature fruits are more firm and have higher mean flesh firmness values than mature fruits. The outer peel color of immature fruits is characterized by light green skin and absence of external latex with the fruit outer segments closely packed and having a rough texture, while for mature fruits, the "fit" stage was characterized by darker green skin with some browning, external dried latex, and fully flattened outer segments. As reported by Ragone (1977), fruits are globose to oblong, with yellowish green to yellow skin when mature with creamy white or pale yellow flesh.

Flesh color of immature fruits has been observed to be ivory white with a strong tendency to brown compared to mature fruits where the flesh color is light cream with little or no browning. On boiling the mature fruits, the taste was good with a starchy, creamy texture. The boiling time was 25–30 minutes. For immature fruits, the boiled slice had a watery texture and the taste was bland. Depending on the intended use of fruit as a vegetable for curry, breadfruits should be harvested when the fruit is pale green in color with latex stains on uniform protruding polygonal segments; free from defects, sunscald, cracks and insect bites; and of uniform shape and firm in texture. If used as desert and in cakes, breadfruits should be harvested when fruit texture is softer and the polygonal segments are smooth with deeper latex stains with a yellow green fruit color. The main internal indices of breadfruit maturity are flesh color and starch content. The flesh of mature unripe breadfruit is ivory white in color and starchy, whereas mature ripe breadfruits have a cream color with a soft texture. The flesh of mature green breadfruit when boiled for 20–25 minutes has a starchy taste, whereas the taste of ripened fruits is sweeter, less starchy due to its conversion of starch to sugar. As reported in the technical bulletin on postharvest care and market preparation information on breadfruit from the New Guyana Marketing Cooperation (May 2004), breadfruit should be harvested when green in color and firm in texture if it is to be used as a starchy vegetable. The fruit peel surface should have protruding segments on the surface that tends to be angular and ridged. If the fruit is to be used as dessert or for roasting or baking, it needs to be harvested when the skin color turns yellow green in color with brown latex stains and flattened expanded polygons of the fruit peel. The principal internal indices are flesh color; the flesh of mature but unripe fruit is white, starchy, and fibrous, while fully ripe fruit has a pale yellow flesh color and is soft and fragrant.

10.3   ACTIVITY 2: SHELF LIFE EVALUATION OF CLING FILMED AND NONCLING FILMED WHOLE BREADFRUIT STORED AT 13°C

10.3.1   MATERIALS AND METHODS

Breadfruits (mature and mature green) were harvested early in the morning from Pamplemousses ES and Reduit CRS at "fit" stage characterized by darker green skin with some browning, external dried latex and fully flattened outer segments as well as at mature green stage. Special care was taken that the fruits (free from pest and disease attack) were not bruised and subjected to mechanical injury. The fruits were then brought to the postharvest laboratory. Fifteen fruits were used per treatment. The fruits were washed under running tap water and then soaked in sterilized javel water (5 mL/L water) for 10 minutes and then drained; the dried fruits were then cling filmed and noncling filmed and stored under ambient and at 13°C. Three batches of trials were carried out for each site. The main objectives of this trial were to assess the shelf-life of whole fresh breadfruits from local germplasm from Reduit and Pamplemousses CRSs when cling filmed and non-cling filmed and stored at 13°C. The parameters assessed and procedures carried out are as mentioned in the following subsections.

10.3.1.1   Percentage Weight Loss

Percentage weight loss was calculated by subtracting the actual weight from the initial weight of fruits per pack and dividing it by the initial weight and multiplying by 100 using a "Sartorius" electronic balance to two decimal places.

10.3.1.2   Disease Incidence

Disease incidence was observed on fruits showing any fungal or bacterial growth. Identification of the disease and pathogen was carried out by the Plant Pathology Division of FAREI (Food and Agricultural Research and Extension Institute, Mauritius).

10.3.1.3   Color

Fruit pulp and peel color were recorded according to the color chart of the Royal Horticultural Society.

10.3.1.4   Firmness

Fruit firmness was measured using an "Atago" penetrometer, and the unit was kg/m^2. Firmness was measured as penetration force required to depress 2.5 mm into the fruit using a penetrometer.

10.3.1.5   Sensory Evaluation

An informal sensory appraisal of the stored fruits by 10 untrained panelists at the Wooton CRS was also carried out from randomly boiled samples.

10.3.3   RESULTS AND DISCUSSION

10.3.3.1   Shelf Life

Under ambient storage, breadfruits stored without any treatment were observed to have ripened within 2 days. Such fruits remained marketable for only 2–3 days, depending on the maturity stage of fruit. Those fruits kept under cling filmed remained marketable for 3–4 days

At 13°C, fruits packed under cling film remained marketable for 14–15 days, while those untreated or soaked in water remained marketable for only 7–10 days for both sites, Reduit CRS and Pamplemousses ES Table 10.2 summarizes the results obtained from the trials.

At 26–28°C, under ambient conditions, the storage life was only 2–3 days (Thompson et al., 1974c; Maharaj and Sankat, 1990a) after which skin browning is noted as well as fruit ripening and softening. As reported in the Journal of National Tropical Botanical Garden, Breadfruit Institute (2012), packaging the fruit in sealed polyethylene bags or plastic wrap can improve shelf life and fruit quality for 5–7 days. The optimum storage temperature is 12–16°C for unpacked fruits, and the shelf-life ranges from 7–10 days. If the fruit is packaged, it can be stored for up to 2 weeks. Packaging influenced positively the shelf-life of stored breadfruits. Storage life in sealed 150-gage polyethylene bags was significantly greater than for unwrapped fruits at both high and low temperatures (Thompson et al., 1974).

10.3.3.2 Weight Loss

Cling filmed fruits lost weight more rapidly under ambient storage than at 13°C for both sites as shown in the table above. Cling filmed fruits from Reduit CRS under ambient storage had an average cumulative weight loss % of 1.7% at 25°C at 3 days storage, whereas those fruits from Pamplemousses ES had a value of 3.1%; control noncling filmed fruits stored under ambient conditions had a cumulative % weight loss of 6–6.1% for

TABLE 10.2   The Results Obtained from the Trials

	25°C cling film	13°C cling film	25°C no cling film	13°C no cling film
Shelf life	3	14	2	7-10
Flesh color	Light cream	Ivory white	Light cream	cream
Taste	starchy	starchy	starchy	starchy
Days to ripening	3-4	14-15	2	7

both sites. Sankat and Maharaj (1993) noted that fruits stored under ambient conditions demonstrated rapid ripening through weight (2.64%) and volume losses (3.46%) (Figure 10.1). As reported by Worrell and Sean Carrington (1997), the use of polyethylene wraps and bags in conjunction with low temperature has proved to be beneficial in maintaining quality and shelf-life for at least 2 weeks, particularly with LDPE (low-density polyethylene) films and HDPE (high-density polyethylene) 40- and 60-micron films. Packaging and refrigeration therefore successfully slow the rate of water loss from the fruits by reducing the water holding capacity of the surrounding air, slowing rates of diffusion, and providing a physical barrier to air currents. Huang and Scott (1985) and Wong et al. (1991) reported that by packing fruit in plastic containers and overwrapping with a semi-permeable membrane, fruit desiccation was reduced with minimum condensation.

Under 13°C storage, the cling filmed fruits had a very low cumulative % weight loss of 2.23% ± SE 1.9 and 0.93% ± SE 1.6 for Pamplemousses ES and Reduit CRS, respectively, over 15 days of storage. Refrigerated storage extends the postharvest life of most produce (Kader, 1992), and

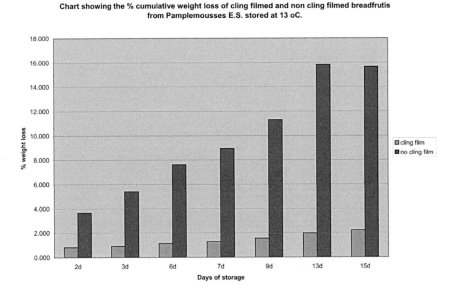

FIGURE 10.1 (See color insert.) Chart showing the % cumulative weight loss of cling filmed and noncling filmed breadfruits from Pamplemousses ES stored at 13°C.

breadfruit being no exception, 12–13°C was found to be optimal (Thompson et al., 1974; Maharaj and Sankat, 1990).

10.3.3.3 Sensory Appraisal

At 13°C, the fruits under cling film retained their flavor and taste at 14 days of storage, while control fruits remained tasty for only 5–7 days. Under ambient conditions, taste of boiled fruits (cling filmed and control) were maintained for only 2–3 days after which they had a slight fermented taste.

10.3.3.4 Color

Color of fruit peel was maintained as bright pale green when cling filmed for a longer period (2 weeks at 13°C) as compared to non-cling filmed fruits, when the fruits assumed a dull green color with brownish tinge after 7 days of storage. Wills et al. (1981) stated that the green color of the fruit is due to the presence of chlorophyll, a magnesium organic compound that degrades in storage principally by pH changes, oxidative systems, and chlorophyllases; along with chlorophyll disappearance, the synthesis of carotenoids is responsible for yellow pigments in fruits. Flesh color of mature green cling filmed fruits were maintained as an ivory cream over the storage time at 13°C. As the fruits ripened, the flesh color changed from ivory to cream.

10.3.3.5 Fruit Firmness

The cling filmed fruits maintained their firmness and ripened gradually over a longer storage period compared to noncling filmed fruits at 13°C. Fully ripened fruits had a lower firmness value (1.5 kg/m^2) compared to nonripened mature fruits at 8–9 kg/m^2. Loss in fruit firmness is caused by the enzyme polygalacturonase (a cell wall-degrading enzyme) that catalyzes upon ripening the hydrolysis of glucosidic linkages of pectic substances in the cell wall and results in weakened texture of fruits (Wills et al., 1981).

10.3.3.6 Postharvest Disease Incidence

The pathogens *Colletotrichum* and *Cladosporium* were identified by the Plant Pathology Division on diseased, cold-stored cling filmed fruits after 2 weeks of storage. As reported by Worell and Carrington (1997), all films delayed both the onset of softening and skin browning by at least a week compared to unwrapped controls. At 13°C, controls began to soften within a week, while wrapped fruit remained firm even after 21 days of storage. One drawback of film wrapping was microbial growth observed from the second week of ambient storage and internal flesh discoloration at 13°C storage. As reported in the technical bulletin on postharvest care and market preparation information on breadfruit from the New Guyana Marketing Cooperation (May 2004), physical damage is the major cause of postharvest decay of unripe fruits, which may be incurred by rough handling, by improper packaging, or during transport. Wounds such as punctures, cuts, abrasions, and cracks provide potential points of entry for decay organisms. Postharvest decay can be adequately controlled by a regular sanitation program involving application of preharvest fungicides and careful harvesting and handling practices and storing at 12.5°C.

No chilling injury was detected in stored fruits at 13°C, but noncling filmed fruits had the fruit peel turning a dull brown color after 7 days of storage. Worell and Carrington (1997) suggested that breadfruit skin browning may be a water loss problem, because water loss from epidermal cells causes cell damage which in turn brings soluble phenolics into contact with polyphenol oxidase, causing browning and ultimately cell death. Recommendations were made to carefully select the fruits at harvest and to minimize damage during handling.

10.4 STORAGE OF MINIMALLY PROCESSED COLD STORED BREAD FRUIT SLICES

10.4.1 INTRODUCTION

Minimal processing involves fruits and vegetables that have been trimmed, peeled, washed, cut, washed, dried, and packed. The purpose of minimal processing is to provide the consumer with a fresh product with an extended shelf life and also ensure food safety and maintain nutritional and sensory quality. In order to produce minimal processing products that are safe, high quality with a long shelf life, the processor has to apply Good Manufacturing Practice in his processing plant. The product flow should be in one direction from the receiving area to the finished packed product to avoid cross-contamination from raw vegetables and fruits to ready-to-eat minimally processed produce. Cleanliness and sanitation are two important factors in quality control and safety for minimally processed products. The main requirements for minimal processing of fruits and vegetables are (a) good quality raw material, (b) efficient and fast operations of cutting, slicing, and trimming, (c) adequate disinfection of the working area and equipment, (d) use of sharp disinfected equipment, (e) provision of clean potable water for washing of produce, (f) maximum draining of produce before packing, (g) proper labeling of packed produce, and (h) provision of optimum temperature (4–5°C) to maintain quality and shelf-life. Special treatments in minimal processing include the use of antibrowning agents such as citric acid and sodium metabisulfite. The quality of minimally processed products should be visually acceptable, have an appealing and fresh appearance, be of consistent quality throughout the package, and be free of defects. Improved shelf-life of fresh-cut breadfruit will encourage local working consumers to adopt breadfruit more often in their routine cooking recipes. The hotel industry will also use this product for exotic tropical cuisine dishes.

10.4.2   MATERIALS AND METHODS

The breadfruits of sound health, free of any disease and pest attack were harvested early in the morning at Reduit CRS and Pamplemousses ES. The fruits were placed in plastic crates and immediately transported to the postharvest laboratory at Wooton CRS. The fruits were washed under running tap water and then peeled to remove the outer skin layer and then placed on a sterilized chopping board and cut longitudinally into regular slices of 2 cm thickness. The cut peeled portions were then washed in sanitized chlorine water (2.0 mL/L of water) to reduce microbial load and rinsed again in tap water (washing after peeling and cutting removes microbes and tissue fluid, thus reducing microbial growth and enzymatic oxidation during storage). The microbiological quality of the washing water must be good, and its temperature should be low. After slicing, soaked in Sodium metabisulfite solution(0.2 g/L of water) for 5–7 minutes and then drained and dried under a fan. The dried segments were then packed in clip-on punnets (100 g per punnet) and also packed under vacuum (85%) in vacuum bags and stored at 5°C. The main objectives were therefore to assess the (a) shelf-life of packed minimally processed treated breadfruit under two packaging at 5°C and (b) to study the physiological disorders and pathological diseases observed in the minimally processed breadfruit during storage trial.

10.4.3   RESULTS AND DISCUSSION

10.4.3.1   Shelf Life

Under clip-on packaging and storage at 5°C, the cut slices packed in clip-on and dipped in sodium metabisulfite retained their color and freshness for 5–7 days irrespective of the site (Reduit CRS and Pamplemousses ES) after which they turned rancid and fermented with the presence of leachate. When vacuum packed at 85% vacuum, the cut slices remained marketable for 18 days irrespective of the site. Vacuum packaging was more advantageous for fresh-cut breadfruits than clip-on barquettes.

10.4.3.1.1   Color

The cut slices maintained ivory white color before turning to a dull yellow color after 5 days storage at 5°C for packed samples in clip-on and sodium metabisulfite. Ivory white color of the cut slices was maintained for 18 days under vacuum packing. Browning of fresh cut was prevented using the anti-browning dip sodium metabisulfite. As reported by Whitaker and Lee (1995), enzymatic browning requires four different components: oxygen, an enzyme, copper, and a substrate. In order to prevent browning, at least one component must be removed from the system. Polyphenol oxidase (PPO)-catalyzed browning of fruits and vegetables can be prevented by factors such as (a) heat or reaction inactivation of the enzyme, (b) exclusion or removal of one or both of the substrates (oxygen and phenols), (c) lowering the pH to two or more units below the optimum, and (d) adding compounds that inhibit PPO or prevent melanin formation.

10.4.3.1.2   Weight loss

Cumulative weight loss % was observed to be minimal for cut slices stored under cold storage at 4–5°C and treated with food-grade sodium metabisulfite, irrespective of the site. Breadfruit slices when vacuum packed at 85% and stored at 5°C were observed to have a lower cumulative % weight loss over storage time compared to those stored in clip-on barquettes. Type of packaging influenced the rate of weight loss of fruits over storage time due to their permeability properties. Sealed packaging type as vacuum-packed products allows for a lower volume of air space compared to clip-on packaging, and hence for a decreased rate of respiration, flow of gases, and water vapor from the vacuum packed produce. For the clip-on barquettes, more freedom of gas exchange and water vapor was allowed through the thin air space between the closed sides of the pack compared to the vacuum-sealed plastic bags. This explains the significantly higher % cumulative weight loss of fresh-cut breadfruit slices (Tables 10.3 and 10.4) when stored in clip-on barquettes over storage time compared to those that were vacuum packed (Figure 10.2).

10.4.3.2   Sensory Appraisal

Boiled samples of stored cut slices were randomly presented at weekly intervals to panelists who affirmed that starchiness, breadfruit taste, texture, and aroma were maintained throughout the storage period until the end of shelf-life.

TABLE 10.3 The Cumulative % Weight Loss of Breadfruit Slices Over 7 Days of Storage in Clip-on Packs at 4–5°C

Site/storage days	3d	7d	S.D	S.E
Reduit Crop Research Station	0.455	1.332	0.620	+/– 0.219
Pamplemousses Experiment Station	0.433	1.161	0.515	+/– 0.143

TABLE 10.4 The Cumulative % Weight Loss of Breadfruit Slices Over 18 Days Storage from Pamplemousses ES when Vacuum Packed and Stored at 4–5°C

Storage period (days)	7d	9d	11d	14d	18d	S.D	S.E
% cumulative weight loss	0.151	0.332	0.599	0.791	0.875	0.305	+/– 0.088

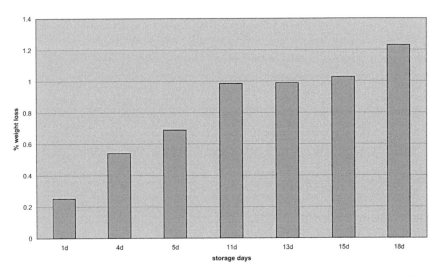

FIGURE 10.2 Chart showing the % cumulative weight loss of fresh-cut vacuum-packed breadfruit slices stored at 5°C.

10.4.3.3 Postharvest Disease Incidence

Postharvest fungi *Rhizopus* was detected on cut slices stored in clip-on barquettes after 7 days of storage. According to Willcox et al. (1994), high humidity and the large number of cut surfaces can provide ideal conditions for the growth of microorganisms. No disease incidence was detected in vacuum-packed samples. Presence of leachate in stored samples at the end of storage life was due to physiological deterioration of cellular tissues. Because minimally processed fresh fruits and vegetables are not heat treated, regardless of additives and packaging, they must be handled and stored at refrigerated temperatures at 5°C or under in order to achieve sufficient shelf-life and microbial safety.

10.5 CONCLUSIONS

The maturity stage at which breadfruits are harvested is of crucial importance for the fruit storability and marketability, and it will depend on immediate end use of the fruit (a) whether for storage for export purposes, (b) for immediate sale in the local markets for curries, or (c) for use for baking or deserts. For export purposes, it is advised to harvest "three quarter" mature green fruits with pale green color with white to brown latex stains but with polygonal segments still tightly bound together. The flesh of mature unripe breadfruit was observed to be ivory white in color and had a starchy taste when boiled. Mature ripe breadfruits had a yellowish green peel color, a cream flesh color with a soft texture, and was sweet to taste when boiled. Handling practices were of utmost importance to minimize mechanical damage as bruised fruits will soften rapidly causing reduction in shelf life for export; thus, fruits should be harvested early morning to avoid heat build-up by using either a fruit picker with mesh bag or a harvesting pole with a cutting scythe and a collecting bag to avoid fruit drop. The stems of the cut fruits of 3–4 cm length should be pointed downward to allow latex flow away from fruit surface and hence prevent brown staining of the peel. Harvested fruits should be placed in clean stackable plastic crates and transported in covered vehicles that provide adequate ventilation and protection (using a cover) to the harvested fruits. Use of a controlled temperature vehicle will be advantageous as it will pre-cool the fruit by removing field heat from the harvested fruit prior to export. It is advised to wash the fruits in sanitized water to remove dust particles and to thoroughly dry them before they are cling filmed prior to storage at 13°C as the fruits retain their peel and flesh color with minimum weight loss as well as maintain firmness taste and flavor over a storage time of 14 days. For increasing marketing opportunities of fresh breadfruit and providing easy access to the working consumer, it is recommended to be vacuum packed at 85% cut slices of breadfruit. That should be dipped in sodium metabisulfite solution (0.2 g/L of water) before being stored at 5°C, resulting shelf life of the cut slices of 18 days during which the color, freshness, flavor, and taste is maintained.

10.6   FUTURE WORK

Future research could involve (a) evaluating the shelf-life of commercial hydro-cooled cling filmed breadfruits in Mauritius, (b) shelf life evaluation of breadfruits using biodegradable type packaging to minimize environmental pollution, and (c) evaluating the use of surface coating based on polysaccharide and sucrose ester-based coatings on cold-stored breadfruits to minimize fruit desiccation.

ACKNOWLEDGMENTS

Grateful to the CEO of FAREI, Mr. Rajkumar, for giving permission to present this paper; Principal Research Scientist, Fruit Division, Mrs Ramburn and staff for their help and cooperation in editing this paper.

KEYWORDS

- breadfruit
- fresh cut
- sensory quality
- shelf life
- storage

REFERENCES

Adel, A. K., (2012). *Breadfruit: Recommendations for Maintaining Postharvest Quality*, Department of Plant Sciences, University of California, Davis CA 95616.

Barbados, Grenada, St Lucia, Dominica, & St Vincent, (2003). *Breadfruit-Postharvest Guidelines.*

Breadfruit Sector Consortium supported by PAEPARD project, Mauritius 2012.

Breadfruit, (2004), *Postharvest Care and Market Preparation*, Ministry of Fisheries, Crops and Livestock, New Guyana Marketing Corporation, National Agricultural Research Institute, with assistance of United States Agency for International Development, Technical Bulletin, No. 24.

Laurila, E., & Ahvenainen, R., (2004). Minimal processing in practice Fresh fruits and vegetables, *Minimal Processing Technologies in the Food Industry.*

Maharaj, R., & Sankat, C. K., (1990). The shelf life of breadfruit stored under ambient and refrigerated conditions. *Acta Horticulturae,* 269:411-424.

Maharaj, R., & Sankat, C. K., (2007). A review of postharvest storage technology of breadfruit. *Acta Horticulturae,* 757:183-192.

Marita Cantwell, (2004). *Postharvest Handling Systems: Minimally Processed Fruits and Vegetables.* University of California Cooperative Extension, Vegetable Research and Information Center. http://vric.ucdavis.edu/selectnewtopic.minproc.htm. April.

National Tropical Botanical garden-Breadfruit Institute, (2012). Tropical Plant Research, Education and Conservation- Breadfruit Institute. http://ntbg/breadfruit/uses/table1.php.

Ragone, D., (1997). *Breadfruit: Artocarus Atilis* (Parkinson) *Fosberg.* International Plant Genetic Resources Institute, Rome.

Ragone, D., (2011). Farm and forestry production and marketing profile for breadfruit, *Speciality Rops for Pacific Island Agroforestry* (http://agroforestry.net/scps).

Thompson, A. K. B. O., & Perkins, C., (1974b). Storage of fresh breadfruit. *Tropical Agri. (Trinidad), 51,* 407–415.

Wills, R. B. H., Lee, T. H., Graham, D., McGlasson, W. B., & Hall, E. G., (1981). *Postharvest: an Introduction to the Physiology and Handling of fruits and Vegetables,* New South Wales University Press, Sydney.

Worell, D. B., & Sean, C. C. M., (1997). Breadfruit, In: Mitra, S. K., (ed.), *Post Harvest Physiology and Storage of Tropical and Sub Tropical Fruits,* edn., Cab International., Oxford, UK, pp. 347–363.

CHAPTER 11

STANDARDIZATION OF DEHYDRATION TECHNIQUES OF SOME ORNAMENTAL FOLIAGES

MOUMITA MALAKAR, SUKANTA BISWAS,
and PINAKI ACHARYYA

*Department of Horticulture, University of Calcutta,
51/2, Ballygunge Circular Road, Kolkata–700019, India,
E-mail: moumitamalakar7@gmail.com*

CONTENTS

ABSTRACT

The eco-friendly dehydrated foliages and plant parts secured much popularity among users and have become key components in floriculture industry. Foliages with highly variable keeping quality are used as filler element in flower vase. Dehydration of foliages has not been studied at large. This investigation was carried out with ornamental foliages of three species, viz., *Araucaria cunninghamii, Thuja orientalis,* and *Juniperus chinensis.* White sand, silica gel, and boric acid were used as embedding materials, and two drying conditions of microwave oven and room drying were adopted for three durational treatments, viz., 10, 20 and 30 s and 4, 8, and 16 days, respectively. In both *Araucaria* and *T. orientalis,* silica gel + microwave oven combination for 30 and 20 s, respectively, exhibited best results in respect of moisture loss (49.23% and 58.33%) and quality. White sand + room condition also caused 61.41% moisture loss in *T. orientalis* when treated for 16 days. In *J. chinensis,* white sand + microwave oven and silica gel + room condition for 20 s and 16 days showed moisture loss of 44.26% and 50.16%, respectively. Boric acid as embedding materials was also found to be effective in dehydration of these species.

All the three species were treated with glycerin:water of 1:1 and 1:3 (vol/vol) for 24, 48, and 96 h, followed by drying in hot air oven at 70–80°C for 5 h and open air of room condition for 24 h. A significant moisture loss of 60.56% to 62.56% was recorded in *T. orientalis* when dehydrated in hot air oven for 96 h.

11.1 INTRODUCTION

Drying and preserving flowers, foliages, and plant materials are a form of artistic expression that was very popular during the Victorian age and has once again gained much popularity in the modern age. The decorative value of dehydrated plant parts led to the use of vase decoration, bouquets, and arrangements of gifts as well as ceremonial decoration of both home and working places. Dried plant materials with decorative value have now been accepted globally as natural, eco-friendly, long lasting, and inexpensive. In the context of international trade of floriculture of India, dried flowers and plant parts are the key segment and constitute 70% of total share of floriculture products exported from this country (Sheela, 2008). The reasons for the development of dry flower industry in India is possibly due to easy availability of wide range of materials throughout the year and manpower for such labor-intensive craft. Zizzo and Fascella (1997) opined that the dried materials can be enjoyed whole year by arranging them in vases, creating arts like candle holders, and in other home decorations. Possibly, the most common use of dried materials is in a wreath or floral arrangement and also in ornament gift packages, masks, hats, and lamp shades. Dried materials often embellish stationery or are used to create unique pictures. Dried plant materials are extra special as they possess the characteristics of novelty, eco-friendly, esthetically near to fresh flowers, flexibility, and year-round availability (Joyce, 1998). In the situation of climatic abnormalities in different parts of the world, which are not congenial for keeping fresh-cut flowers in vases, dried flowers have established tremendous potentiality, which is very much observed during the last decades (Dhatt et al., 2007).

Moisture holding of dried flowers and foliages influence their quality and longevity (Singh et al., 2003). Gill et al. (2002) studied the efficacy of various methods of drying, viz., microwave oven, embedded in a desiccant, solar drying, press drying, and preservation in glycerin for drying fern asparagus (*Asparagus* sp.) and silver oak (*Grevillea robusta*) leaves for commercial and domestic purpose, and observed that embedding in silica gel for 30 h was the best method for drying of ferns for commercial purposes, whereas drying by embedding in a silica gel for 36 h and press drying for 60 h was best for domestic purpose. In the light of the above, the present investigation was undertaken with three perennial species to standardize suitable method for developing dry leaves of these species. Cut leaves of these species are widely used for vase decoration and other decorations. Different methods, viz., embedding in different desiccants and drying either in microwave oven or in room condition was attempted along with preservation in glycerin.

11.2 MATERIALS AND METHODS

Fresh leaves of three different ornamental species of conifers, viz., *Araucaria cunninghamii,* Sweet (Hoop pine), *Juniperus chinensis*, Linn. (Chinese juniper) belonging to the family Aracariaceae, and *Thuja orientalis*, Linn., (Chinese thuja) of the family Cupressaceae were used for this study. Mature fresh young green foliages with more or less uniform in size and shape were collected in the morning hours followed by bagging in polyethylene bag to avoid further desiccation. Fresh weight and size of foliages were recorded. The experiment was conducted in the laboratory of the Department of Horticulture, Institute of Agricultural Science, University of Calcutta.

Foliages were embedded in pure drying media, viz., silica gel, boric acid, and white sand followed by drying in microwave oven (600 wt) for 10, 20, and 30 s and open air in room condition for 4, 8, and 16 days. Leaf samples were placed horizontally on desiccants at the bottom of microwave-resistant glass containers followed by covering of about a depth of 2 inches with the same media to avoid further moisture absorption from air, and the lid was tightened. Post drying characteristics were recorded 2 h after taking out from the microwave oven condition and keeping in ambient condition. For glycerin preservation, foliages were submerged in glycerin:water solution of 1:1 and 1:3 for 24, 48, and 96 h using boiling water in which glycerin was added slowly, gently, and then dried in hot air oven at 70–80°C for 5 h and open air of room condition for 24 h. To discard excess glycerin:water solution, the materials after taking out of from the glycerin were kept in hanging position.

In both the experiments, each treatment was replicated thrice considering one specimen as replication and the average data of each parameter are presented. The data were analyzed using completely randomized block design with factorial concept (Panse and Sukhatme, 1985).

11.3   RESULTS AND DISCUSSIONS

Results revealed (Table 11.1) that accelerated moisture loss is proportional with the increase of duration in all three desiccants and species.

Highest moisture loss was recorded in silica gel. *T. orientalis* and *A. cunninghamii* foliage materials recorded moisture loss of 58.33% in 20 s and 49.23% in 30 s under microwave oven condition. Tandon (1982), Bhutani (1990), and Datta (1999) established silica gel as the best embedding material followed by borax and sand. *J. chinensis,* on the other hand, recorded 44.2% moisture loss in white sand + microwave oven condition for 20 s treatment (Table 11.1). Electronically produced microwaves liberate moisture from organic substances by agitating the water molecules;

TABLE 11.1 Effect of Different Embedding Media in Drying of Three Different Foliages under Different Drying Conditions

Embedding Materials	Duration of treatments (sec.)	*Araucaria cunninghamii*			*Thuja orientalis*			*Juniperus Chinensis*		
		Average moisture loss (%)	Average size reduction (%)	Chlorophyll content (mg/gm.)	Average moisture loss (%)	Average size reduction (%)	Chlorophyll content (mg/gm.)	Average moisture loss (%)	Average size reduction (%)	Chlorophyll content (mg/gm.)
Microwave oven condition										
Silica gel	10	39.48c	2.64c	0.26a	56.66a	1.10b	0.35a	35.40c	1.20b	0.44a
	20	46.37b	3.32b	0.20b	58.33a	1.30a	0.30a	40.28b	1.50b	0.37b
	30	49.23a	3.88a	0.16c	57.07a	1.40a	0.23b	41.45a	2.10a	0.24c
White sand	10	23.47c	3.82b	0.13a	29.02c	1.20b	0.35a	40.44b	3.58b	0.43a
	20	32.30b	4.01b	0.07b	45.63b	2.19a	0.29b	44.26a	4.51a	0.35b
	30	39.53a	5.92a	0.05b	49.91a	2.23a	0.16c	43.83a	4.64a	0.22c
Boric acid	10	29.19c	2.67c	0.14a	35.04c	1.07b	0.38a	33.60b	0.12b	0.38a
	20	34.80b	4.02b	0.09b	53.92a	3.75a	0.28b	35.75b	0.13b	0.29b
	30	40.88a	4.89a	0.06b	50.86b	4.27a	0.17c	40.40a	3.33a	0.21c
Room condition										
Silica gel	4	49.72a	5.90a	0.40a	44.07b	3.27a	0.25a	31.48b	2.23c	0.36a
	8	53.40a	6.03a	0.25b	57.86a	3.85a	0.22a	33.09b	3.53b	0.31a
	16	56.16a	6.19a	0.21b	60.36a	3.86a	0.21a	50.16a	4.42a	0.28b

TABLE 11.1 (Continued)

Embedding Materials	Duration of treatments (sec.)	*Araucaria cunninghamii*			*Thuja orientalis*			*Juniperus Chinensis*		
		Average moisture loss (%)	Average size reduction (%)	Chlorophyll content (mg/gm.)	Average moisture loss (%)	Average size reduction (%)	Chlorophyll content (mg/gm.)	Average moisture loss (%)	Average size reduction (%)	Chlorophyll content (mg/gm.)
White sand	4	27.64b	4.50a	0.37a	22.58c	3.14b	0.23a	20.39c	1.71c	0.24a
	8	50.09a	5.50a	0.28b	48.73b	3.26b	0.20a	26.16b	2.27b	0.15b
	16	50.21a	5.53a	0.22b	61.41a	4.22a	0.19b	45.90a	3.11a	0.11b
Boric acid	4	12.38c	2.06a	0.30a	19.75b	1.04b	0.21a	16.47c	2.00c	0.29a
	8	31.59b	2.15a	0.24b	56.20a	3.04a	0.18b	36.72b	2.93b	0.23b
	16	37.04a	2.62a	0.18c	60.68a	3.90a	0.16b	49.07a	4.90a	0.23b

this is the principle underlying the quickest microwave oven drying (Bhu-tani, 1990a). In all the three species, the effect of white sand and boric acid as desiccants was almost at par. Size reduction of leaves was increased with the increase in treatment duration. Bhalla et al. (2006) observed that average foliage size reduction in chrysanthemum was more with white sand and boric acid than silica gel, but maximum size reduction of flowers was noted while embedded in silica gel and dried in microwave condi-tion. In *Araucaria cunninghamii* and *Juniperus chinensis* size reduction of 5.92% and 4.64%, respectively, was recorded while drying under white sand + microwave for 30 s. In *T. orientalis*, a reduction of size by 4.27% was recorded in boric acid +microwave for 30 s.

Dried leaves with better chlorophyll content could be obtained with minimum duration of treatment, as it is the main pigment content of foliages and play a vital role for the determination of dried plant part's attractive-ness. In regard to chlorophyll restorability, silica gel-embedded materials showed maximum chlorophyll recovery with 0.26 and 0.44 mg/g in 10 s treatment in *A. cunninghamii* and *J. chinensis*. Boric acid represented as best desiccant for *T. orientalis* with the maximum chlorophyll recovery of 0.38 mg/g in 10 s treatment. Meman and Barad (2009) observed higher pigment reduction at higher temperature, which was consistent with our result. No marked variation in the shape and texture of post-drying foli-ages was observed.

In room drying, increased duration caused increased moisture reduc-tion. The activity of all three desiccants was more or less the same under room drying, but among them, silica gel showed good result by reduc-ing 56.16% of moisture in 16 days treatment, followed by white sand in *A. cunninghamii*. Moisture reduction of *T. orientalis* leaves was similar under all embedding materials with an extent of losing 60.82% moisture. Silica gel and boric acid caused 50.16% and 49.07% moisture loss from *J. chinensis* leaves, respectively, but white sand failed to show any effective result. Misra (2002) established the fact that dried flowers and leaves with a specific moisture level can be stored for a very long period without los-ing their appearance and decorative value.

With regard to size reduction of leaves through room drying, the reduction varied with the desiccants and species. Size reduction was little with boric acid in *A. cunninghamii* and *T. orientalis* (2.06% and 1.04%, respectively). Color retention was much better in room drying than in microwave oven + sand media. It is noteworthy to mention that the main principle of drying is to diminish moisture content to a point at which the biochemical changes are minimized but maintaining cell structures, pigment level, and actual shape of flowers or foliages (Dana and Larner, 2001).

In another part of experiment, foliages of the three species were allowed to absorb in two concentrations of glycerin:water solution, viz., 1:1 and 1:3 for 24, 48, and 96 h. Lee et al. (2003) exclaimed that dye with glycerin gave the flowers and foliages a more vibrant color in addition to its soft and pliable feel (Table 11.2).

It is evident from Table 11.2 that moisture loss from materials increased with the imbibition time of glycerin:water, irrespective of drying techniques. Highest moisture loss was recorded with leaves of all three species treated for 96 h in both combination of glycerin used. Insignificant variation of moisture loss from leaves due to the use of different concentrations of glycerin were also observed. Among three species, *T. orientalis* leaves showed a moisture loss of 62.5% under hot air oven drying after 96 h absorption in 1:1 glycerin:water solution, while the same trend was also noted under room drying. Minimum size reduction of leaf was recorded with 1:1 glycerin:water solution (24 h) in *T. orientalis* and *J. chinensis* leaves both under hot air oven and room drying. In *A. cunninghamii* foliages, the 1:3 glycerin:water solution showed a reduction of 1.20–2.04% in different drying conditions. Chlorophyll deterioration trend was prominent with the increase in treatment duration of all the species. Campbell et al. (2001) and Sohn et al. (2003) experimented and proved glycerol as an appropriate preservative for *Eucalyptus cinerea* and *Magnolia grandiflora* for retaining natural appearance as it replenishes the natural moisture of the leaf with a substance that maintains leaf fall, texture, color, and sugar level.

11.4   CONCLUSION

Dehydration is an important post-harvest technology for enhancing ornamental keeping quality of flowers and foliages as it quickly

TABLE 11.2 Effect of Glycerin Preservation of Three Different Foliages under Different Drying Conditions

Treatments	Duration of treatments (hrs.)	Araucaria cunninghamii			Thuja orientalis			Juniperus Chinensis		
		Average moisture loss (%)	Foliage size reduction (%)	Chlorophyll content (mg/gm.)	Average moisture loss (%)	Foliage size reduction (%)	Chlorophyll content (mg/gm.)	Average moisture loss (%)	Foliage size reduction (%)	Chlorophyll content (mg/gm.)
Hot-air-oven condition										
Glycerin-water (1:1)	24	13.81c	2.01b	0.24a	24.48c	3.44c	0.25a	26.79c	2.05c	0.36a
	48	24.62b	2.90a	0.17b	42.23b	4.02b	0.21a	32.12b	2.38b	0.25b
	96	33.33a	3.06a	0.12b	62.50a	4.47a	0.15b	33.95a	4.23a	0.20c
Glycerin-water (1:3)	24	22.98b	1.20c	0.25a	24.40c	3.49a	0.22a	24.60c	3.19c	0.31a
	48	23.44b	2.62b	0.16b	36.81b	3.51a	0.19b	30.50a	4.20b	0.22b
	96	30.08a	2.79a	0.11b	60.56a	3.59a	0.13b	31.50a	5.30a	0.18c
Room condition										
Glycerin-water (1:1)	24	15.10c	2.56c	0.22a	27.20b	0.10b	0.23a	13.49c	1.33a	0.30a
	48	25.50b	2.86b	0.16b	28.30b	0.11ab	0.16b	20.50b	1.48a	0.22b
	96	34.40a	3.38a	0.14b	29.60a	0.12a	0.13b	22.80a	1.56a	0.15c
Glycerin-water (1:3)	24	23.50c	2.04b	0.20a	26.50b	0.14b	0.21a	10.00c	2.36b	0.28a
	48	25.50b	3.08a	0.14b	27.60ab	0.20a	0.14b	19.07b	2.58ab	0.21b
	96	31.50a	3.18a	0.09c	28.50a	0.23a	0.12b	21.65a	2.92a	0.13c

reduced the moisture content to a point at which there is little biochemical change, while keeping cell structure and shape unaffected. Dry foliages can be stored for unlimited period if they are well secured from the damage of atmospheric high humidity. Since the last three decades, scientists are putting their efforts to standardize the dehydration techniques for flowers and foliages that require special care to protect their shape, sizes, and color, while other plant parts like branches, cones, and barks can be dried with little care. With this aim, this investigation succeeded to point out the efficacy of various desiccants and drying conditions for dehydrating the selected foliages. Drying in microwave oven showed quicker result as compared to room drying, as increased rate of moisture loss due to more conduction and convection of heat to foliage tissue and water evaporation from the surface might have caused rapid drying in microwave oven (Singh and Dhaduk, 2005). Irrespective of media, moisture reduction was accelerated with the increased duration in all species of plants. Silica gel caused maximum moisture loss by attracting moisture following the phenomenon known as physical absorption and capillary condensation irrespective of drying condition, followed by boric acid and white sand. In microwave oven drying, *Araucaria* sp. foliages reduced maximum in size, while silica gel reduced maximum size in room condition. Sand media in microwave drying proved best for *J. chinensis.* Foliage color in all species were bleached to some extent, while the fading effect of boric acid has also been reported by Pamela (1992) and subsequently confirmed by Singh and Dhaduk (2005). Embedding methods were found to be appropriate for maintaining shape of foliages as it recommended for drying to get three dimensional views (Singh and Dhaduk, 2005). It was observed through our experiment that glycerin uptake was increased with the increase in absorption duration. Highest moisture loss percentage was obtained in *T. orientalis* leaves in hot air oven drying after 96 h absorption in 1:1 glycerin:water solution. Similar findings were also reported by Joyce and Dubious (1992). Thus, owing to great availability of variety of plants, flowers, and other artistic raw materials, India has got enormous scope to propagate this field, as dry flower market is still small in comparison to that of the fresh flower market, which might be due to its more recent introduction in the floricultural trade.

KEYWORDS

- **dehydrated**
- **eco-friendly**
- **embedding**
- **foliage**
- **moisture**

REFERENCES

Bhalla, R., Moona, S., Dhiman, R., & Thakur, K. S., (2006). Standardization of drying techniques of Chrysanthemum (*Dendranthemum grandiflorum* Tzvelev). *Journal of Orna- mental Horticulture*, *9*(3), 159–163.

Bhutani, J. C., & Tandon, R. K., (1982). Sukhe phoolon se sajawat kijiye. *Phal-Phool.*, *5*(3), 3–5.

Bhutani, J. C., (1990). Dry rose craft. *Rose News*, *9*(11), 8–9.

Buzarbarua, A., (2000). *A Textbook of Practical Plant Chemistry*. S., Chand and Company, New Delhi, pp. 83–87.

Campbell, S. J., Ogle, H. J., & Joyce, D. C., (2001). *Glycerol Uptake Preserves Cut Juvenile Foliage of Eucalyptus Cinera-School of Land and Food*, 3rd edition., Australia, vol. *15*, pp. 492.

Dana, M. N., & Larner, B. R., (2002). *Preserving Plant Materials-* University Cooperative Extension Service. 2nd edition., West Lafayette. Department of Horticultural Flowers, vol. *19*, pp.102.

Datta, S. K., (1999). Dehydrated flowers, foliage and floral craft. In: *Floriculture and Landscap-ing*. Naya Prakash, Kolkata, pp. 696–703.

Dhatt, K. K., Singh, K., & Kumar, R., (2007). Studies on methods of dehydration of rose buds. *Journal of Ornamental Horticulture*, *10*(4), 264–267.

Gill, S., Bakshi, S., & Arora, S., (2002). Standardization of drying methods for certain cut flowers. *Journal of Ornamental Horticulture*, *9*(3), 159–163.

Joyce, D. C., & Dubious, P., (1992). Preservation of fresh cut ornamental plant material with glycerol. *Postharvest Biology and Technology*, *2*(2), 145–153.

Joyce, D. C., (1998). Dried and preserved ornamental plant materials- not new but often over-looked and underrated. *Acta. Horticulture*, *45*(4), 133–145.

Lee, W. Y., Mijeong, Y., Chunho, P., & Beyounghwa, K., (2003). Effects of various drying methods for wild flowers. *Korean Journal of Horticulture Science and Technology*, *21*(1), 50–56.

Meman, A., & Barad, V., (2009). Study on dry leaf production of asparagus. *Journal Horticulture and Forestry*, *11*(3), 43–47.

Pamela, W., (1992). *The Complete Flower Arranger*. Annes Publishing Limited, London, pp. 255.

Panse, V. G., & Sukhatme, P. V., (1985). *Statistical Methods for Agricultural Workers*. ICAR, New Delhi, pp. 55–70.

Ranjan, J. K., & Mishra, S., (2002). Dried flowers: A way to enjoy their beauty for a long period. *Indian Horticulture*, *10*, 32–33.

Sharma, G. K., Semwal, A. D., & Arya, S. S., (2000). Effect of processing treatments on carotenoids composition of dehydrated carrots. *Journal of Food Science Technology*, *37*(2), 196–200.

Sheela, V. L., (2008). Dry flowers: a profitable floriculture industry. In: *Flower for Trades*. 3rd edition., New India Publishing Agency, New Delhi., vol. 10, pp. 65–67.

Singh, A., Dhaduk, B. K., & Shah, R. R., (2003). Effect of dehydration on post harvest life and quality of zinnia flowers. *Journal of Ornamental Horticulture*, *6*(2), 141–142.

Singh, A., Dhaduk, B. K., & Shah, R. R., (2005). Effect of dehydration techniques of some selected flowers. *Journal of Ornamental Horticulture*, *6*(2), 155–156.

Sohn, K., Kwon, H., & Kim, E., (2003). A optimum drying temperature to maintain sea sand dry color of roses. *Korean Journal of Horticulture Science and Technology*, *21*(2), 141–145.

Swarnarupa, R., & Jayasekar, M., (2008). Dry flower production. *Journal of Tropical Agriculture*, *10*(2), 45–49.

Zizoo, G., & Fascella, G., (1997). How to obtain dried flowers. *Colture Protette*, *24*(10), 51–60.

CHAPTER 12

DEPENDENCE ON NON-TIMBER FOREST PRODUCTS FROM A COMMUNITY FOREST AS A SAFETY NET FOR LIVELIHOOD SECURITY AMONG THE VILLAGERS OF MAMIT DISTRICT, MIZORAM

K. LALHMINGSANGI and U. K. SAHOO

Department of Forestry, School of Earth Sciences and Natural Resource Management, Mizoram University, Aizawl–796004, India, E-mail: uksahoo_2003@rediffmail.com

CONTENTS

ABSTRACT

Non-timber forest products (NTFPs) provide an important source of livelihood security for the inhabitants living adjacent to forest areas. A survey was conducted in five villages under Mamit district of Mizoram, and it was found that NTFPs act as safety nets by collecting available NTFPs from the community forest (Village Forest Development Committee (VFDC) plantation areas). The main aim of the study is to focus on the dependence of NTFPs from the community forest of the surveyed villagers. For this, participatory rural appraisal (PRA) was done along with questionnaires, personal interviews, and group discussion with the villagers. The weaker section of the society gets benefited by providing food security in times of unavailability of agricultural cash crops, which was one of the basic needs for their livelihood. Besides the home consumption of NTFPs, they were sold to meet the cash requirements especially by the widow and landless farmers in all the surveyed villages. Such is the case; it gave them a kind of natural insurance and security for future needs. Bamboo pole, broom stick, fruits, wild foods, and fuelwood are the main NTFPs that they have collected from the community forest area. Among the various NTFPs, fuelwood (56%) has the highest percentage of household involved in harvesting alone among the studied villages, followed by wild food (46%), broomstick (35%), bamboo pole (31%), fruits (8%), and medicinal plants (5%). The potential of NTFPs in community forest are well benefited by fulfilling their daily food requirements, building materials, meeting cash, and saving money by exploiting NTFPs and reduce the yearly expenditure.

12.1   INTRODUCTION

Non-timber forest products (NTFPs) are considered to be important for sustaining rural livelihood, reducing rural poverty, biodiversity conservation, and facilitating rural economic growth (Global NTFP partnership, 2005). Several opportunities for improved rural development are linked to NTFPs (Adepoju et al., 2007) as the villagers are traditionally dependent on forest resources for their food, shelter, and income through collection and marketing of NTFPs (Sahoo et al., 2010). If the villagers were not extracting wild food plants from the forest, their needs of cash income would double to maintain their daily meal. Household may trade in NTFP's in times of hardship, with this activity forming an important safety net (Takasaki et al., 2004).

In recent years, with the advent of globalization, there are indications of a rapid increase in the extraction mostly from natural populations. In India alone, it is estimated that over 50 million people are dependent on NTFPs for their subsistence and cash income (National Centre for Human Settlements and Environment, 1987; Hegde et al., 1996). Recent studies estimated that 275 million poor rural people in India equal to 27% of the total population depend on NTFPs for at least part of their subsistence and cash livelihoods (Bhattacharya and Hayat, 2009; Malhotra and Bhattacharya, 2010). The women of mountainous regions have maximum work inputs for agricultural activities and domestic responsibilities, particularly in rural areas (Salam et al., 2005).

Forest resources are also important to battle the global warming and climate change problems such as carbon sink (IIPCC, 2001). It has also been realized that this will not be possible without active co-operation of the inhabitants living adjacent to forest areas. This situation demands a more coherent system for sustainable development of forest resources in the Joint Forest Management program, which should be based on regulating management for capacity building of community. Introduction of forestry practices for maximum production of fuel wood, fodder, fruits, timber, fiber, etc., must be encouraged (Rawat and Sharma, 2010). However, when there is no market for the NTFPs, the methods of evaluation are not as straight forward. While environmental economists and ecological economists have done extensive research on hypothetical markets, the techniques they have developed are not always suitable to estimate the value of nonmarketed NTFPs. In some communities, NTFPs might be exchanged for marketed products, rather than sold directly. NTFPs that are consumed rather than sold in the market can be considered income in kind rather than in cash. Thus, ignoring the role of NTFPs consumption in the livelihoods of rural populations gives a very distorted view of the importance of NTFPs and of their economic values (Chopra, 1993).

The Joint Forest Management program is a co-management regime for protection, regeneration, and development of degraded forests where the role of NGOs' has been very useful in bridging the gap between state and the people dependent on forests. Under this program, the forest area under consideration is protected and managed jointly by the local community and forest department, while the ownership of the forest lies with the government. The people get in turn the forest products free of cost from the community-managed forests. This paper analyzes the extent of dependence on NTFPs by the people from the jointly managed forest area.

12.2 MATERIALS AND METHODS

12.2.1 STUDY SITE

The study was carried out in five villages of Mamit district, viz., Dapch-huah, Chhipui, Chungtlang, Tuahzawl, and Lengte Village Forest Development Committee (VFDC) plantation areas during the year 2013–2015 (Figure 12.1). Even though the occupation of the people from these five villages is mainly *jhum* cultivation, they are exploiting NTFPs from VFDC plantation areas to supplement the agricultural crops and meet their daily needs. These VFDC plantation areas are mainly situated 2–5 km away from the village in the community land areas and are monitored by the villagers along with the Forest Department, which provide a sense of ownership to the villagers.

12.2.2 METHODOLOGY

Participatory rural appraisal (PRA) was adopted for the field study. Primary and secondary data were collected using questionnaire, personal interview, and interaction with the villagers. Semi-structured questionnaire was given to approximately 10% of the household from each village to provide information on to what extent they were involved in the exploitation of NTFPs as well as to know the different benefits they got through the community forest, i.e., VFDC plantation areas. A small group interaction was done along with the village leaders including females to avoid gender division and to get a real picture discussing the positive and negative aspect in harvesting, processing, and marketing of NTFPs from the plantation sites. Socioeconomic survey, land use pattern, value addition of NTFP, and dependency on forest as well as marketing strategy were also taken into consideration. Major emphasis was given on the collection of data by direct interaction with those individuals who were actually engaged in utilization of NTFPs.

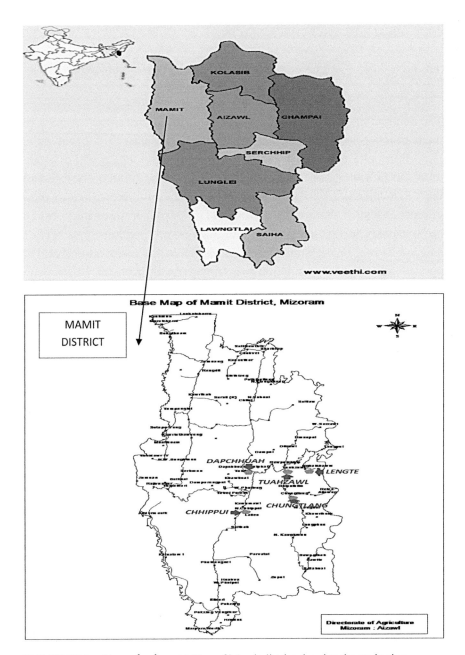

FIGURE 12.1 **(See color insert.)** Map of Mamit district showing the study sites.

12.3   RESULTS AND DISCUSSION

12.3.1   SOCIOECONOMIC PROFILE OF THE STUDIED VILLAGE

The occupation of most of the villagers is agriculture in which most of them practice shifting cultivation, and approximately 9–11% of them have horticulture farm, 1–2% hold government jobs, and 1–2% are involved in trading. NTFPs are exploited from the community forest of the VFDC plantation area and from other community forests nearby their villages. Most of them are part-time NTFP exploiter since these NTFP are seasonal; besides marketing, they harvest and exploit those NTFPs as a substitute for their cash crop. As shown in the table, among all the studied villages, the population was highest in Dapchhuah village with 1130 total population and 235 households, followed by Chhippui village with 950 total population and 186 households, Lengte village with 530 population and 135 households, Tuahzawl village with 520 population and 100 households, and the least was in Chungtlang village with 476 total population and 100 households. Literacy rate was high in all the studied villages and varied between 97% and 99%. Tuahzawl village has the highest literacy rate of 99%, followed by Chhippui, Chungtlang, and Lengte village constituting 99% literacy rate and the least was in Dapchhuah village at 97%.

12.3.2   INFRASTRUCTURE

As shown in Table 12.1, Anganwadi centers and primary and middle schools were located in all the studied villages, whereas high school was present only in Dapchhuah and Chhippui villages; these schools are mainly run by the state government. Health subcenters and bank facilities were absent in all the five villages, and the villagers, in case of health emergency, have to go to their nearby town. One community hall was present in Dapchhuah, Tuahzawland, and Lengte village each. Four to five churches as well as spring water sources were present in each village.

TABLE 12.1 Socioeconomic Profile of the Five Surveyed Villages under Mamit district in Mizoram

Attributes	Surveyed VFDC Villages				
	Dapchhuah	Chhippui	Tuahzawl	Chungtlang	Lengte
No. of house-hold	235	186	100	100	135
Male	510	550	288	271	250
Female	620	400	232	205	280
Total population	1130	950	520	476	530
Matriculation	15	36	41	20	39
Graduate	2	20	20	7	15
Literacy rate %	97	98	99	98	98
Average land holding (ha)	1	1	1	1	1
Horticulture garden (HH)	30	15	12	10	27
Jhum cycle (year)	7	6	8	6	7
Anganwadi Centre	3	4	3	2	3
Primary school	1	1	1	1	1
Middle school	1	1	1	1	1
High school	1	1	-	-	-
Health Sub centre	-	-	-	-	-
Bank	-	-	-	-	-
Community hall	1	-	1	-	1
Church	4	5	5	4	5
Spring water	4	4	5	4	4

12.3.3 USE OF VARIOUS NTFPS

From the surveyed villages, 31% of household are involved in harvesting and processing of bamboo pole, whereas 35.4% household in broom grass, 6% in fruits, 45.8% in wild foods, 56.4% in fuel wood and 4.6% in medicinal plants. Fuel wood has the highest percentage of household involvement among all the NTFPs mainly because it is in high demand by most of the households all throughout the year. Least involvement of the villagers in the selected NTFPs was found in medicinal plants due to less abundance in all the areas, and youths are more interested in other medicines and therefore are lesser drawn to it (Table 12.2).

12.3.3.1   Bamboo Pole

The most common bamboo species used for bamboo pole in the studied villages are *Bambusa tulda* (Rawthing), *Melocanna baccifera* (Mautak), *Dendrocalamus longispathus* (Rawnal), *Schizostachylum dullooa* (Rawthla), *Bambusa vulgaris* (Vai rua), *Dendrocalamus giganteus* (Rawpui), *Schizostachyum fuchsiamum* (Raw te), *Dendrocalamus hookeri* (Rawlak), *Schzostachyum mannii* (Raw te), and *Dendrocalamus hamiltonii* (Phul rua), which meet the requirement of bamboo poles; however, people do also harvest bamboo poles from the adjoining forest areas. People have a strong believe that for all species of bamboo, harvesting mature culms and in the right season sustains their productivity. The number of household using bamboo pole is high in most of the villages, with the highest in Chhippui at 60%, followed by Chungtlang at 30%, Dapchhuah at 30%, Lengte at 25%, and the least in Tuahzawl at 10%.

Most of them harvest for their own consumption, except in Chungtlang and Chhippui villages. *Dendrocalamus longispathus* is mainly harvested for weaving local carrier (em), and for this one, mature bamboo of 15 ft is sufficient to complete the carrier (Figure 12.2 and 12.3). *Bambusa tulda* is preferred for weaving local winnowing fan, and for this, 1 m bamboo with four nodes is usually harvested. *M. baccifera* is used for making different instrument handles. The local carrier made from *D. longispathus* is in high demand all throughout the year, especially in towns and villages. Weaving is done according to the demands and order.

12.3.3.2   Broom Grass (*Thysanolaena maxima*)

The inflorescence of broom grass is harvested on maturity when the panicles become tough and their color change to light green or red during

TABLE 12.2 Use of Different NTFPs in the Five Studied Villages

NTFPs		Surveyed villages				
		Dapchhuah	Chhippui	Tuahzawl	Chungtlang	Lengte
Bamboo Poles	% of HHs involved/year	30	60	10	30	25
	Quantity harvested (kg/hh/yr)	100	200	20	50	30
	Home consumption (kg/hh/year)	45	70	20	50	30
	Quantity sold (Kg/hh/year)	55	120	-	-	-
	Income generated (Rs/hh/year)	300	1400	-	-	-
Broom Grass	% of HHs involved/year	7	80	50	20	20
	Quantity harvested (kg/hh/yr)	28	20	10	10	5
	Home consumption (kg/hh/year)	10	20	10	10	5
	Quantity sold (Kg/hh/year)	18	-	-	-	-
	Income generated (Rs/hh/year)	280	-	-	-	-
Fruits	% of household involved/year	15	3	9	-	3
	Quantity harvested (kg/hh/yr)	12	5	5	-	5
	Home consumption (kg/hh/year)	12	5	5	-	5
	Quantity sold (Kg/hh/year)	-	-	-	-	-
	Income generated (Rs/hh/year)	-	-	-	-	-

TABLE 12.2 (continued)

NTFPs		Surveyed villages				
		Dapchhuah	Chhippui	Tuahzawl	Chungtlang	Lengte
Wild food	% of household involved/year	60	70	20	50	29
	Quantity harvested (kg/hh/yr)	104	149	30	35	50
	Home consumption (kg/hh/year)	20	120	10	30	10
	Quantity sold (Kg/hh/year)	84	29	20	5	40
	Income generated (Rs/hh/year)	1050	240	200	50	650
Fuel Wood	% of household involved/year	57	60	50	60	55
	Quantity harvested (kg/hh/yr)	66	50	35	45	30
	Home consumption (kg/hh/year)	16	50	35	45	30
	Quantity sold (Kg/hh/year)	50	-	-	-	-
	Income generated (Rs/hh/year)	150	-	-	-	-
Medicinal plants	% of household involved/year	6	9	4	2	2
	Quantity harvested (kg/hh/yr)	5	4	4	3	3
	Home consumption (kg/hh/year)	5	4	4	3	3
	Quantity sold (Kg/hh/year)	-	-	-	-	-
	Income generated (Rs/hh/year)	-	-	-	-	-

FIGURE 12.2   Weaving of local carrier (Pai em) from *Dendrocalamus longispathus*.

FIGURE 12.3   Different weaving products of *Dendrocalamus longispathus*.

winter season from January to March. It is harvested by hand pulling of the culms. It is further processed by exposure to sunlight for few days and then cleaning off the seeds. It is then wrapped and cut in a proper manner to add the value. Thatch grass is harvested in the month of November to April when the grass reach its highest length and maturity and further processed by exposure to direct sunlight for about a month. Even though broom grass is used by all the households, harvesting is done mainly for their own consumption and not for market purpose. Further, 80% of households are involved in harvesting of broom grass in Chhippui, followed by 50% in Tuahzawl, 20% in Chungtlang and Lengte, and the least 7% in Dapchhuah. Even though Dapchhuah has the least percentage of household involvement in harvesting, they are the only one who market broom grass to local markets.

12.3.3.3 Wild Fruits

Garcinia lanceifolia Roxb, *Artocarpus heterophyllus*, and *Emblica officinalis* fruits are harvested by the villagers. They harvest only for their own consumption. Children below the age of 15 years are mostly involved in harvesting of these fruits. Some adult exploiters when they cross by the plant also harvested it. Highest consumption of wild fruits was seen in Dapchhuah village constituting 15% of household; this is mainly because the community forest is easily accessible to them, and they need to cross that area frequently. This was followed by Tuahzawl at 9%, Chhippui and Lengte at 3%, and Chungtlang village at 0%. There are no market benefits from the fruits harvested from the studied villages (Figure 12.4).

FIGURE 12.4   *Garcinia lanceifolia* fruit harvest in Tuahzawl village.

12.3.3.4   Wild Food

Amorphophallus paeonifolius, *Homalomena aromatica*, mushroom, *Solanum torvum* Sw, bamboo shoots (*M. baccifera* and *D. longispathus),* *Eurya acuminatea,* and *Amomum dealbatum* Roxb. are the main foods harvested from the community forest of the studied villages (Figure 12.5). In the case of elephant foot yam (*A. paeonifolius*), only the male tuber weighing about 4–5 kg is harvested by digging the ground with spade or any other equipment without harming the tuber. Value addition is done by slicing the tuber and then boiling it with water for 1 h; subsequently, the water is drained, the peel is taken off, and the tuber is boiled again with local-made sodium carbonate. It is then smashed and wrapped in a banana leaves. *Solanum torvum* young fruits are also harvested and cooked along with other vegetables and serve as important vegetables. A common medicinal plant *Homalomena aromatic* leaves is harvested and

FIGURE 12.5 (1) *Amorphophallus paeonifolius* before processing. (2) *Amomum dealbatum* in local market. (3) Bamboo shoots (*Dendrocalamus longispathus*) in local market. (4) *Eurya acuminatea* in its habitat.

consumed by the villagers of Chhippui villages as a vegetable. *Tremellomycetes fuciformis* (pa sawntlung), *Schizzophylum commune* (pasi), and other varieties of mushrooms are available for consumption during the month of July to October and are harvested by direct hand plucking, which provide good income. However, mushroom once available in plenty are very scanty now due to the death of bamboo (owing to bamboo flowering) a few years back and because of insufficient decomposition of bamboo base that produce mushroom (Figure 12.6). Lesser people are involved in harvesting of mushroom because mushrooms need a good timing to harvest them and also because it is less abundant in the plantation area. The marketing rate of mushroom is more or less high in all the villages. Local market price of mushroom was 100–150 Rs/kg. Bamboo shoots are one of the most important vegetables and have a high market demand as well. The shoots of *M. baccifera* and *D. longispathus* are mainly consumed by the villagers. They are consumed in fresh form, and the cooked shoots are even sun dried and preserved for off season. Bamboo shoots are one the most important substituents of agricultural cash crops and constitute the highest percentage of household involved among all the foods collected from the community forest. *Eurya acuminatea* leaves are also harvested and used as vegetables from the surveyed villages. *Amomum dealbatum* young shoots and buds are cooked and used as vegetables. Dapchhuah at 60% and Chhippui at 70% have the highest percentage of household involved in harvesting different available wild foods from the community forest areas and also the highest income as compared to the other villages (Chungtlang 50% and Lengte 29% and Tuahzawl 20%).

FIGURE 12.6   Processing of bamboo shoot at a roadside market.

12.3.3.5   Fuel Wood

In all the studied villages, all the households are using fuelwood in addition to LPG. Consumption of fuelwood is high in most of the villages; the more interior the place, the higher is consumption of fuelwood. Even though most of them have an LPG connection, they cannot fully rely on that due to scarcity of gas and the high cost. They are more comfortable using fuelwood; therefore, all the household somehow harvest fuelwood from the nearby forest as well as from the VFDC plantation area. Common fuelwood species harvested by the villagers are *Quercus pachyphylla, Anoglissus acumulata, Mesua ferrea, Schima wallichi, Bischofia javanica,* and *Callicarpa arborez. M. baccifera* is also harvested as a substitute to these species in times of scarcity of fuelwood. A total of 56.4% of household are involved in harvesting fuelwood from the community forest, and the remaining households collect fuelwood from their own land. Among all the NTFPs, household participation in the collection of fuelwood from the community forest is highest in fuel wood as compared to other NTFPs. Chungtlang and Chhippui villages have the highest density of fuelwood species (Figure 12.7), covering 50% of the plantation area. Most of them are naturally grown. The highest consumption of fuelwood from the area is also from these two villages, with 60% of the households involved in the harvesting of these fuelwood species from the plantation area. This was followed by Dapchhuah at 57%, Lengte at 55% and Tuahzawl at 55%.

FIGURE 12.7   Fuelwood in Dapchhuah village.

Besides the density of fuelwood species, the distance of community forest from the village also matters; the nearer the community forest, the more is the collection seen in most of the villages. Dapchhuah is the only village that sells the fuelwood and gets income from it. This is mainly because Dapchhuah is located on national highway, which gives a good chance of marketing the fuelwood.

In ancient Mizo culture, every family is expected to have a fuelwood stock. This is mainly because when some people pass away within the villages, the Young Mizo Association (YMA) (which was one of the biggest nongovernmental organization (NGO) in Mizoram) used to collect 1–2 fuelwood for that family as a helping hand. This is still practiced in all the studied villages, which make them to increase the rate of harvesting from the plantation areas as well as from the nearby forest besides their own consumption.

Mature *Gmelina arborea* is also used for making the wheels of local-made vehicle (Figure 12.8), which is one of the most important means of transportation, especially for carrying goods from the field to home, for carrying water, etc., within the village.

12.3.3.6 Medicinal Plants

There is no specific season for harvesting of medicinal plants. The villagers harvest different medicinal plants and consume them in fresh form only when needed. Leaves are mostly harvested. *Litsea monopetala* (Roxb.) (Figure 12.9) Pers leaves are harvested by the villagers of Chhippui for

FIGURE 12.8   Local-made vehicle with wheels made of mature *Gmelina arborea.*

FIGURE 12.9   *Litsea monopetala*, an important medicinal plant in the area.

the treatment of cattle sores. They add the leaves along with the cattle food so as to heal the sores. Leaves of *Mikania micrantha* Kunth, a quick growing climber, is commonly used as an antiseptic and applied to fresh cuts. *Clerodendrum colebrookianum* Walp. leaves are also harvested by the villagers for reducing high blood pressure, in addition to being used as vegetables. Plant medicines remain one of the most affordable and easily accessible sources of treatment in primary healthcare in the studied villages. The beneficiaries at Chhippui village (9% household) are making the best use of the medicinal plants from the plantation area, followed by Dapchhuah village (6% house hold) while least utilization on medicinal plants was found on Chungtlang and Lengte village (2% household each).

12.3.4   MARKET FLOW OF NTFPS

In the present study, from the five villages, the localities did all the harvesting, processing and even marketing of NTFPs. Marketing of selected NTFPs are mainly done in their own village, neighboring villages and towns, roadside, junction selling points, and markets in the nearest urban centers and cities. Most of the NTFP exploiters are part-time exploiter as they are seasonal and occasional. However, approximately 3% household among the NTFP producers are full time with varying products and locations. Medicinal plants are not sold by the villagers; they use it only for their own consumption. Few years back, *Homalomena aromatica* rhizomes were harvested in a large amount and were sold in national market through a middleman from Chhippui village; the rhizomes are now preserved, and collection of such plants in a large amount was prohibited by the local leaders (Figure 12.10–2.15).

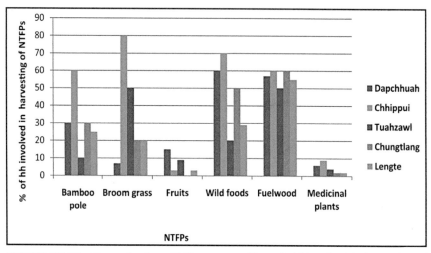

FIGURE 12.10   (See color insert.) Percentage of household involved in harvesting of NTFPs.

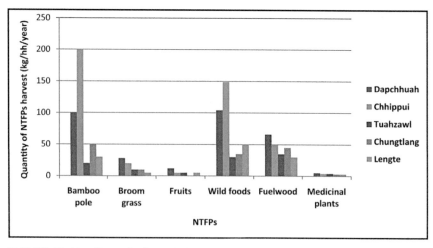

FIGURE 12.11   (See color insert.) Quantity of NTFPs harvested in the studied villages (kg/hh/year).

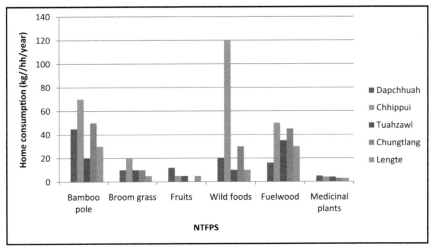

FIGURE 12.12   (See color insert.) Amount of NTFPs used for home consumption in different villages (kg/hh/year).

For a period of time, prohibition of harvesting of bamboo shoots by the Forest Department created a problem in marketing of bamboo shoots. Broom grass is the only NTFP which have its marketing channel at national level. A number of stakeholders are involved in the marketing of broom grass as it is a lengthy channel. Though, local trade has an advantage which allow participation by the poor.

12.3.5   SOCIAL BENEFITS FROM ENTRY POINT ACTIVITIES (EPA)

Various incentive works of community development was done with maximum benefits to the village people. This led to draw the attention and attitude of village inhabitants toward the importance of conservation and propagation of the forest. Minor infrastructural benefits like construction of community information center, community hall, public urinal, bazaar set, vegetable warehouse, public water tank, approach road, step, and funds for playground were also obtained. Distribution of torch light (for aged persons), pressure cooker, and LPG gas connection was also done to reduce the dependence of the villagers on fuelwood alone, which is a means of conservation of forest resources. They even got an

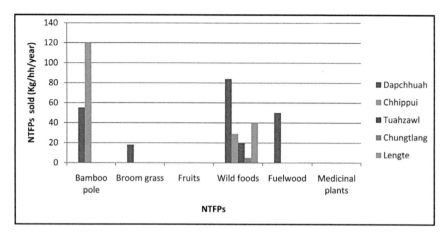

FIGURE 12.13 **(See color insert.)** Amount of NTFPs sold in different villages.

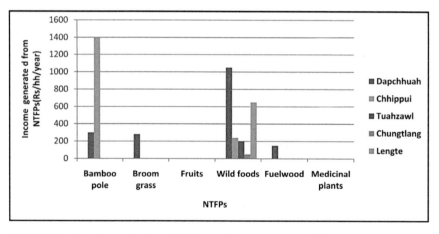

FIGURE 12.14 **(See color insert.)** Estimated income generated from NTFPs in different villages.

opportunity to work in the community forest and received payment from the Forest Department (Figure 12.16).

12.4 CONCLUSIONS

It is important to explore potential of different NTFPs, which can bring in more economic benefits to local communities if their potential is har-

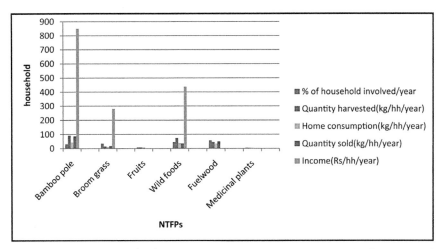

FIGURE 12.15 **(See color insert.)** Mean of all the NTFPs uses in the five surveyed villages.

nessed properly. The NTFPs provides a security net by providing the basic requirements for the villagers in one way or the other. Hence, it is very important to protect them and harvest in a sustainable way, and more emphasis has been placed on this. While unsustainable and sustainable practices are seen in harvesting forest fruits, it is important to have awareness among the villagers who are always in touch with the forest resources. Villagers have to put extra efforts to bridge a wide gap between the large demand and less supply of NTFP resources. Efforts should be made to introduce the quick profitable crops (cash crops, agricultural, horticultural, herbal, etc.) and multipurpose tree species for fulfilling the basic requirements of the villagers, which will reduce the pressure on natural resources.

FIGURE 12.16 Entry point activities (EPA) in Lengte and Tuahzawl village.

KEYWORDS

- bamboo
- broom grass
- forest product
- fuel wood
- livelihood
- wild fruits

REFERENCES

Anderson, A., (1992). 'Land use strategies for successful extractive economies in Amazonia.' *Advances in Economic Botany, 9*, 67–77.

Chaluvaraju, Singh, D. B., Rao, M. N., Ravikanth, G., Ganeshaiah, K. N., & Uma S. R., (2001). Conservation of bamboo genetic resources in Western Ghats: status, threats and strategies. In: *Forest Genetic Resources: Status, Threats and Conservation Strategies,* New Delhi, India: Oxford-IBH Publications, pp. 97–113.

Chopra, K., (1993). The value of non-timber forest products: an estimation for tropical deciduous forests in India. *Economic Botany, 47*(3), 251–257.

Hegde, R., Suryaprakash, S., Achoth, L., & Bawa, K. S., (1996). Extraction of NTFPs in the forests of BR Hills 1. Contribution to rural income. *Economic Botany, 50*, 243–250.

Homma, A., (1992). 'The dynamics of extraction in Amazonia: A historical Perspective.' *Advances in Economic Botany, 9*, 23–31.

Maikhuri, R. K., & Gangwar, A. K., (1991). Fuel wood use by different tribal and nontribal communities in North East India. *Natural Resources Forum, 15*, 162–165.

National Centre for Human Settlements and Environment, (1987). *Documentation of Forest and Rights. vol. 1.* New Delhi, India, National Centre for Human Settlements and Environment.

Prasad, S., Chellam, R., & Krishnaswamy, J., (2001). Fruit removal patterns and dispersal of Emblica officinalis (Euphorbiaceae) at Rajaji National Park, India. In: *Tropical Ecosystems: Structure, Diversity and Human Welfare*, New Delhi, India. Oxford-IBH Publications, 513–516.

Salam, M. A., Noguchi, T., & Koike, M., (2005). Factors influencing the sustained participation of farmers in participatory forestry: a case study in central Sal forests in Bangladesh. *J., of Environment and Management, 74*, 43–51.

Takasaki, Bahram, B. L., & Coomes, O. T., (2004). Risk coping strategies in tropical forests: floods, illness and resource extraction. *Environment and Development Economics, 9*, 203–2224.

Uma Shaanker, R., Ganeshaiah, K. N., & Rao, M. N., (2001). Genetic diversity of medicinal plant species in deciduous forest of South India: impact of harvesting and other anthropogenic pressures. *Journal of Plant Biology, 28*, 91–97.

Yashwant, S., Rawat, Chandra, M.,& Sharma. (2010). Sustainable development and management of forest resources: a case study of site specific microplan preparation and joint forest management (JFM) Implementation in district rudraprayag, Central Indian Himalaya. *Eviron. We. Int. J., Sci. Tech., 5*, 1–12.

CHAPTER 13

EFFECTS OF BLACK ROT ON THE ANTIOXIDANT PROPERTIES OF MORRIS AND SARAWAK PINEAPPLE (ANANAS COMOSUS)

HASVINDER KAUR BALDEV SINGH, VICKNESHA SANTHIRASEGARAM, ZULIANA RAZALI, and CHANDRAN SOMASUNDRAM

Institute of Biological Sciences, Faculty of Science and the Centre for Research in Biotechnology for Agriculture (CEBAR), University of Malaya, 50603, Kuala Lumpur, Malaysia, E-mail: hasvin_gill@yahoo.com

CONTENTS

Black rot is a postharvest disease that is currently affecting the pineapple industry of Malaysia. The pineapple cultivars that are susceptible to this postharvest disease include Morris and Sarawak. The objective of this study was to determine the antioxidant properties of healthy and infected

Morris and Sarawak pineapples. The methanolic extract of the pineapples was assessed for total polyphenol content, total flavonoid content, total antioxidant capacity, and DPPH radical scavenging activity. The total polyphenol content was assayed using the Folin-Ciocalteu method. The total flavonoid content was detected using the aluminum chloride colorimetric method. Antioxidant activities were determined using the total antioxidant capacity and DPPH radical scavenging assay. Both the infected pineapple cultivars were found to contain a higher content of polyphenols and flavonoids. Similarly, both the antioxidant assays conducted showed a higher level of antioxidant activity in infected pineapples. The increased level of antioxidant properties in pineapples infected by black rot is related to the secondary response of the plant defense mechanism.

13.1　INTRODUCTION

Agriculture plays an important role in the local and export industries of Malaysia. Pineapple is one of the industry that contributes to the country's economic status. Based on statistics, Malaysia's production of pineapples was estimated to be around 309,331 metric tons, 314,405 metric tons, and 315,977 metric tons in the year 2011, 2012, and 2013 respectively (FAO, 2016). Pineapple is a short, herbaceous perennial plant that is grown on peat and mineral soils. Ranking as the third tropical fruit in the world, pineapple is the only edible member of the Bromeliaceae family and can be consumed fresh, cooked, or preserved (Bartholomew et al., 2003; Chan, 2010; Kudom and Kwapong, 2010).

Pineapple has an extraordinary flavor, aroma, and juiciness. This queen of fruits also contains a rich amount of nutrients such as vitamin C, calcium, potassium, magnesium, phosphorus and crude fiber. Some minerals of pineapple such as manganese is essential for the activation of certain enzymes in the human body as well as for the formation of bones. Other beneficial trace elements like copper helps to promote iron absorption and aids in regulating heart rate and blood pressure. Besides these benefits, pineapple is used as a nutritional supplemented fruit for health purpose as it contains a high amount of ascorbic acid, sugar, and moisture content. The high level of vitamin C in pineapple is helpful in preventing the risk of gastrointestinal cancer. Pineapple also contains bromelain, which is a proteolytic enzyme that acts as an anti-inflammatory and digestive agent. Pineapple fruit also has a rich amount of antioxidants such as flavonoids, ascorbic acid, and other types of phenolic compounds that are associated with antioxidant activities and are significant indicators of fruit standards in the global market (Hemalatha and Anbuselvi, 2003; Brat et al., 2004; Mhatre et al., 2009; Sabahelkhier et al. 2010; Debnath et al., 2012).

In Malaysia, the most prominent pineapple cultivars are Morris, Sarawak, Gandul, Maspine, and Yankee. Josapine and N36 pineapple hybrids are also produced in Malaysia. However, as reported by the Malaysian Agricultural Research and Development Institute (MARDI), the pineapple industry of Malaysia is facing difficulties with the Morris and Sarawak cultivars as they are susceptible to a destructive pineapple disease called black rot.

The pineapple black rot disease is caused by a facultative parasitic fungus called *Thielaviopsis paradoxa* (De Seyen.). This postharvest disease is a common fresh fruit problem and usually occurs when the fruit is wounded during harvest. This fruit disease is characterized by a water-soaked lesion and browning of the infected tissue. During an early infection, the tissue becomes soft and watery. When the fruit further decays, the tissue will be enclosed with a black coating composed of macrospores of the fungus near to the core of the fruit (Bratley & Mason, 1939; Paull & Reyes, 1996; Mardi, 2015). Thus far, there are no studies done to evaluate the changes in antioxidant activities in pineapples infected by the black rot disease. Hence, the objective of this study was to determine the antioxidant properties of healthy and infected Morris and Sarawak pineapples.

13.2 MATERIALS AND METHOD

13.2.1 PLANT MATERIAL

Morris and Sarawak pineapples were used as samples for the experiments. Fresh and infected fruit were obtained from the local market located 3.7 km from the Postharvest Physiology and Biotechnology Laboratory, University of Malaya. Both the Morris and Sarawak pineapple were selected visually by picking those that have the same size and harvest time. The fresh fruits were ensured not to have any defects as indicated by the firmness and physical appearance; the infected fruits were picked based on external defects.

13.2.2　SAMPLE EXTRACTION

The pineapple peels were removed, and the flesh was cut into small chunks. The small chunks were then homogenized using liquid nitrogen with a mortar and pestle set. The extraction method was performed according to Xu et al. (2008) with slight modifications. Equal parts of the homogenized sample (10 g) was mixed with 10 mL of 80% methanol to purify the sample (ratio 1:1). The mixture was placed in a shaking incubator (Shellab Orbital Shaking Incubator S14, OR, USA) at 250 rpm for 30 min at room temperature and was then centrifuged. The supernatant was used for the analysis of antioxidant activity.

13.2.3　DETERMINATION OF ANTIOXIDANT PROPERTIES

13.2.3.1　Total Polyphenol Content

Total polyphenol content of sample extracts was determined using the Folin-Ciocalteu assay (Singleton et al., 1965) modified to a microscale (Bae et al., 2007). A standard curve of gallic acid ($y = 0.00566x$, $r^2 = 0.9955$) was prepared, and the results were reported as milligrams of gallic acid equivalent (GAE) per 100 mL sample extract.

13.2.3.2　Total Flavonoid Content

Flavonoid content of the samples was determined using a colorimetric method described by Sakanaka et al. (2005). A standard curve of (+)-catechin ($y = 0.0135x$, $r^2 = 0.9943$) was prepared, and the results were reported as milligrams of catechin equivalent (CE) per 100 mL sample extract.

13.2.3.3 Total Antioxidant Capacity

Total antioxidant capacity of the samples was determined using the phosphomolybdenum method described by Prieto et al. (1999). A standard curve of ascorbic acid ($y = 0.0018x$, $r^2 = 0.9981$) was prepared, and the results were reported as micrograms of ascorbic acid equivalent (AAE) per milliliter sample extract.

13.2.3.4 DPPH Radical Scavenging Activity

DPPH assay was carried out as described by Oyaizu (1986) and Bae and Suh (2007). A standard curve of ascorbic acid ($y = 10.145x$, $r^2 = 0.9907$) was prepared, and the results were reported as micrograms of ascorbic acid equivalent (AAE) per milliliter of the extract. The radical scavenging activity was calculated accordingly:

$$\% \text{ DPPH inhibition} = (A_{control} - A_{sample}/A_{control}) \times 100$$

where $A_{control}$ is absorbance reading of control and A_{sample} is absorbance reading of the sample.

13.2.4 STATISTICAL ANALYSIS

Statistical analysis was carried out using SPSS 19.0 software (SPSS Inc., IBM), and the data were represented by mean values ± standard deviation (SD). The differences found between the mean values of the samples were obtained through analysis of variance (one-way ANOVA) by referring to Tukey's honestly significant difference (HSD) test. This was done at $p < 0.05$ significance level.

13.3 RESULTS AND DISCUSSION

In this study, four antioxidant assays were performed on Morris and Sarawak pineapples. The antioxidant assays were total polyphenol content, total flavonoid content, total antioxidant capacity, and DPPH radical scavenging activity (Figure 13.1).

As can be seen in the results above, Figure 13.1 showed that the infected Morris and Sarawak pineapples possessed highest total polyphenol content of 64.46 ± 3.16 mg GAE/mL and 29.46 ± 5.89 mg GAE/mL, respectively. The healthy Morris and Sarawak pineapples had much lower polyphenol content of 44.11 ± 1.64 mg GAE/mL and 24.82 ± 9.41 mg GAE/mL than the infected pineapples (Figure 13.2). Higher values for the total polyphenol content were observed in both healthy and infected Morris pineapples than those observed in Sarawak pineapples. Nevertheless, a previous study reported three-fold higher increase in polyphenol content in the infected tissue of plum leaves. It is reported mentioned that the polyphenols can act as defense-related metabolites in the infected plants and that antioxidative substances are also related to the stress factors of the plants (Favali and Pressacco, 2000).

In plants, flavonoids act as a physiological regulator and chemical messenger other than having a role in the plant flower coloration. Flavonoids also defend plants against the attack of herbivores and pathogens. Figure 13.2 above shows the total flavonoid content estimated using a catechin equivalent. The infected Morris pineapple showed the highest total flavonoid content with 20.00 ± 0.92 mg CE/100 mL, while

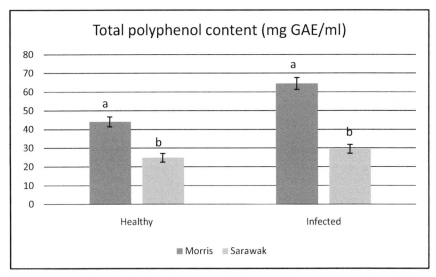

FIGURE 13.1 Total polyphenol content of Morris and Sarawak pineapples. [1]Values followed by different letters are significantly different (p < 0.05) (n = 9).

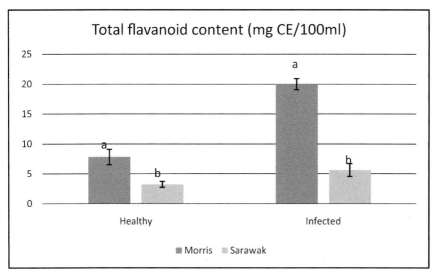

FIGURE 13.2 Total flavonoid content of Morris and Sarawak pineapples. [1]Values followed by different letters are significantly different (p < 0.05) (n = 9).

the infected Sarawak pineapple showed total flavonoid content of 5.63 ± 1.08 mg CE/100mL. Both the healthy pineapples of Morris and Sarawak were found to have a lower amount of total flavonoid content than the infected pineapples. Healthy Morris pineapple had 7.80 ± 1.28 mg CE/100 mL flavonoid content, while the healthy Sarawak pineapple had a lower flavonoid content of 3.22 ± 0.49 mg CE/100 mL. The increase in flavonoids is associated with its function as a secondary antioxidant defense system when a plant is exposed to the conditions of abiotic and biotic stresses. The results were supported by several studies showing the protective effects of flavonoids against pathogenic infections (Rice-Evans et al., 1995; Cook and Samman, 1996; Kumar et al., 2003; Pandey, 2007; Agati et al., 2012).

Total antioxidant capacity was determined according to Prieto et al. (1999). In Figure 13.3, the antioxidant capacity was found to be highest in the infected Morris and Sarawak pineapples. However, the total antioxidant capacity was higher in the infected Sarawak pineapple (1132.2 ± 50.22 µg AAE/mL) than in the infected Morris pineapple (1099.40 ± 20.55 µg AAE/mL). Both the healthy Morris and Sarawak pineapples

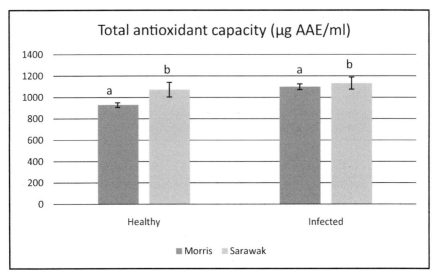

FIGURE 13.3   Total antioxidant capacity of healthy and infected fruits of both Morris and Sarawak pineapples. [1]Values followed by different letters are significantly different ($p < 0.05$) ($n = 9$).

had a lower total antioxidant capacity of 928.90 ± 22.50 µg AAE/mL and 1072.20 ± 37.88 µg AAE/mL, respectively.

DPPH is a stable free radical that is commonly used to determine the radical scavenging activity of natural compounds. With the reduction by an antioxidant, the absorption of DPPH reduces because of the formation of DPPH-H, which is a nonradical form. Figure 13.4 shows that the infected pineapples of Morris and Sarawak had significantly higher radical scavenging activity than the healthy pineapples. In comparison, the infected Morris pineapple had higher amount of radical scavenging activity of 6.01 ± 0.09 µg AAE/mL than the infected Sarawak pineapple, which had only 3.03 ± 0.22 µg AAE/mL. Similarly, the healthy Sarawak pineapple showed a higher radical scavenging activity of 4.2 ± 0.08 µg AAE/mL than healthy Morris pineapple, which had only 2.42 ± 0.39 µg AAE/mL. In a study done by Nahiyan and Matsubara (2012), the infected plant showed an increase in DPPH radical scavenging activity, polyphenols, and ascorbic acid content. Thus, the disease tolerance is linked to the antioxidative substance of a plant.

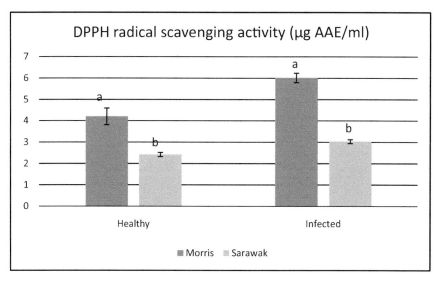

FIGURE 13.4 DPPH radical scavenging activity of healthy and infected fruits of both Morris and Sarawak pineapples. [1]Values followed by different letters are significantly different ($p < 0.05$) (n = 9).

13.4 CONCLUSION

Results obtained revealed a significant increase in antioxidant properties of the Morris and Sarawak cultivars during the infection of black rot. Overall, Morris pineapple showed a higher total polyphenol content, total flavonoid content, and radical scavenging activity than the Sarawak pineapple. The Sarawak pineapple only showed higher amount in total antioxidant activity than Morris pineapple. All the four assays showed increase in antioxidant properties during the infection. More studies are needed to investigate in detail the biochemistry and physiology of these actions.

ACKNOWLEDGMENT

The authors would like to thank the University of Malaya and MOSTI E-Science grant 02-01-03-SF1019 for supporting this research.

KEYWORDS

- **antioxidants**
- **black rot**
- **flavonoid**
- **morris**
- **pineapple**
- **polyphenol**
- **Sarawak**

REFERENCES

Agati, G., Azzarello, E., Pollastri, S., & Tattini, M., (2012). "Flavonoids as antioxidants in plants: location and functional significance." *Plant Science,* vol. *196*, pp. 67–76.

Bae, S. H., & Suh, H. J., (2007). Antioxidant activities of five different mulberry cultivars in Korea. *LWT-Food Sci. Technol., 40*, 955–962.

Bartholomew, D. P., Paul, R. E., & Rorbach, K. G., (2003). *The Pineapple 'Botany,* Production and Uses'. Department of tropical plant and soil science, CTAHR. University of Hawaii.

Brat, P., Thi-Hoang, L. N., Soler, A., Reynes, M., & Brillouet, J. M., (2004). Physicochemical characterization of a new pineapple hybrid (FLHORAN41 Cv.). *J., Agric. Food Chem., 52*, 6170–6177.

Bratley, C. O., & Mason, A. S., (1939). *Control of Black Rot of Pineapples in Transit* of US. Department of Agriculture. vol. *511*, of USDA Circular. pp. 12.

Chan, Y. K., (2000). Status of pineapple industry and research and development in Malaysia. Proc. III International pineapple symposium (Subhadrabandhu, S., & P., Chairidchai, P., eds.), *Acta. Hort., 529*, ISHS. DOI: 10.17660/ActaHortic.2000.529.7

Cook, N. C., & Samman, S., (1996). "Review: flavonoids-chemistry, metabolism, cardio protective effects and dietary sources," *Journal of Nutritional Biochemistry, 7*(2), pp. 66–76.

Debnath, P., Dey, P., Chanda, A., & Bhakta, T., (2012). A survey on pineapple and its medicinal value. *Scholars Academic J., Pharm, 1*(1).

FAO, (2015). Food and Agricultural Commodities Production. http://faostat3.fao.org. Obtained on 25th Feb., 2015.

Favali, M. A., & Pressacco, L., (2000). Histopathology and polyphenol content in plants infected with phytoplasmas. *Cytobios, 102*, 133–147.

Hemalatha, R., & Anbuselvi, S., (2013). Physicochemical constituents of pineapple pulp and waste. *J., Chem. Pharm. Res., 5*(2), 240–242.

Kudom, A. A., & Kwapong, P. K., (2010). Floral visitors of *Ananas comosus* in Ghana: A preliminary assessment. *Journal of Pollination Ecology, 2*(5), 27–32.

Kumar, S., Gupta, A., & Pandey, A. K., (2013). "Calotropis procera root extract has capability to combat free radical mediated damage," *ISRN Pharmacology*, vol. *2013*, Article ID 691372, pp. 8.

Paull, R. E., & Reyes, M. E. Q., (1996). Preharvest weather conditions and pineapple fruit translucency. *Scientia Hortic., 66*, 59–67.

Malaysian Agricultural Research and Development Institute (MARDI). Pejabat Besar Pos, 50774 Kuala Lumpur, Malaysia.

Mhatre, M., Tilak-Jain, J., De, S., & Devasagayam, T. P. A., (2009). Evaluation of the antioxidant activity of non-transformed and transformed pineapple: A comparative study. *Food Chem. Toxicol, 47*, 2696–2702.

Nahiyan, A. S. M., & Matsubara, Y., (2012). Tolerance to fusarium root rot and changes in antioxidative ability in mycorrhizal asparagus plants. *Hortscience, 47*(3), 356–360.

Oyaizu, M., (1986). Studies on products of browning reactions: antioxidative activities of products of browning reaction prepared from glucosamine. *Jpn. J. N.,Utr., 103*, 413–419.

Pandey, A. K., (2007). "Anti-staphylococcal activity of a pan-tropical aggressive and obnoxious weed *Parihenium histerophorus*: an in vitro study," *National Academy Science Letters, 30*(11–12), pp. 383–386.

Prieto, P., Pineda, M., & Aguilar, M., (1999). Spectrophotometric quantitation of antioxidant capacity through the formation of a phosphomolybdenum complex: Specific application to the determination of vitamin E., *Analytical Biochemistry, 269*, 337–341.

Rice-Evans, C. A., Miller, N. J., Bolwell, P. G., Broamley, P. M., & Pridham, J. B., (1995). "The relative antioxidant activities of plant derived polyphenolic flavonoids," *Free Radical Research, 22*(4), pp. 375–383.

Sabahelkhier, K. M., Hussain, A. S., & Ishag, K. E. A., (2010). Effect of maturity stage on protein fractionation, in vitro protein digestibility and anti-nutrition factors in pineapple (*Ananas comosus*) fruit grown in Southern Sudan. *Afr. J., Food Sci., 4*(8), 550–552.

Sakanaka, S., Tachibana, Y., & Okada, Y., (2005). Preparation and antioxidant properties of extracts of Japanese persimmon leaf tea (kakinoha-cha). *Food Chemistry, 89*, 569–575.

Singleton, V. L., Joseph, A., & Rossi, Jr. J., (1965). Colorimetry of total phenolic with phosphomolybdic-phosphotungstic acid reagents. *Am. J., Enol. Vitic., 16*, 144–158.

Xu, G., Liu, D., Chen, J., Ye, X., Ma, Y., & Shi, J., (2008). Juice components and antioxidant capacity of citrus varieties cultivated in China. *Food Chem., 106*, 545–551.

PART II

PROTECTION OF HORTICULTURE CROPS

CHAPTER 14

ESSENTIAL OILS AS GREEN PESTICIDES FOR PLANT PROTECTION IN HORTICULTURE

MURRAY B. ISMAN and RITA SEFFRIN

Faculty of Land and Food Systems, University of British Columbia, Vancouver V6T1Z4, Canada

CONTENTS

ABSTRACT

Interest in, and research on, bioactivities of plant essential oils to insects has exploded in the past 15 years, according to a recent bibliometric analysis. However, commercial exploitation of this knowledge is being realized much more slowly although essential oil-based pesticides have begun to establish a market presence at least in the USA. The volatility of essential oils makes them especially suitable as fumigants in protected environments and for protection of stored products, but they also have demonstrated utility for protection of horticultural crops. Many essential oils and their major constituents, monoterpenes and sesquiterpenes, have contact toxicity to insects and mites, but their utility is broadened owing to their sublethal behavioral effects as deterrents and repellents. These bioactivities result from neurotoxicity of the terpenes, with at least two distinct mechanisms of action identified thus far. One intriguing aspect of the toxicity of some essential oils in insects is synergy among particular terpenes within an oil, thus enhancing bioactivity. Although many essential oils display bioactivity against insects when tested in the laboratory, only a few commodity oils – those used extensively in the flavor and fragrance industries – have been developed for use as pesticides. These include certain oils from the families Lamiaceae, Lauraceae, Myrtaceae, and Poaceae. Their potential as protectants for horticultural crops is discussed.

14.1 INTRODUCTION

Our global population has doubled in the last 45 years. If the present growth rate of 1.3% per year persists, the population will double again within a mere 50 years according to World Watch Magazine (2004). There are huge pressures to provide food at low cost. Synthetic pesticides have been an important component of industrialized agriculture throughout the world since the 1950s. The "second generation" pesticides were very effective at killing pests and thus boosting crop yields, and being relatively inexpensive, their use quickly spread over the globe. Over time, we discovered that many of these chemicals were extremely pervasive in our environment as a result of their widespread and repeated use and, in some cases, their environmental persistence. Some chemicals take an extremely long time to degrade, such that even those banned decades ago, including dichlorodiphenyltrichloroethane (DDT) and its secondary products, can still be found in the environment today. Organochlorine, organophosphate, carbamate, and pyrethroid pesticides were introduced between the late 1940s and the mid of 1970s, and they helped to usher in industrial agriculture or the "Green Revolution." More recently, other types of pesticides (e.g., neonicotinoids) have been introduced into the world market and industrial agriculture has come to rely more and more on the use of synthetic chemical pesticides to protect crops from pests.

In recent decades, there has been a steady increase in the amount of pesticides used in agriculture. In the European Union alone, more than 200,000 tons of pesticides (active ingredients) are used annually (Eurostat Statistical, 2007). In developing countries, the effects of acute poisoning due to exposure to dangerous pesticides in food are far more severe than in industrialized countries. For example, an estimated 70,000 tons of active ingredient of pesticides per year are imported annually in Central America, and the main groups of insecticide include organophosphates, carbamates and pyrethroids (Castillo et al., 2010). Pesticides can be hazardous to humans and their residues can accumulate in food chains, damaging birds, fish, and other forms of wildlife. In many cases, these side effects are not immediately apparent, but may show up later, for example, in the abnormal eggs of birds that have eaten pesticide-laded insects.

The recent increase in organic farming practices in Europe demonstrates that farming without synthetic pesticides is feasible, scalable, economically viable, and environmentally safe. Land under organic cultivation increased from 5.7 million hectares in 2002 to 9.6 million hectares in 2011, and includes horticultural and orchard crops as well as animal sectors (European Commission, 2013). Investigators have long searched for new, highly selective, and biodegradable pesticides to solve the problem of long-term toxicity to mammals and, environmental persistence of pesticides and develop techniques that can be used to reduce overall pesticide use while maintaining crop yields. Plant essential oils have been used since antiquity for many purposes, including pest control in agriculture and against nuisance pests such as mosquitoes, flies, and ticks. In recent years, consumers have increasingly expressed interest in purchasing organically grown foods as well as using natural and naturally derived materials to eradicate pests in their lawns, gardens, and homes. Pesticides based on plant essential oils or their constituents have demonstrated efficacy against many agricultural pests. In fact, pesticides derived from plant essential oils can have several important benefits. Because of their volatile nature, there is a much lower level of risk to the environment than with current synthetic pesticides. Predator, parasitoid, and pollinator insect populations will be less impacted because of the minimal residual activity, making essential oil-based pesticides compatible with integrated pest management programs. Additionally, resistance to essential oil-based pesticides may develop more slowly if at all, owing to the complex mixtures of constituents that characterize many of these oils.

14.2 BIOLOGICAL ACTIVITY OF ESSENTIAL OILS: A HISTORICAL PERSPECTIVE

Interest has increased in the use of oils to replace conventional chemical pesticides in pest control. Many plant essential oils show a broad spectrum of activity against pest insects owing to their repellent, fumigant, insecticidal, attractant, antifeedant, oviposition deterrent, and growth regulatory activities. These oils also have a long tradition of use in the protection of stored products. Citronella oil, discovered in 1901, was the most widely used personal repellent before the 1940s, and is still used today in many formulations (Brown and Hebert, 1997). The fumigant toxicity of essential oils owing to their high volatility may have first attracted the attention of researchers. The fumigant toxicity of essential oils and their main constituents, the volatile monoterpenes, was first reported in the 1960s (Smelyanets and Kuznetsov, 1968).

Many studies of the insecticidal activity of essential oils were published in the 1970s, especially against stored product pests. Patchouli, *Pogostenmon heyneanus* (Solanaceae) and sweet basil, *Ocimum basilicum* (Lamiaceae), essential oils showed insecticidal activity against *Sitophilus oryzae* (Coleoptera: Curculionidae), *Stegobium paniceum* (Coleoptera: Anobiidae), *Tribolium castaneum* (Coleoptera: Tenebrionidae) and *Bru-

chus chinensis (Coleoptera: Bruchidae) (Deshpande et al., 1974; Deshpande and Tipnis, 1977). Pulegone, linalool, and limonene were effective fumigants against rice weevil, *Sitophilus oryzae*, while *Mentha citrata* oil containing linalool and linalyl acetate also exhibited significant fumigant toxicity to these weevils (Singh et al., 1989). Oviposition inhibition and ovicidal activities have been reported for carvacrol, carveol, geraniol, linalool, menthol, terpineol, thymol, verbenol, carvones, fenchone, menthone, pulegone, thujone, verbenone, cinnamaldehyde, citral, citronellal, and cinnamic acid against house fly, *M. domestica* eggs (Rice and Coats, 1994). Cinnamyl alcohol, 4-methoxy-cinnamaldehyde, cinnamaldehyde, geranylacetone, and α-terpineol are attractive to adult corn rootworm beetles, *Diabrotica* sp. (Hammack, 1996; Petroski and Hammack, 1998) and have been used in traps for these pests.

According to Regnault-Roger (1997), the examination of patents involving essential oils showed that a majority of the inventions focused on household uses, and several formulations were proposed to control mosquitoes and flies; some of them in combination with pyrethroids (Liang, 1988; Kono et al., 1993). Isman (2000) mentioned several investigations confirming that certain plant essential oils not only repelled insects but had contact and fumigant insecticidal actions against specific pests. As part of an effort aimed at the development of reduced-risk pesticides based on plant essential oils, toxic and sublethal effects of some essential oil terpenes and phenols were investigated using the tobacco cutworm (*Spodoptera litura*) and the green peach aphid (*Myzus persicae*) as model pest species.

Survey of the scientific literature on the biopesticidal potential of essential oils from the year 2000 onwards indicates that plants of the families Myrtaceae, Lamiaceae, Asteraceae, Apiaceae, and Rutaceae are important sources for natural pesticides effective against pests in the insect orders Lepidoptera, Coleoptera, Diptera, Isoptera, and Hemiptera.

The mode of action of essential oils also received attention from the research community. Enan (2001, 2005) has provided evidence that many essential oil constituents poison insects by blocking octopamine receptors. Octopamine, synthesized from tyramine, is a neurotransmitter and neuromodulator in arthropods and may have neurohormonal influences as well. The rapid action against some pests is indicative of a neurotoxic mode-of-action and there is some evidence for interference with gamma-aminobutyric acid (GABA)-gated chloride channels as well (Priestley et al., 2003).

According to Isman and Machial (2006), the majority of studies on essential oil bioactivity to insects between 2005 and 2008 were aimed at stored products pests with 78 essential oils tested on *Sitophilus oryzae* and 22 on *Acanthoscelides obtectus* with fumigant toxicity evaluated. They also noted that since 2005, essential oils from 30 plant families were tested against coleopterans pests of stored grain. Essential oils from 22 species belonging to the family Lamiaceae, 17 species of Asteraceae, and 10 species of Myrtaceae were shown to provide control of coleopteran pests of the families Bostrichidae, Bruchidae, Chrysomelidae, Cucujidae, Curculionidae, Dermestidae, and Tenebrionidae (Pérez et al., 2010).

Much research of this type has focused on greenhouse experiments (Regnault-Roger, 2012). The repellence of essential oils from fruit skins of laranja pera (*Citrus sinensis* Osbeck var. pera) and laranja lima (*Citrus aurantium* L.) to the two-spotted spider mite, *Tetranychus urticae* Koch, on string was evaluated in greenhouse experiments. Although both oils, rich in *d*-limonene, showed similar repellent effects in laboratory bioassays, lima oil prevented the movement of mites between plants across oil-treated strings for 1 week in the greenhouse (Camara et al., 2015). In 2011, around 60,500 pounds of *d*-limonene (the dominant monoterpene from *Citrus* peels) were applied as a pesticide in California, although 71% of that total was for non-agricultural uses, primarily for structural pest control (Isman, 2014).

14.3 MEDICINAL PLANTS USED TO PRODUCE ESSENTIAL OIL AS PESTICIDES AND THEIR CONSTITUENTS

Not all terrestrial plants produce essential oils, but many aromatic plants, including some used as culinary herbs and spices, have been used since antiquity as folk medicine and as preservatives in foods (Christaki et al., 2012). Certain of these source plants have been traditionally used for the protection of stored commodities, especially in the Mediterranean region and in Southern Asia, but interest in the oils for pest control was renewed with demonstrations of their fumigant and contact insecticidal activities to a wide range of pests in the 1990s (Isman, 2000).

Many herbs and spices, including rosemary, oregano, sage, thyme, peppermint, and garlic, can be found worldwide, with many originating from the Mediterranean area (Bampidis et al., 2005; Kadri et al., 2011). They contain a wide range of chemical substances such as polyphenols, alkaloids, polypeptides, or their oxygen-substituted derivatives (Cowan, 1999; Perumalla and Hettiarachchy, 2011). Most spices belong to the families Lamiaceae, Lauraceae, Myrtaceae, and Poaceae. Common members from which essential oils are obtained are described in more detail in the following sections.

14.3.1 LAMIACEAE FAMILY

Rosmarinus officinalis, commonly known as rosemary, is a garden plant grown around the world. Aerial parts of rosemary are used as flavoring agent in foods, beverages, and cosmetic preparations and have various traditional uses in ethnomedicine including analgesic, anti-inflammatory, anti-rheumatic, and spasmolytic applications (Minaiyan et al., 2011). Native to the Mediterranean, rosemary grows freely in large areas of southern Europe and is cultivated worldwide. Leading regions of rosemary production are the Mediterranean countries, Northern Africa, England, Mexico, and the USA. For commercial purposes, there are three main types based on their predominant chemical constituent and geographical origin: camphor-borneol (Spain), 1,8-cineole (Tunisia), and verbenone (France). To obtain essential oil of the highest quality, plants should be in bloom and only the flowering tops should be harvested for distillation. With mechanical harvesting, it is better to cut frequently because yields are higher from rapid regrowth. Isman et al. (2008) explored the relationship between chemical composition and insecticidal activity of 10 commercial samples of rosemary oil, based on laboratory bioassays with two agricultural pests, the armyworm *Pseudaletia unipuncta* Haworth (Noctuidae) and the cabbage looper *Trichoplusia ni*. Hübner (Noctuidae). Nine major terpenoid constituents of rosemary oil were quantified in the samples by gas chromatography–mass spectrometry (GC-MS). The major constituents were 1,8-cineole, α-pinene, β-pinene, and camphor; on average 1,8-cineole made up 52% of the oil by weight. Rosemary oil repels female *Thrips tabaci* (Koschier and Sedy, 2003) and is toxic to the two-spotted spider mite *Tetranychus urticae* (Miresmailli et al., 2006). In a fumigant toxicity study on stored grain pests, both rosemary oil and linalool proved to be highly effective in controlling the lesser grain borer, *Rhyzopertha dominica* (Shaaya et al., 1991).

14.3.1.1   Mentha × Piperita

The three main types of mint are peppermint, spearmint, and cornmint. While peppermint and spearmint are used as flavorings in their own right, cornmint is primarily used as a source of menthol. The plant is aromatic, stimulant, and used for relieving chest and nasal congestion. Its oil is one of the most widely used essential oils in food products, cosmetics, pharmaceuticals, dental preparations, mouthwashes, soaps, and alcoholic liquors (Sujana et al., 2013). Over the past 20 years, India has come to dominate the global production of mint oils. Peppermint and spearmint command higher prices than cornmint. Major constituents of mint oil are menthol and menthone and vary according to species. Koschier and Sedy (2003) tested repellence and oviposition effects of mint oil, along with several other essential oils, against onion thrips (*Thrips tabaci*). Mint oil concentrations of 0.1% and 1% significantly reduced egg laying by females on leaves (0.8 eggs per leaf disc treated with 1% mint oil compared to 7.2 eggs per leaf disc on untreated controls). The volatile components of peppermint oil are primarily menthol (29–48%), menthone (20–31%), menthyl acetate (31%), menthofuran (1–7%), and limonene (Khan and Abourashed, 2010). Peppermint was included in a study by Choi et al. (2003) that tested the efficacy of 53 essential oils on 3 life stages of the greenhouse whitefly *Trialeurodes vaporariorum*. Peppermint oil was ranked as one of the eight most toxic oils on all three life stages. In a separate study by Choi et al. (2004), peppermint oil was considered highly toxic in a diffusion bioassay to two-spotted spider mites (*Tetranychus urticae*). Harwood et al. (1990) reported that peppermint monoterpenes resulted in reduced growth and molting abnormalities in cutworms. Peppermint oil was also highly effective on the red flour beetle (*Tribolium castaneum*) as a fumigant in a study evaluating toxicity of essential oils and constituents against four stored-grain pests. Of the constituents tested, 1, 8-cineole, which is commonly found in peppermint oil and is also known as eucalyptol, was one of the most toxic components (Shaaya et al., 1991).

14.3.1.2 *Thymus vulgaris*

Common thyme, is a shrubby, woody plant native throughout the Mediterranean region (Spain, France, and Italy) (Stahl-Biskup and Sáez, 2002). It is also cultivated in some European and New World countries such as Brazil. It is a very popular aromatic herb used as a condiment in many dishes (Jakiemiu et al., 2010). Oil of thyme is the important commercial product obtained by distillation of the fresh leaves and flowering tops of *T. vulgaris*. Thyme oil consists mainly of the phenols carvacrol and thymol (20–80%), with thymol typically being the dominant compound present. In some varieties, up to 51% monoterpene hydrocarbon content has been reported, being made up largely of *p*-cymene and γ-terpinene. Alcohols such as linalool, α-terpineol and thujan-4-ol are also present (Khan and Abourashed, 2010). Choi et al. (2003) tested efficacy against greenhouse whitefly adults, nymphs, and eggs. At the highest rate of 9.3×10^{-3}, thyme oil produced 100% mortality to adults and 88% mortality to eggs. In a study by Shaaya et al. (1991), thyme oil had high fumigant toxicity against the stored-grain pest *Oryzaephilus surinamensis*. Among constituents tested in the same study, carvacrol, linalool, and α-terpineol were also highly toxic. Machial et al. (2010) tested 17 essential oils for toxicity to two lepidopteran species, the oblique-banded leafroller (*Choristoneura rosaceana*), and the cabbage looper (*Trichoplusia ni*), in which thyme oil was the second most toxic to 1st instar *C. rosaceana* larvae.

14.3.2 *LAURACEAE FAMILY*

14.3.2.1 *Cinnamomum* Species

Cinnamomum is a large genus, many species of which yield a volatile oil on distillation. The most important *Cinnamomum* oils in world trade are those from *C. verum* (cinnamon bark and leaf oils), *C. cassia* and *C. camphora*. Cinnamon is one of the most important spices used daily by people all over the world. Cinnamon oil primarily contains cinnamaldehyde, cinnamic acid, and cinnamic alcohol. In addition to being an antioxidant, anti-inflammatory, antidiabetic, antimicrobial, and lipid-lowering compound, cinnamon has also been reported to have activities against neurological disorders, such as Parkinson's and Alzheimer's diseases (Visweswara Rao and Hua Gan, 2014). Given the large number of *Cinnamomum* species that exist, their widespread distribution in Asia, and the number still not characterized in terms of essential oil content and composition, the genus has much potential for providing new tree crops in developing countries. *Cinnamomum camphora* is native to China south of the Yangtze River, Taiwan, southern Japan, Korea, and Vietnam, and has been introduced to many other countries. The top four cinnamon-producing countries are Indonesia, China, Sri Lanka, and Madagascar. In a study conducted by Eun-Jeong et al. (2008), cinnamon oil was found to have an LD_{50} value of 0.016 mg/cm^2 on rice weevil (*Sitophilus oryzae*), a common destructive pest of stored grains. This study also tested fumigant activity of (E)-cinnamaldehyde and 41 structurally related compounds, of which allyl cinnamate was the most toxic, with 83% mortality using the closed container fumigant method at a 0.013 mg/cm^3 dosage.

14.3.3 POACEAE FAMILY

Cymbopogon species are commonly known as lemongrass. Some species are commonly cultivated as culinary and medicinal herbs by people in many countries because of their scent, resembling that of lemons. In Brazil, for example, the tea, infusion, and extracts of *C. citratus*, which are prepared with fresh or dry leaves, are often used in popular medicine as a restorative, digestive, effective drug against colds, and as an analgesic, antihermetic, anticardiopatic, antithermic, and anti-inflammatory (Negrelle and Gomes, 2007). Citronella oil is one of the essential oils obtained from the leaves and stems of different species of *Cymbopogon*. The oil is used extensively as a source of perfumery chemicals such as citronellal, citronellol, and geraniol. Supply is dominated by India and Guatemala. There are two main types of citronella oil, referred to as Ceylon and Java. Both types contain citronellal, citronellol, and geraniol as the major components, but the proportions of these vary greatly depending on source and type, with Java having a higher percentage made up of these 3 components. The Ceylon type contains a higher percentage of monoterpenes than the Java type, and West Indian lemongrass oil contains 65–85% citral, whereas Cameroonian *C. citratus* contains geranial as its major component, comprising 33% of the oil. Other compounds that may be present in lemongrass oil include myrcene (12–20%), dipentene, methylheptenone, β-dihydropseudoionone, neral, β-pinene, linalool, methylheptenol, α-terpineol, geraniol, nerol, farnesol, citronellol, and volatile acids such as isovaleric, geranic, caprylic, citronellic and others (Khan and Abourashed, 2010).

Lemongrass oil is also known for its insecticidal properties and was included, among 52 other essential oils, in a test against greenhouse whitefly by Choi et al. (2003). The oil was the most effective on the egg stage (98% mortality). A fumigant toxicity study of essential oils and constituents on four stored-product beetle species showed that lemongrass oil had little to no toxic effect. However, linalool and terpinen-4-ol, found in lemongrass oil as minor constituents, were found to be highly effective (Shaaya et al., 1991). In a study by Machial et al. (2010) on the toxicity of essential oils on two lepidopteran species, lemongrass oil was the second most toxic to 1st instar cabbage looper larvae, with 53% mortality at a concentration of 5 µL/mL, an LC_{50} of 7.2 µL/mL, and an LD_{50} of 60.5 µL/mL. A GC-MS analysis found the major constituents of this lemongrass oil sample to be citral (47.1%), *trans*-verbenol (32.1%), and camphene (10.7%).

14.3.4   MYRTACEAE FAMILY

Clove (*Syzygium aromaticum*) is one of the most valuable spices that has been used for centuries as a food preservative and for many medicinal purposes. Clove is native to Indonesia but is now cultivated in several parts of the world including Brazil. This plant represents one of the richest source of phenolic compounds and has great potential for pharmaceutical, cosmetic, food, and agricultural applications (Cortés-Rojas, 2014). A major component of clove taste is imparted by the chemical eugenol. It is widely used in agricultural applications to protect foods from microorganisms during storage, which might have an effect on human health, and as a pesticide and fumigant (Kamatou et al., 2012). The production of all types of clove oil – leaf, stem, and bud – is dominated by Indonesia. Madagascar, the largest exporter of cloves, also exports some oil, typically clove leaf oil. Clove bud oil typically contains 60–90% eugenol, 2–27% eugenol acetate, and 5–12% β-caryophyllene. Constituents found in clove leaf and clove stem oils are very similar, but may differ in ratio. Naphthalene may be present in leaf and stem oils, but does not occur in the bud oil (Khan and Abourashed, 2010). In Choi et al. (2003), clove oil provided 98% mortality to greenhouse whitefly nymphs at 9.3×10^{-3} concentration, and clove (bud) oil produced 94% mortality of eggs at the same concentration. Among the 53 essential oils included in this study, only 8 (including clove), were considered highly effective against all three stages (adult, nymph, and egg) of greenhouse whitefly. Eugenol is a phenylpropanoid found in many essential oils, but in highest concentrations in clove and cinnamon oils. Eugenol is known for its herbicidal, insecticidal, and antifungal activity. Enan (2005) determined eugenol LD_{50} values of 1.9 µg/insect for fruit fly (*Drosophila melanogaster*). Isman (2000) compared the toxicity of eugenol with α-terpineol and terpinen-4-ol. Eugenol was 7–9 times more toxic than the two terpenes in the western corn rootworm beetle (*Diabrotica virgifera virgifera*).

14.3.4.1 Eucalyptus Species

Some *Eucalyptus* species have attracted attention from horticulturists, global development researchers, and environmentalists because of desirable traits such as fast-growing sources of wood, and production of oil that can be used for cleaning and as a natural insecticide.

Eucalyptus oil is the generic name for distilled oil from the leaf of *Eucalyptus*, a genus of the plant native to Australia and cultivated worldwide. Eucalyptus oil has a history of wide application. The oil is antiseptic and is used in infections of the upper respiratory tract and certain skin diseases (Kumar et al., 2013). Production of eucalyptus oil is dominated by China. A range of secondary sources include Brazil, India, Australia, and South Africa. There are a range of eucalyptus oil types – medicinal (1,8-cineole-rich); perfumery (citronellal-type); industrial (piperitone-type) – and care is therefore needed in interpreting prices and volumes.

The insecticidal effects of *Eucalyptus dundasii* Maiden essential oil was studied on the adults of the lesser grain borer, *Rhyzopertha dominica* (F.), and the saw-toothed grain beetle, *Oryzaephilus surinamensis* (L.). Chemical analysis indicated that 1,8-cineole (54.15%), *p*-cymene (12.41%), α-thujene (11.37%), and β-caryophyllene (6.7%) were the major constituents. *E. dundasii* essence was repellent for both insects (Aref et al., 2005). Jemâa et al. (2012) investigated seasonal variation in chemical composition of essential oils isolated from leaves of *Eucalyptus camaldulensis, E. astringens, E. leucoxylon, E. lehmannii,* and *E. rudis* and assessed their fumigant activity against three stored-date moth pests: *Ephestia kuehniella, Ephestia cautella,* and *Ectomyelois ceratoniae.* The five essential oils contained 1,8-cineole, α-pinene, and α-terpineol as the major constituents. Of the other major constituents, β-pinene and *p*-cymene were only present in *E. rudis* essential oil. In addition, *o*-cymene was specific only to *E. camaldulensis* and *E. rudis* essential oils. Results demonstrated that fumigant toxicity varied with season, insect species, essential oil concentration and exposure time. *E. camaldulensis* essential oil was more toxic against *E. cautella* and *E. kuehniella.* LC_{50} values were, respectively, 11.07 and 26.73 mL/L air. However, for *E. ceratoniae, E. rudis* essential oil was more effective, with an LC_{50} of 31.4 mL/L air.

14.3.5   GERANIACEAE FAMILY

The essential oil of geranium is extracted through steam distillation of stems and leaves of the geranium plant *Pelargonium* spp. Essential oils derived from these aromatic plants have demonstrated biological properties and can be used to prevent and treat human systemic diseases, including infectious diseases (Carmen and Hancu, 2014). Production of geranium oil is dominated by Egypt and China. Very small quantities of "rose" geranium oil (bourbon oil) are produced in a number of African countries – South Africa, Madagascar, Rwanda – and this oil commands a significant price premium. Geranium oil consists of 60–70% alcohols, primarily citronellol and geraniol, with linalool and phenethyl alcohol also present. The esters geranyl tiglate, geranyl acetate, citronellyl formate and citronellyl acetate comprise 20–30% of the oil (Khan and Abourashed, 2010). Geranium oil was tested for toxicity against three life stages of greenhouse whitefly, among 53 other essential oils by Choi et al. (2003). All three stages were effectively controlled (>90% mortality) by geranium oil, but at the highest dose only. Joen et al. (2009) found that the acaricidal effects of geraniol on storage food mites (*Tyrophagus putrescentiae*) were more effective than the industry standard benzyl benzoate, with LD_{50} values of 1.27 μg/cm^3 and 1.95 μg/cm^3, respectively.

14.3.6   AMARYLLIDACEAE FAMILY

With a history of human use of over 7,000 years, *Allium sativum*, commonly known as garlic, is native to central Asia. Garlic has been cultivated for around 4,000 years and has not only been used in food preparation but also as a medicine and crop protection product. Garlic is known for its positive effects on health, particularly the prevention of cardiovascular diseases and certain digestive cancers (Lalla et al., 2013). This long history gave garlic a head start in recent efforts to gain EU regulatory approval for its use in agriculture. A vast body of literature was available to drawn upon when compiling the required data, for example, data on toxicity, residues, ecotoxicity and fate and behavior in the environment. Most published literature cites that diallyl disulfide is the main compound in garlic oil, at 60%, with diallyl thiosulfinate, allylpropyl disulfide, diallyl disulfide, and diallyl trisulfide also being major components. Non-sulfur compounds that may be found in garlic oil include citral, geraniol, linalool, and β-phellandrene (Khan and Abourashed, 2010). Machial et al. (2010) reported garlic oil to be the most effective of 17 tested essential oils on 1st instar cabbage looper larvae (*Trichoplusia ni*), with LC_{50} = 3.3 μL/mL and LD_{50} = 22.7 μg/insect. It ranked 5th out of 17 in terms of its toxic effect to the oblique-banded leaf roller (*Choristoneura rosaceae*), with 22% mortality of 1st instar larvae at 5 μL/mL. The major constituents of the garlic oil sample used in that study were 35.2% diallyl disulfide, 26.2% di-2-propenyl trisulfide, 20.7% 3,3 thiobis-1-propene, 6.6% methyl 2-propenyl trisulfide, and 4.5% methyl 1-propenyl disulfide.

14.4   ESSENTIAL OILS SYNERGY VERSUS ISOLATED CONSTITUENTS AND THEIR EFFECTS ON AGRICULTURAL PESTS

Essential oils have received attention in recent years, in part owing to concerns about synthetic pesticides and for their potential to reduce the development of resistance to pesticides. They possess various biological properties. The wide range of biological activities showed by an essential oil can be related to its qualitative and quantitative composition. They can consist of terpenoids (monoterpenes, sesquiterpenes, and diterpenes in the form of hydrocarbons, alcohols, aldehydes, ketones, ethers, esters, peroxides, and phenols), aromatic compounds (C6-C3 and C6-C1 compounds; less frequent but characteristic of certain essential oils), and low-molecular-weight aliphatic compounds (hydrocarbons, alcohols, acids, aldehydes, esters, and lactones) with different physical, chemical, and pharmacological properties, responsible for the activity of whole essential oil (Blázquez, 2014). Many formulations include a blend of essential oils to create a product with a broad spectrum of action and multiple modes of action. In fragrances and flavors, mixing essential oils or blending is considered part art and part science. When a formulation is prepared, it is necessary to consider: (1) the chemistry of the oil to determine its volatility, viscosity, and other physical properties and (2) the desired action as blending correctly allows for a synergistic effect within the blend. This means the action of the oil is increased by mixing several oils together; in some cases, the sequence in which the oil is blended can also be a factor. Changing the sequence can change the properties.

Isolated compounds from essential oils can also be effective as pesticides, and in some cases, higher toxicity is achieved when the compound is removed from the context of the parent oil. Certain essential oils, commonly used in fragrances and as flavors, are exempt from pesticide registration in the USA. Specifically, those essential oils (and constituents) on FIFRA (Federal Insecticide, Fungicide, and Rodenticide Act) List 25B ("Exempt Active Ingredients") have been used to create insecticides, fungicides, and herbicides since 1998. The following essential oils and compounds are exempt from pesticide registration in the USA: cinnamon, citronella, clove, garlic, geranium, lemongrass, mint, peppermint, rosemary, thyme, eugenol, and geraniol. Australia has a similar list of exempt products that includes many essential oils.

Considerable research has been conducted in Canada to evaluate synergy and compare toxicity of individual compounds to their parent essential oils. Contact and fumigant toxicities of thymol, citronellal, eugenol, and rosemary oil were tested on the wireworm *Agriotes obscurus*. Thymol was the best contact toxin (LD_{50} = 196.0 μg/Larva), whereas citronellal and eugenol were less toxic (LD_{50} = 404.9 and 516.5 μg / larva, respectively). Rosemary oil did not show any significant contact toxicity, even at 1,600 μg /larva. In terms of fumigant toxicity, citronellal was the most toxic to wireworm larvae (LC_{50} = 6.3 μg /cm³) followed by rosemary oil (LC_{50} = 15.9 μg /cm3), thymol (LC_{50} = 7.1 μg /cm3), and eugenol (LC_{50} = 20.9μg /cm3) (Waliwitiya et al., 2005). Monoterpenoids (terpenes and biogenically related phenols) commonly found in plant essential oils were tested for acute toxicity *via* topical application to tobacco cutworms (*Spodoptera litura* Fab.), the most toxic among 10 compounds being thymol (LD_{50} = 25.4 μg /larva) from garden thyme, *Thymus vulgaris*. The compounds were then tested for sublethal effects, specifically inhibition of larval growth after topical application of low doses. Because minor constituents in complex essential oils have been suggested to act as synergists, binary mixtures of the compounds were tested for synergy vis à vis acute toxicity and feeding deterrence. *Trans*-anethole synergized with thymol, citronellal, and α-terpineol in terms of both acute toxicity and feeding deterrence. On the basis of these findings, several complex mixtures were developed and tested as leads for effective control agents. Candidate mixtures demonstrated good synergistic effects (Hummelbrunner and Isman, 2001). Bioassays of rosemary (*Rosmarinus officinalis* L.) essential oil and blends of its major constituents were conducted using host-specific strains of the two-spotted spider mite, *Tetranychus urticae* Koch, on bean and tomato plants. Two constituents tested individually against a bean host strain and five constituents tested individually against a tomato host strain accounted for most of the toxicity of the natural oil. Toxicity of blends of selected constituents indicated a synergistic effect among the active and inactive constituents, with the presence of all constituents necessary to equal the toxicity of the natural oil (Miresmailli et al., 2006). Tak et al. (2015) reported a strong synergistic interaction between 1,8-cineole and camphor, the major constituents of rosemary oil, against the cabbage looper *Trichoplusia ni* and the mechanism of synergy is through enhanced penetration of the cuticle by camphor when admixed with 1,8-cineole (Tak and Isman, 2015).

14.5   ESSENTIAL OIL-BASED PESTICIDES-FROM THE DATA TO THE CROP FIELD AFTER 40 YEARS OF RESEARCH

A literature search encompassing the past 40 years with the keywords "essential oil" and "insects" yielded no less than 2,000 scientific papers. Most papers document the immediate effects as acute toxicity or repellence on arthropods (Regnault-Roger et al., 2012). In spite of widespread public concern for long-term health and environmental effects of synthetic pesticides, especially in Europe and North America, natural pesticides of plant origin have not had much impact in the marketplace thus far. Many conventional insecticides upon which growers had depended for decades (e.g., organophosphates and carbamates) in the USA. were dramatically restricted in use by the Food Quality Protection Act of 1996. In turn, this act created a market opportunity for alternative products, in particular "reduced-risk" pesticides that are favored by the Environmental Protection Agency. Some American companies took advantage of the regulatory exemption and have been able to bring essential oil-based pesticides to market in a far shorter time period than would normally be required for a conventional pesticide.

The most attractive aspect of using essential oils and/or their constituents as crop protectants (and in other contexts for pest management) is their favorable mammalian toxicity (Isman, 2000). Since 2000, EcoSMART Technologies Inc. has been a world leader in essential oil-based pesticides. An insecticide/miticide containing rosemary and peppermint oils as active ingredients was introduced for use on horticultural crops under the name EcoTrol EC. These and other EcoSMART pesticides have also been approved for use in organic food production. They are currently sold for home and garden use under the EcoSMART brand name (a division of Kittrich Corporation, Pomona, CA, USA), and for crop protection as Ecotec EC (Brandt Consolidated, Springfield, IL, USA) and Ecotrol

EC (KeyPlex, Winter Park, FL, USA). Several smaller companies in the USA and the UK developed garlic-oil based pest control products and in the USA. There are also consumer insecticides for home and garden use that contain mint oil as the active ingredient. Menthol was approved for use in North America for the control of tracheal mites in beehives, and a product produced in Italy (Apilife VAR™) containing thymol and lesser amounts of 1,8-cineole, menthol and camphor has been used to control *Varroa* mites in honeybee hives since 1996. Cinnamite™ insecticide/miticide was developed for greenhouse use by Mycotech Corporation in 1999. Based on cinnamon oil, this pesticide was labeled for use against mites and aphids, but is no longer produced.

Although several plant essential oils are exempt from registration in the USA, many more oils are not, and few other countries currently provide for such exemptions. Accordingly, regulatory approval continues to be a barrier to commercialization and will likely continue to be a barrier until regulatory systems are adjusted to better accommodate these products (Isman and Machial, 2006).

Requiem®, an essential oil-based pesticide, was registered in USA in 2008. It was the first botanical registered in the USA since 1990 (Chiasson et al., 2004) and is based on terpene constituents of *Chenopodium ambrosioides*. It is currently marketed by Bayer Cropscience, both in the USA and EU. Plants of the genus *Chenopodium*, notably the species *C. quinoa, C. album, and C. ambrosioides*, have traditionally been used in agriculture as food staples or as natural insecticides (Quarles, 1992). Eco-oil® is an Australian botanical miticide/insecticide designed to control a wide range of insects such as scale, mites, aphids, whitefly, and leafminers. It is a blend of essential oils and contains 2% tea tree oil (*Melaleuca alternifolia*), eucalyptus oil, and canola oil.

Topia™ insecticide was registered in 2009. It is the first completely organic aerosol product from FMC, utilizing geraniol as the active ingredient to eliminate a wide variety of household pests including ants, bed bugs, cockroaches, silverfish, and stink bugs. Even more recently (2014), the EPA approved Captiva®, an insecticide based on *Capsicum* oleoresin and garlic oil, produced by Ecoflora Agro in Colombia, and Prev-Am® (orange peel oil) was approved in the European Union. Other formulations based on garlic oil (Biorepel®) or mixed with clove oil (Pest Out®;

SaferGro) are miticide/insecticides that provide control of mites, thrips, and aphids.

Some recently introduced botanical pesticides based on essential oils were registered in Mexico (Isman, 2014), including Akabrown based on cinnamon, oregano, mint, and clove oils and Ebioluzion with 5% terpenoids. EcoVia™ EC (Rockwell Labs Ltd), based on thyme oil and rosemary oil, was developed for the control of mosquitoes, ticks, sucking insects as well as numerous other flying and crawling insect pests on flowering ornamental plants, trees and turf.

14.6 ADVANTAGES AND DISADVANTAGES OF ESSENTIAL OIL-BASED PESTICIDES FOR PEST MANAGEMENT IN HORTICULTURE

The environmental problems caused by overuse of synthetic pesticides have been the matter of concern for both scientists and the general public in recent decades. Their high toxicity and residues in soil, water resources, and crops that affect public health led researchers to search for new highly selective and biodegradable pesticides. Natural products are one viable alternative to synthetic pesticides as a means to reduce negative impacts to human health and the environment. The move toward green chemistry and the continuing need for developing new crop protection tools with novel modes of action makes discovery and commercialization of natural products as green pesticides an attractive and profitable pursuit that is commanding attention (Koul et al., 2008). The concept of "Green Pesticides" refers to all types of nature-oriented and beneficial pest control materials that can contribute to pest population suppression and increased food production. They are more compatible with the environmental than many synthetic pesticides (Isman and Machial, 2006). Use of essential oils or their components are consistent with this natural concept owing to their volatility and limited persistence under field conditions, and several are exempt from registration in the USA. In general, essential oil-based pesticides are considered safe, and they offer some advantages when compared with synthetic pesticides:

1. Essential oils have been widely used for medicinal and clinical proposes and in addition, they show low toxicity to other verte-

brates including fish and birds because they do not persist in soil and water (Isman, 2000).

2. Essential oils can be applied as tank mixtures with synthetic insecticides that can lessen the quantities of synthetic pesticides used. Further, essential oils can be applied in rotation with conventional products to mitigate the development of insecticide resistance in pest population or for early season application in conjunction with augmentative biological control when pest pressures are low. Nattudurai et al. (2013) tested two synthetic volatile compounds (benzaldehyde and propionic acid) and two volatile oils (camphor and eucalyptus). They were screened individually and in combinations against different life stages of *Tribolium castaneum*. The individual treatments of camphor and eucalyptus oils were less effective, but combinations of benzaldehyde–camphor oil were found to be effective. Benzaldehyde–propionic acid combination recorded 99.3% adult mortality inside a 1 m^3 wooden cage after 15 days, and this mixture can be used as a fumigant in store houses.

3. They can be compatible with biocontrol because of the lack of foliar residues. The essential oils from leaves of *Schinus molle* var. *areira*, *Aloysia citriodora*, *Origanum vulgare*, and *Thymus vulgaris* have shown potential as insecticides against the green stink bug *Nezara viridula*. Their toxicological and behavioral effects on the parasitoid *Trissolcus basalis*, a biological control agent of this pest insect, were also evaluated. The essential oils from *O. vulgare* and *T. vulgaris* proved to be highly selective when used as fumigants and did not change parasitoid behavior. After 1 week, the residues of these oils were harmless and did not show sublethal effects against *T. basalis*. Based on these results, essential oils have potential applications for the integrated management of *N. viridula* (Werdin González, 2013).

4. Owing to their volatility, the oils and their constituents are environmentally non-persistent, with outdoor half-lives of <24 h on surfaces, in soil and in water. There has not been any report of biomagnification of essential oils through the food chain (Regnault-Roger et al., 2012).

5. Limited toxicity to pollinators. Although essential oils have fewer nontarget effects on natural enemies, direct contact on beneficial

insects such as pollinating bees can cause mortality. White et al. (2009) showed that wintergreen oil applied as a fumigant on crop pollinators, such as *Osmia cornifrons* (Radoszkowski) (Hymenoptera: Megachilidae) to control parasitic mites, required >2,473.5 ppm to cause bee mortality. However, when wintergreen oil was topically applied to bees, 353.4 ppm of wintergreen oil caused bee mortality within 10 min. On the other hand, Ebert et al. (2007) focused on *Apis mellifera* adult toxicity when testing 10 products: 1,8-cineole, clove oil, formic acid, marjoram oil, menthol, oregano oil, oxalic acid, sage oil, thymol, and wintergreen. Each product was tested at several concentrations in a sugar syrup fed to bees over several days, and dead bees were counted daily. Menthol and 1,8-cineole had mortality levels no different from controls fed plain syrup after 8 days of treatment. At 14 days of treatment, wintergreen oil was the least toxic. These results indicate that the tested products could all be used safely for treating bees orally if dose is carefully managed in the hive.

6. In terms of green pesticide technology, using oil-in-water microemulsions as a nano-pesticide delivery system to replace traditional emulsifiable concentrates (oil), in order to reduce the use of organic solvents and increase dispersion, wettability and penetration of the droplets is being developed. The advantages of using pesticide oil-in-water microemulsions for improving the biological efficacy and reducing the dosage of pesticides would be a useful strategy in green pesticide technology (Koul et al., 2008).

7. No toxic residues remain on harvested produce. Many oils are obtained from medicinal herbs and have been used as food preservation and flavor for centuries. They are also safe for applicators and field workers.

8. They can be cost competitive and some are approved for organic agriculture in the USA and EU.

Despite many advantages, the market for essential oil-based insecticides has a number of major challenges, and is very small at present. However, at least based on data from California, it is growing at a much faster rate than that for synthetic insecticides.

1. Since essential oils tend to evaporate quickly from surfaces, the spice-based pesticides need to be applied to crops more frequently than conventional pesticides. Some last only a few hours, compared to days or even months for conventional pesticides. Because these natural pesticides are much less toxic than conventional pesticides, they will likely be applied in higher concentrations. Therefore, coverage is important, and reapplications may be necessary to achieve acceptable control.

2. Some essential oils can be phytotoxic if very high rates are applied although phytotoxicity depends on plant species' susceptibility. No phytotoxic symptoms were observed on grape leaves treated with citrus essential oils, and low phytotoxicity was caused by the essential oils of lavender, thyme-leaved savory, and mint, whereas the highest phytotoxicity was observed when basil oil was used (Karamaouna et al., 2013). Vapors of lavender oil, lemon balm, oregano, and thyme caused desiccation of cayenne plants at 2 μL/L, and the same concentration of oregano killed broad bean plants (Digilioa et al., 2008).

3. Many essential oils have best efficacy against small, soft bodied pests (mites, thrips, aphids, mealybugs) and as fumigants for coleopteran stored product pests, but less efficacy against more robust foliar-feeding lepidopterans and coleopterans.

14.7 CONCLUSION

In recent years, consumers have increasingly expressed interest in purchasing organically grown foods, as well as using natural and naturally derived materials to eradicate pests in their lawn, garden and homes. Also, there has been a steady increase in the amount of pesticides marketed for organic food production, demonstrating that farming without synthetic pesticides is entirely feasible, scalable, economically profitable, and environmentally safe. Pesticides based on plant essential oils or their constituents have demonstrated efficacy against many agricultural pests. Due to their volatile nature, there is a much lower level of risk to the environment than with current synthetic pesticides, and they can be compatible with integrated pest management programs. The volatility of essential oils makes them especially suitable as fumigants in protected environments and for protection of stored products, but they also have demonstrated utility for protection of horticultural crops.

KEYWORDS

- **essential oil**
- **fumigant**
- **green pesticide**
- **medicinal plants**
- **plant protection**
- **toxicity**

REFERENCES

Aref, S. P., Valizadegan, O., & Farashiani, M. E., (2015). *Eucalyptus dundasii* Maiden essential oil, chemical composition and insecticidal values against *Rhyzopertha dominica* (F.) and *Oryzaephilus surinamensis* (L.). *J. Plant Prot. Res.*, *55*, 35–41.

Bampidis, V. A., Christodoulou, V., Christaki, E., Florou-Paneri, P., & Spais, A. B., (2005). Effect of dietary garlic bulb and garlic husk supplementation on performance and carcass characteristics of growing lambs. *Anim. Feed Sci. Technol.*, *121*, 273–283.

Blázquez, M. A., (2014). Role of natural essential oils in sustainable agriculture and food preservation. *J. Sci. Res. Rep.*, *3*, 1843–1860.

Brown, M., & Herbert, A. A., (1997). Insect repellents: An overview. *J. Am. Acad. Dermat.*, *36*, 243–249.

Camara, C. A. G., Akhtar, Y., Isman, M. B., Seffrin, R. C., & Born, F. S., (2015). Repellent activity of essential oils from two species of *Citrus* against *Tetranychus urticae* in the laboratory and greenhouse. *Crop Prot.*, *74*, 110–115.

Carmen, G., & Hancu, G., (2014). Antimicrobial and antifungal activity of *Pelargonium roseum* essential oils. *Adv. Pharm. Bull.*, *4*(2), 511–514.

Castillo, L. E., Ruepert, C., & Ugalde, R., (2010). Ecotoxicology and pesticides in Central America. In: *Fundamentals of Ecotoxicology*, Michael, C., Newman (ed.), CRC Press, USA, *3*, 47–57.

Chiasson, H., Vincent, C., & Bostanian, N. J. (2004). Insecticidal properties of a *Chenopodium*-based botanical. *J. Econ. Entomol.*, *97*, 1378–1383.

Choi, W. I., Lee, E. H., Choi, B. R., Park, H. M., & Ahn, Y. J. (2003). Toxicity of plant essential oils to *Trialeurodes vaporariorum* (Homoptera: Aleyrodidae). *Hort. Entomol.*, *96*, 1479–1484.

Choi, W. I., Lee, E. H., Choi, B. R., Park, H. M., & Ahn, Y. J. (2004). Toxicity of plant essential oils to *Tetranychus urticae* (Acari: Tetranychidae) and *Phytoseiulus persimilis* (Acari: Phytoseiidae). *J. Econ. Entomol.*, 97, 553–558.

Christaki, E., Bonos, E., Giannenas, I., & Florou-Paneri, P., (2012). Aromatic plants as a source of bioactive compounds. *Agric.*, 2, 228–243.

Cortés-Rojas, D. F., Souza, C. R. F., & Oliveira, W. P., (2014). Clove (*Syzygium aromaticum*): a precious spice. *Asian Pac. J. Trop. Biomed.*, 4, 90–96.

Cowan, M. M., (1999). Plant products as antimicrobial agents. *Clin. Microbiol. Rev.*, 12, 564–582.

Deshpande, R. S., & Tipnis, H. P., (1977). Insecticidal activity of *Ocimum basilicum* Linn. *Pest.*, 11, 11–12.

Deshpande, R. S., Adhikary, P. R., & Tipnis, H. P., (1974). Stored grain pest control agents from *Nigella sativa* and *Pogostemon heyneanus*. *Bull. Grain Tech.*, 12, 232–234.

Digilioa, M. C., Mancinib, E., Votob, E., & Feob, V., (2008). Insecticide activity of Mediterranean essential oils. *Journal of Plant Interactions*, 3(1), 17–23.

Ebert, T. A., Kevan, P. G., Bishop, B. L., Kevan, S. D., & Downer, R. A., (2007). Oral toxicity of essential oils and organic acids fed to honey bees (Apis mellifera). *J. Apic. Res.*, 46(1), 220–224.

Enan, E., (2001). Insecticidal activity of essential oils: octopaminergic sites of action. *Comp. Biochem. Physiol.*, 130, 325–337.

Enan, E., (2005). Molecular and pharmacological analysis of an octopamine receptor from American cockroach and fruit fly in response to plant essential oils. *Arch. Insect Biochem. Physiol*, 56, 161–171.

Eun-Jeong, L., Kim, J. R., Choi, D. R., & Ahn, Y. J. (2008). Toxicity of cassia and cinnamon oil compounds and cinnamaldehyde-related compounds to *Sitophilus oryzae* (Coleoptera: Curculionidae). *J. Econ. Entomol.*, 10, 1960–1966.

European Commission, (2013). Facts and figures on organic agriculture in the European Union. http:// ec. europa. eu/agriculture/markets-and-prices/more-reports/pdf/organic-013_en. pdf (accessed Jan 21, 2016).

Eurostat Statistical Books, (2007). The use of plant protection products in the european union – Data 1992–2003, Luxembourg. http://ec. europa. eu/eurostat/documents/3217494/5611788/KS-76–06–669-EN. P.,DF/36c156f1–9fa9–4243–9bd3-f4c7c3c8286a?version=1.0 (access Jan 21 2016).

Hammack, L., (1996). Corn volatiles as attractants for northern and western corn rootworm beetles (Coleopteran: Chrysomelidae: Diabrotica sp.). *J. Chem. Ecol.*, 22, 1237–1253.

Harwood, S. H., Moldenke, A. F., & Berry, R. E., (1990). Toxicity of peppermint monoterpenes to the variegated cutworm (Lepidoptera: Noctuidae). *J. Econ. Entomol.*, 83, 1761–1767.

Hummelbrunner, L., & Isman, M. B., (2001). Acute, sublethal, antifeedant, and synergistic effects of monoterpenoid essential oil compounds on the tobacco cutworm, *Spodoptera litura* (Lep., Noctuidae). *J. Agric. Food Chem.*, 49, 715–720.

Isman, M. B. J. A., Bradbury, R., (2008). Insecticidal activities of commercial rosemary oils (*Rosmarinus officinalis.*) against larvae of *Pseudaletia unipuncta* and *Trichoplusia ni* in relation to their chemical compositions. *Pharm. Biol.*, 46, 82–87.

Isman, M. B., & Machial, C. M., (2006). Pesticides based on plant essential oils: from traditional practice to commercialization. In: *Advances in Phytomedicine: Naturally Occurring Bioactive Compounds.* Rai, M., & Carpinella, M. C., (eds). Elsevier, New York, *3*, 29–44.

Isman, M. B., (2000). Plant essential oils for pest and disease management. *Crop Protection, 19,* 603–608.

Isman, M. B., (2014). Botanical insecticides: A global perspective. In: *Biopesticides: State of the Art and Future Opportunities*, Gross, A. D., Coats, J. R., Duke, S. O., & Seiber, J. N., eds., *American Chemical Society Symposium Series, 1172,* 21–31.

Jakiemiu, E. A. R., Scheer, A. P., Oliveira, J. S., Côcco, L. C., Yamamoto, C. I., & Deschamps, C., (2010). Estudo da composição e do rendimento do óleo essencial de tomilho (*Thymus vulgaris L.*) *Semina-Ciencias Agr., 3,* 683–88.

Jemaa, J. M. B., Haouel, S., Bouaziz, M., & Khouja, M. L., (2012). Seasonal variations in chemical composition and fumigant activity of five *Eucalyptus* essential oils against three moth pests of stored dates in Tunisia. *J. Stored Prod. Res., 48,* 61–67.

Joen, J. H., Lee, C. H., & Lee, H. S., (2009). Food protective effect of geraniol and its congeners against stored food mites. *J. Food Prot., 72,* 1468–1471.

Kadri, A., Zarai, Z., Chobba, I. B., Bekir, A., Gharsallah, N., Damak, M., & Gdoura, R., (2011). Chemical constituents and antioxidant properties of *Rosmarinus officinalis* L., essential oil cultivated from south-western Tunisia. *J. Med. Plants Res., 5,* 5999–6004.

Kamatou, P. G., Vermaak, I., & Viljoen, A. M., (2012). Eugenol - from the remote Maluku Islands to the international market place: A review of a remarkable and versatile molecule. *Molecules, 17,* 6953–6981.

Karamaouna, F., Kimbaris, A., Michaelakis, A. D. M., Papatsakona, P., & Tsora, E., (2013). Insecticidal activity of plant essential oils against the vine mealybug, *Planococcus ficus. J. Insect Sci., 13*(142), 1–13.

Khan, I. A., & Abourashed, E. A., (2010). *Leung's Encyclopedia of Common Natural Ingredients- Used in Food, Drugs, and Cosmetics,* 3rd ed, Wiley, New York.

Kono, M., Ono, M., Ogata, K., Fujimori, M., Imai, T., & Tsucha, S., (1993). *Control of Insects with Plant Essential Oils and Insecticides.* Patent JP 93–70745 930308.

Koschier, E. H., & Sedy, K. A., (2003). Labiate essential oils affecting host selection and acceptance of *Thrips tabaci* lindeman. *Crop. Prot., 22,* 929–934,.

Koul, O. W., Alia. S. W., & Dhaliwal, G. S., (2008). Essential oils as green pesticides: potential and constraints. *Biopestic. Int., 4*(1), 63–84.

Kumar, M., Faheem, M., Singh, S., Shahzad, A., & Bhargava, A. K., (2013). Antifungal activity of the *Eucalyptus australe* important medicinal plant. *Int. J. Eng. Sci., 2,* 27–30.

Lalla, F. D., Ahmed, B., Omar, A., & Mohieddine, M., (2013). Chemical composition and biological activity of *Allium sativum* essential oils against *Callosobruchus maculatus. J. Environ. Sci. Toxicol. Food Technol., 3,* 30–36.

Liang, K., (1988). *Insecticidal Composition Containing Pyrethrinoids for Domestic* Use. Patent CN 88–105917 88071.

Machial, C. M., Shikano, I., Smirle, M., Bradbury, R., & Isman, M. B., (2010). Evaluation of the toxicity of 17 essential oils against *Choristoneura rosaceana* (Lepidoptera: Tortricidae) and *Trichoplusia ni* (Lepidoptera: Noctuidae). *Pest Manag. Sci., 66,* 1116–1121.

Minaiyan, M., Ghannadi, A. R., Afsharipour, M., & Mahzouni, P., (2011). Effects of extract and essential oil of *Rosmarinus officinalis* L., on TNBS-induced colitis in rats. *Res. Pharm. Sci., 6*, 13–21.

Miresmailli, S., Bradbury, R., & Isman, M. B., (2006). Comparative toxicity of *Rosmarinus officinalis* L., essential oil and blends of its major constituents against *Tetranychus urticae* Koch (Acari: Tetranychidae) on two different hostplants. *Pest Manag. Sci., 62*, 366–371.

Nattudurai, G., Paulraj, M. G., & Ignacimuthu, S., (2012). Fumigant toxicity of volatile synthetic compounds and natural oils against red flour beetle *Tribolium castaneum* (Herbst) (Coleopetera: Tenebrionidae. *J. of King Saud University - Sci, 24*(2), 153–159.

Negrelle, R. R. B., & Gomes, E. C., (2007). *Cymbopogon citratus* (D. C) Stapf: chemical composition and biological activities. *Rev. Bras. Pl. Med., 9*, 80–92.

Pérez, S. G., Ramos-López, M. A., Zavala-Sánchez, M. A., & Cárdenas-Ortega, N. C., (2010). Activity of essential oils as a biorational alternative to control coleopteran insects in stored grains. *J. Med. Plant Res., 4*, 2827–2835.

Perumalla, A. V. S., & Hettiarachchy, N. S., (2011). Green tea and grape seed extracts - potential applications in food safety and quality. *Food Res., 44*, 827–839.

Petroski, R. J. & Hammack, L., (1998). Structure activity relationships of phenyl alkyl alcohols, phenyl alkyl amines and cinnamyl alcohol derivatives as attractants for adult corn root worm (Coleoptera: Chrysomelidae: *Diabrotica* sp.). *Environ. Entomol., 27*, 688–694.

Priestley, C. M., Williamson, K. A., & Wafford, D. B., (2003). Sattelle. Thymol, a constituent of thyme essential oil, is a positive allosteric modulator of human GABA receptors and homo-oligomeric GABA receptors from *Drosophila melanogaster. Br. J. Pharmacol., 140*, 1363–1372.

Quarles, W., (2002). Botanical pesticides from *Chenopodium. The IPM Pract., 14*, 2, 1–11.

Regnault-Roger, C., (1997). The potential of botanical essential oils for insect pest control. *Integ. Pest Manag. Rev., 2*, 25–34.

Regnault-Roger, C., Vincent, C., & Arnason, J. T., (2012). Essential oils in insect control: Low-risk products in a high-stakes world. *Ann. Rev. Entomol, 57*, 405–424.

Rice, P. J. & Coats, J. R., (1994). Insecticidal properties of several monoterpenoids to the housefly (Diptera: Muscidae), red flour beetle (Coleoptera : Tenebrionidae) and southern corn root-worm (Coleoptera : Chrysomelidae). *J. Econ. Entomol., 87*, 1172–1179.

Shaaya, E., Ravid, U., Paster, N., Juven, B., Zisman, U., & Pissarev, V., (1991). Fumigant toxicity of essential oils against four major stored-product insects. *J. Chem. Ecol., 17*, 499–504.

Singh, D., Siddiqui, M. S., & Sharma, S., (1989). Reproductive retardant and fumigant properties in essential oils against rice weevil in stored wheat. *J. Econ. Entomol., 82*, 727–733.

Smelyanets, V. P., & Kuznetsov, N. V., (1968). Toxicity of some terpene compounds. *Khimija y Selskom Khozjajste, 6*(10), 754–755.

Stahl-Biskup, E., & Sáez, F., (2002). *Thyme - the Genus Thymus*, Taylor & Francis, New York.

Sujana, P., Sridhar, T. M., Josthna, P., & Naidu, C. V., (2013). Antibacterial activity and phytochemical analysis of *Mentha piperita* L. (peppermint) — an important multipurpose medicinal Plant. *Am. J. Plant Sci., 4*, 77–83.

Tak, J. H., & Isman, M. B., (2015). Enhanced cuticular penetration as the mechanism for synergy of insecticidal constituents of rosemary essential oil in *Trichoplusia ni*. *Scientific Reports, 5*, 12690.

Tak, J. H., Jovel, E., & Isman, M. B., (2016). Comparative and synergistic activity of *Rosmarinus officinalis* L., essential oil constituents against the larvae and an ovarian cell line of the cabbage looper, *Trichoplusia ni* (Lepidoptera: Noctuidae). *Pest Manag. Sci., 72*, 474–480.

Visweswara Rao, P., & Gan, S. H., (2014). *Cinnamon: A Multifaceted Medicinal Plant. Evidence-Based Complementary and Alternative Medicines*, pp. 12.

Waliwitiya, R., Isman, M. B., & Riseman, A., (2005). Insecticidal activity of selected monoterpenoids and rosemary oil to *Agriotes obscurus* (Coleoptera: Elateridae). *J. Econ. Entomol., 98*, 1560–1565.

Werdin González, J. O., Laumann, R. A., Da Silveira, S., Moraes, M. C., Borges, M., & Ferrero, A. A., (2013). Lethal and sublethal effects of four essential oils on the egg parasitoids *Trissolcus basalis*. *Chemosph, 92*(5), 608–615.

White, J. B., Park, Y. L., West, T. P., & Tobin, P. C., (2009). Assessment of potential fumigants to control *Chaetodactylus krombeini* (Acari: Chaetodactylidae) associated with *Osmia cornifrons* (Hymenoptera: Megachilidae). *J. Econ. Entomol., 102*(6), 2090–2095.

World Watch Magazine, (2004). 17, 5 (http://www.worldwatch.org/node/552) (accessed Jan 23, 2016).

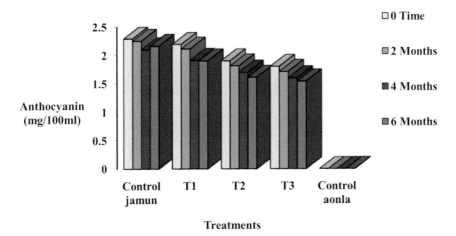

FIGURE 6.1   Changes in anthocyanin content during storage of blended jamun-aonla ready-to-serve beverages.

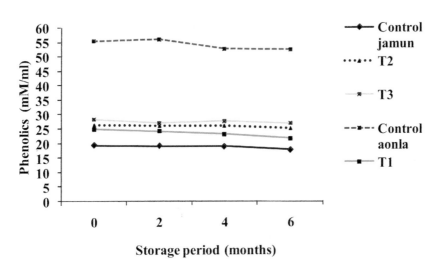

FIGURE 6.2   Changes in phenolics in blended jamun–aonla ready to-serve beverages during storage.

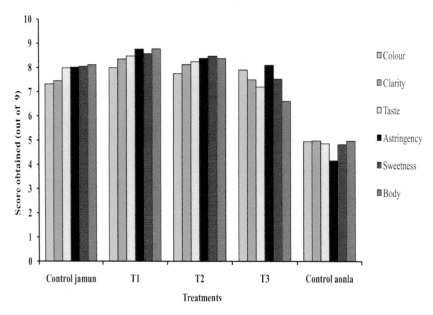

FIGURE 6.3 Sensory evaluation of blended jamun–aonla ready-to-serve beverages during storage.

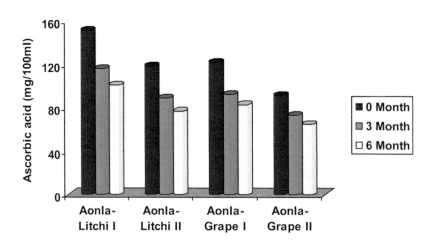

FIGURE 7.1 Changes in ascorbic acid content of blended aonla squash during storage.

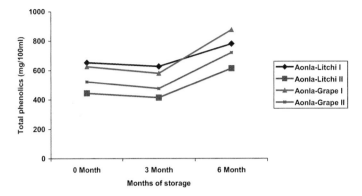

FIGURE 7.2   Changes in total phenolic content of blended aonla squash during storage.

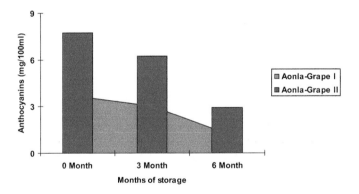

FIGURE 7.3   Changes in anthocyanin content of aonla–grape squash during storage.

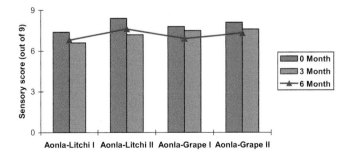

FIGURE 7.4   Changes in sensory scores of blended aonla squash during storage.

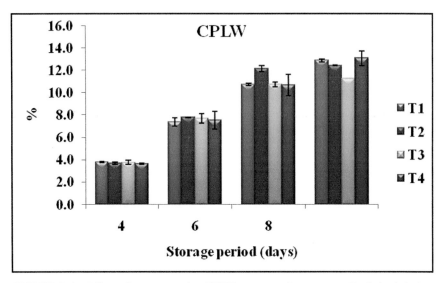

FIGURE 8.1 Effect of yeasts on the CPLW percent of mango cv. Dashehari during storage. T_1 (baker's yeast), T_2 (industrial yeast), T_3 (*Saccharomyces cerevisiae*), and T_4 (control).

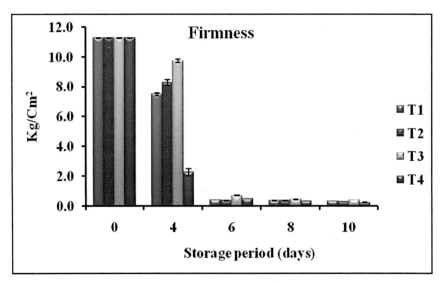

FIGURE 8.2 Effect of yeasts on the firmness (kg/cm²) of mango cv. Dashehari during storage. T_1 (baker's yeast), T_2 (industrial yeast), T_3 (*Saccharomyces cerevisiae*), and T_4 (control).

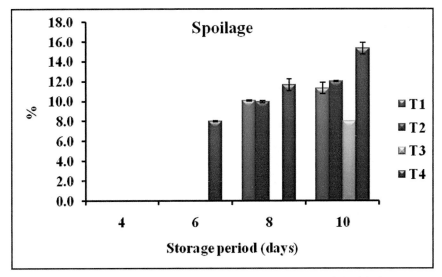

FIGURE 8.3   Effect of yeasts on the spoilage (%) of mango cv. Dashehari during storage. T_1 (baker's yeast), T_2 (industrial yeast), T_3 (*Saccharomyces cerevisiae*), and T_4 (control).

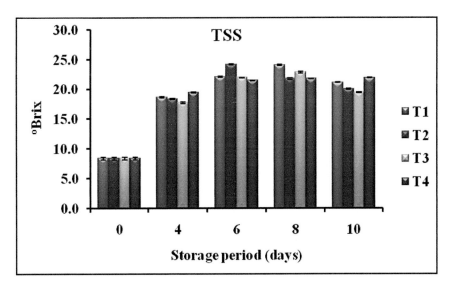

FIGURE 8.4   Effect of yeasts on the TSS (°Brix) of mango cv. Dashehari during storage. T_1 (baker's yeast), T_2 (industrial yeast), T_3 (*Saccharomyces cerevisiae*), and T_4 (control).

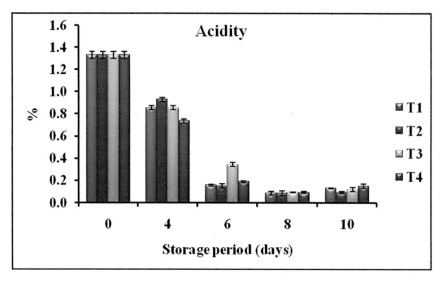

FIGURE 8.5   Effect of yeasts on the titratable acidity (%) of mango cv. Dashehari during storage. T_1 (baker's yeast), T_2 (industrial yeast), T_3 (*Saccharomyces cerevisiae*), and T_4 (control).

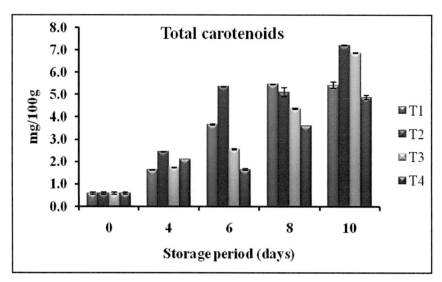

FIGURE 8.6   Effect of yeasts on the total carotenoids (mg/100 g) of mango cv Dashehari during storage. T_1 (baker's yeast), T_2 (industrial yeast), T_3 (*Saccharomyces cerevisiae*), and T_4 (control).

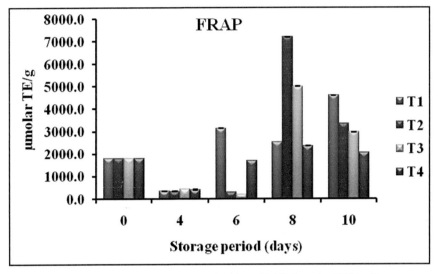

FIGURE 8.7 Effect of yeasts on the antioxidant FRAP (μmolar TE/g) of mango cv. Dashehari during storage. T_1 (baker's yeast), T_2 (industrial yeast), T_3 (*Saccharomyces cerevisiae*), and T_4 (control).

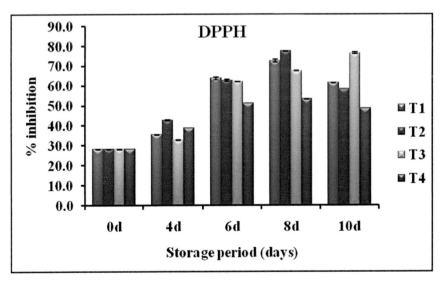

FIGURE 8.8 Effect of yeasts on the antioxidant DPPH (% inhibition) of mango cv. Dashehari during storage. T_1 (baker's yeast), T_2 (industrial yeast), T_3 (*Saccharomyces cerevisiae*), and T_4 (control).

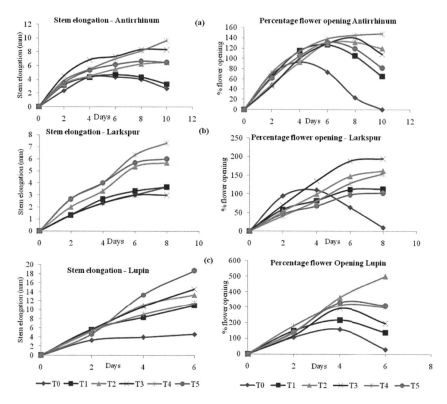

FIGURE 9.1 Effect of treatments on stem elongation and percent flower opening of (a) antirrhinum, (b) dimorphotheca, (c) larkspur, (d) lupin, and (e) Sweet William.

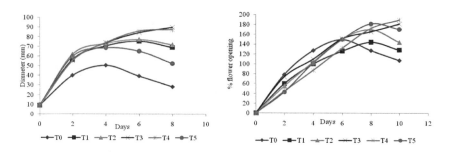

FIGURE 9.2 (a) Effect of treatments on flower diameter (mm) of dimorphotheca and (b) effect of treatments on percent flower opening of Sweet William.

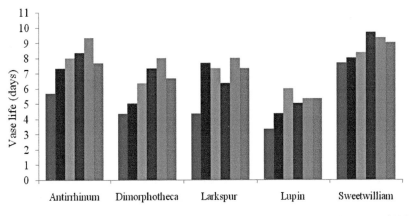

■ Control ■ Sucrose ■ Sucrose + HQC ■ Sucrose + AOA ■ Sucrose + BA ■ Sucrose+ Al2SO4

FIGURE 9.4   Effect of treatments on vase life of winter annuals.

Chart showing the % cumulative weight loss of cling filmed and non cling filmed breadfrutis from Pamplemousses E.S. stored at 13 oC.

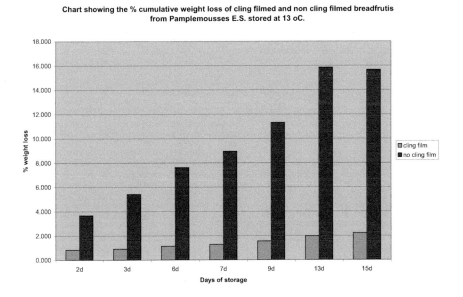

FIGURE 10.1   Chart showing the % cumulative weight loss of cling filmed and noncling filmed breadfruits from Pamplemousses ES stored at 13°C.

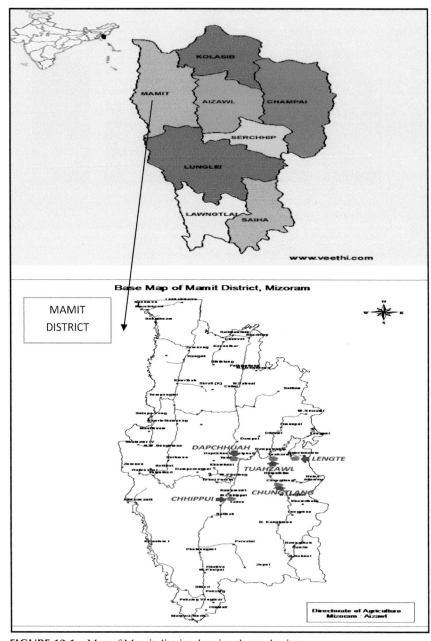

FIGURE 12.1   Map of Mamit district showing the study sites.

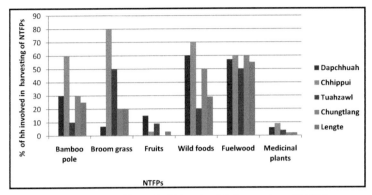

FIGURE 12.10   Percentage of household involved in harvesting of NTFPs.

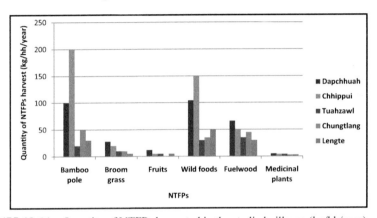

FIGURE 12.11   Quantity of NTFPs harvested in the studied villages (kg/hh/year).

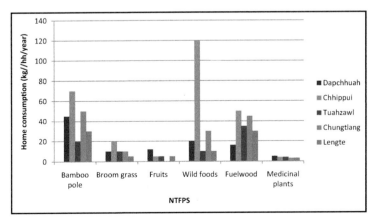

FIGURE 12.12   Amount of NTFPs used for home consumption in different villages (kg/hh/year).

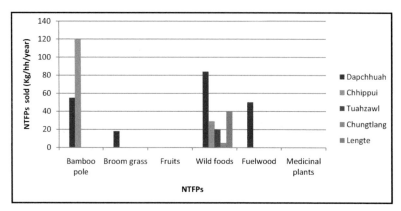

FIGURE 12.13 Amount of NTFPs sold in different villages.

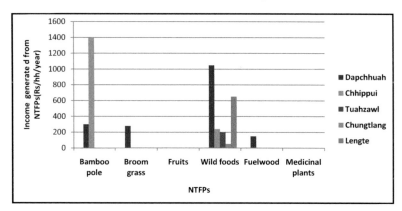

FIGURE 12.14 Estimated income generated from NTFPs in different villages.

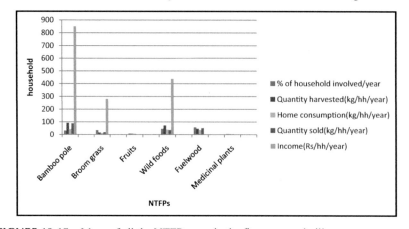

FIGURE 12.15 Mean of all the NTFPs uses in the five surveyed villages.

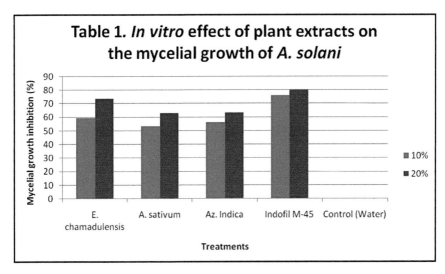

FIGURE 18.1   In vitro effect of plant extracts on the mycelial growth of *Alternaria solani*.

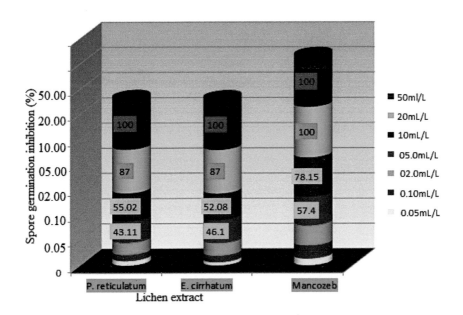

FIGURE 20.1   Antifungal efficacy (%) of the test samples against C. *capsici* – A graphical representation.

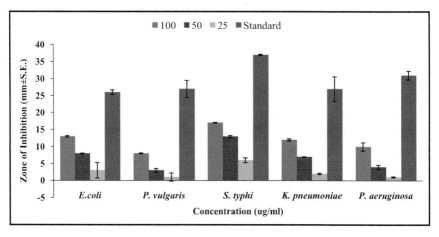

FIGURE 22.1 Effect of essential oil showing zone of inhibition against bacterial pathogens.

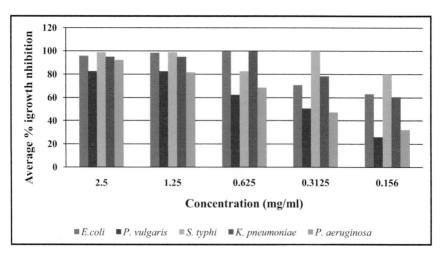

FIGURE 22.2 Average percent growth inhibition of peppermint oil against bacterial pathogens.

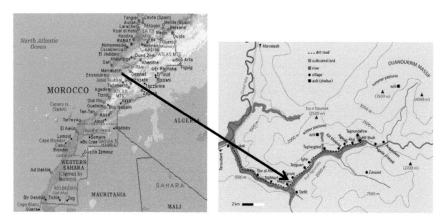

FIGURE 24.1   Map of Morocco and the study area.

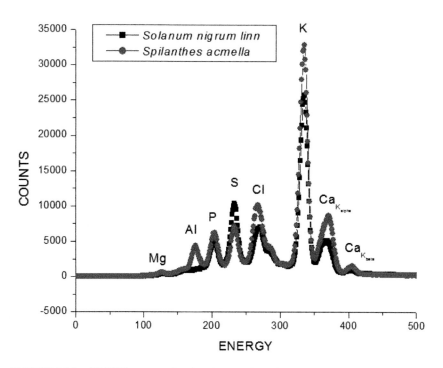

FIGURE 26.2   EDXRF spectra of major elements in *Solanum nigrum* linn and *Spilanthes acmella*.

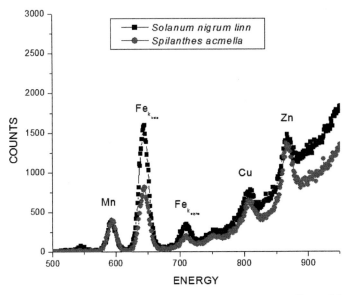

FIGURE 26.3   ED-XRF spectra of minor elements in *Solanum nigrum* linn and *Spilanthes acmella*.

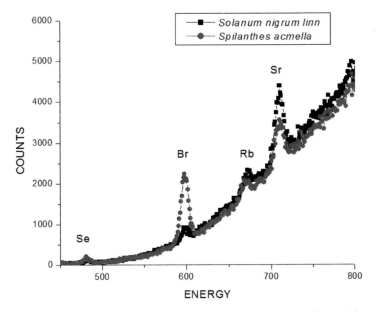

FIGURE 26.4   ED-XRF spectra of earth elements in *Solanum nigrum* linn and *Spilanthes acmella*.

CHAPTER 15

INTEGRATED HEALTH MANAGEMENT IN HORTICULTURAL CROPS

C. CHATTOPADHYAY[1] and AJANTA BIRAH[2]

[1]Uttar Banga Krishi Viswavidyalaya, Pundibari, Coochbehar–736165, West Bengal, India, E-mail: chirantan_cha@hotmail.com

[2]ICAR-National Research Centre for Integrated Pest Management, New Delhi–110012, India

CONTENTS

ABSTRACT

Plant protection is one of the key issues in the overall gamut of Indian agriculture. In view of indiscriminate use of chemical pesticides, environmental safety vis-á-vis sustaining crop yields, threats to farm bio-security, and crop health in the era of globalization, the situation has become rather challenging. Integrated pest management (IPM) follows the principles of understanding of the crop, pest, and the environment and their interrelationships to enable advanced planning with emphasis on routine monitoring of crop and pest conditions and balancing of cost/benefits of all management practices. Agriculture and allied activities are the main source of livelihood for the people of northeastern region of our country, and any attempt to reduce poverty as well as to place the region in developmental paradigm should have a system-based eco-regional planning for agricultural development. IPM is a complex process, and farmers lack understanding of the biological processes of pests, their predators, and methods of application of new components. There are a number of IPM practices that work best when applied by the entire community and in a synchronized mode. Thus, an integrated decision support system for crop protection services may be required to be devised centrally to monitor the pest dynamics through e-pest surveillance, analysis of pest risks, and provision of pest forecasts along with mobile-based dissemination of advisories keeping in view prevailing weather and changes in climate. This would necessitate networking of all stakeholders so that they could contribute effectively in a cohesive manner.

15.1 INTRODUCTION

Crop yield losses in India due to pests (includes all biotic stresses, viz., weeds, insect pests, diseases, nematodes, rodents, etc.) range 15–25% (varied estimates by different sources). Losses due to damages caused by pests in terms of quantity are almost to an extent such that if that loss could be saved, India could meet its needs of 2020 domestically even with the present levels of crop productivity as well as keeping in view the stagnation in yields for certain crops vis-á-vis impacts of climate change. Lack of knowledge/awareness about eco-friendly methods of pest management apart from unavailability of inputs required for the same makes the job further difficult. However, efforts of researchers in crop protection have been able to make sizable dent on reducing losses due to biotic stresses with the use and up-scaling of eco-friendly technological packages, thereby improving crop productivity and livelihood of farming community. There has been quite some time since we faced any acute epidemic or epizootic, and relevant technologies have also played their roles in minimizing such incidences. There has been considerable head way in the area of technology development and its implementation at the field level – for instance, the National Food Security Mission has a rich component of plant protection in overall implementation of the scheme at the national level. Therefore, plant protection currently holds the key to redeem much of the losses due to pests through convergence of stakeholders.

15.2 INTEGRATED HEALTH MANAGEMENT

IPM technology envisages to integrate different management techniques to keep crop plant stress due to different deleterious organisms below economic injury/threshold level and thereby targets to maintain good crop health. The components of host resistance (manipulation of crop to withstand or tolerate pests) include use of clean certified seed/indexed planting material of the variety recommended for the targeted region, cultural management (to optimize growing conditions for the crop or anything that increases a crop's competitive edge to result in increased tolerance to pests often resulting in reduced pesticide use or create unfavorable conditions for the pest), sanitary management that avoids introducing pest to crop field (clean field equipment, removal of the affected plant/its part), and natural and biological management that enhances beneficial organisms or releases predators, parasites, etc. Economic injury level (EIL) is the cost of control that equals value of damage caused by the pest, which is determined through extensive research and is the information necessary to develop an economic threshold (ETL), which is used by crop advisors.

The northeast (NE) region comprising eight states, viz., Assam, Arunachal Pradesh, Meghalaya, Manipur, Mizoram, Nagaland, Tripura, and Sikkim, has a total geographical area of 262180 km^2, which is nearly 8% of the total area of the country with more than 39 million people. The total area under horticultural crops is around 822.5 thousand hectares, which is around 3.14% of the total geographical area of the region, which provides a production of 6818.4 thousand tons. Out of the total area under different fruit crops in the NE region, ~60.6 thousand hectares is only under banana cultivation. Among the biotic stresses, various fungal, bacterial, and viral diseases starting from nursery to postharvest stage cause considerable loss. As the diseases are either planting material-borne or soil-born, they are very difficult to manage. Only early diagnosis and IPM practices can be effective in managing these problems. Food grain production and requirement scenario indicates a deficit existing in all the NE states, varying from 10% for Tripura/Sikkim to 69% for Mizoram, except Nagaland that has some surplus food grains. While planning this, the strength of the farming system approach to judicious utilization and conservation of natural resources of the region with concurrent policy and research support to increase production, add value to the produce, and their disposal/sale management shall be of paramount importance (Vision 2030, ICAR NEH).

Green revolution, one of the greatest success stories of India, was due to intensive agriculture (IA). However, factors involved in IA like genetic uniformity of crops, dense plant population, mono-cropping, higher fertilization and irrigation, inappropriate cropping systems, practices (chemicalization) with immediate profit motives favoured insect pests and diseases incidence. IPM and the use of biotic agents to minimize the unsystematic and indiscreet use of chemical pesticides will be the fundamental theory of plant protection. Biopesticides represent only 2.89% (as on 2005) of the overall pesticide market, with an expected annual growth rate of 2.3% (as in 2006). Neem-based pesticides, *Bt*, NPV (*Ha/Sl*), and *Trichoderma*, are the major biopesticides produced and used in India. Although only 0.6 kg/ha chemical pesticide is used in India compared to 14 kg/ha in China and 12 kg/ha in Japan, the use is highly injudicious in some pockets that have led to several issues of pest resurgence, development of resistance in pests to chemical pesticides apart from human health problems [Chemical pesticides were responsible for 49% lower sperm count in men eating raw fruits (Sheiner et al., 2003); they are among the factors responsible for neurological problems (Ascherio et al., 2006; Baldi et al., 2010), neurodevelopmental disorders (Beseler et al., 2008; Jurewicz and Hanke 2008), birth defects (Winchester et al., 2009). Organochlorine pesticides were linked to preterm deaths, reduced baby weight, and ovarian cancer in north India (Mustafa et al., 2015; Sharma et al., 2015; Tyagi et al., 2015), and around 2.2 million people die annually of cancer related to chemical pesticide poisoning (McCauley et al, 2006; Gilden et al., 2010). Large use of chemical pesticides leads to several diseases (April 7, 2015, Times of India; WHO, CSAUAT). Chemical pesticides are linked to increased risk of diabetes, and exposure to them significantly increases the risk of type 2 diabetes by nearly 60% (Fotini Kawoura, September 25, 2015, Medscape Medical News, UK).] and disturbance to environment and biodiversity. This also brought significant shift in the insect population dynamics and change in the status of several insect pests. The cost of plant protection on various crops ranged 7–40% of the total crop production cost. Though IPM has been advocated for the past two decades, the number of farmers who adopted IPM practices in various crops in India is debatable. IPM research in the past decade brought out changes in the farmers' attitude in pest management, which resulted reduction in pesticide use in different crops. To be more effective, readdressing the policies for encouraging eco-friendly options and strengthening extension, involving farmers should be considered as high priority.

In most cases, IPM consists of scouting with timely application of a combination of strategies and tactics. These may include site selection and preparation; utilizing resistant cultivars; altering planting practices; modifying the environment by drainage, irrigation, pruning, thinning, shading, etc.; and applying pesticides, if necessary. But in addition to these traditional measures, monitoring environmental factors (temperature, moisture, soil pH, nutrients, etc.), disease forecasting, and establishing economic thresholds are important to the management scheme. These measures should be applied in a coordinated and harmonized manner to maximize the benefits of each component. For example, balancing fertilizer applications with irrigation practices helps promote crop health.

India has successfully reduced pesticide consumption without adversely affecting the agricultural productivity. This was facilitated by appropriate policies that discouraged pesticide use and favored IPM application. Despite such efforts, adoption of IPM is low owing to a number of socioeconomic and other constraints. Though many technology programs are based on community approach, they do not have any proper policy to sustain the group approach. The IPM policy should also provide incentives to farmers to adopt IPM as a cardinal principle of plant protection.

The IPM is knowledge intensive on crop, pest(s), environment, and their inter-relationships; holistic in approach; and requires expert advice, timely decision-making, and immediate actions for solutions to pests. Basic principles of IPM involve advanced planning, balancing of cost and benefit from any management interventions, and routine monitoring of crop and pest conditions. Farmers needs involve pest (including pathogen, weed, etc.) identification (diagnostics) during crop surveillance, pest surveillance, monitoring preferably accomplished by e-pest surveillance at the national/global level, pest forecasting and dissemination of expert information on pest management for quick action to solution. Availability of appropriate inputs of IPM also impedes following principles of the same. Thus, an important need of the hour is to improve levels of awareness on IPM among field functionaries and farmers.

Since 1985, the Government of India (GoI) has enabled farmers to adopt IPM practices to bring down losses due to pests and also provide ways to reduce use of chemical pesticides. In view of the same, the National Research Centre for Integrated Pest Management (NCIPM) was set up in VII Plan (1988) under the Indian Council of Agricultural Research. In the National Agricultural Policy announced by the GoI in 2001, para 24 emphasizes IPM and use of biological control agents to minimize indiscriminate and injudicious use of chemical pesticides as a cardinal principle for crop protection. During >25 years of existence, NCIPM has achieved successes in validating and harmonizing IPM technologies in different crops, viz., Astha village (Maharashtra) for cotton, Bambawad (UP) in basmati rice, etc. In 1992, Central Integrated Pest Management Centres (CIPMCs) were established by the GoI by merging all Central Plant Protection Stations (CPPS), Central Surveillance Stations (CSS), and Central Biological Control Stations (CBCS). Presently, there are 31 CIPMCs in 28 states and 1 UT, who are engaged in pest monitoring and field release of biological control agents, conduct farmers' field schools (FFSs), and train extension officers and master trainers. CIPMCs are in turn linked with 98 state biocontrol laboratories.

To facilitate popularizing the IPM approach among farming community under the Central Sector Scheme "Promotion of Integrated Pest Management" of the Department of Agriculture and Cooperation, GoI, an information system for IPM has been created, which helps in efficient reporting and dissemination of information on pest surveillance, rearing of host culture, production and release of biological control agents in the field, conservation of naturally occurring biological control agents for the control of crop pests, and transfer of innovative IPM skills/methods/techniques to extension workers and farmers through conduct of training and FFSs in all states by CIPMCs of the Directorate of Plant Quarantine and Storage, Faridabad.

Biological control is also a very effective component of crop protection. Due to public awareness about the hazards related to the use of chemical pesticides, there has been a lot of interest generated for the use of eco-friendly strategies targeted at the management of crop pests. For this purpose, biopesticides could be a cost-effective, eco-friendly, and sustainable option, when proven source of host resistance/tolerance against several pests is not available. However, the quality, quantity, application method, and timeliness play a significant role in determining the level of success of biological control. There are several success stories of biological control doing a commendable job in the field of crop protection. Successful biological management of papaya mealy bug and sugarcane wooly aphid alone have saved >Rupees 2.5 thousand crore (>4100 m US$) in 2 years for the nation. The dreaded weed *Mikania micrantha* is being successfully managed in southwest and NE India by *Puccinia spegazzinii* (rust fungus). Credibility of the bioformulations *Kalisena* [technology developed by Indian Agricultural Research Institute (IARI), New Delhi and Indian Council of Agricultural Research (ICAR), and transferred for marketing in Asia, Africa, North and South America to M/s Cadila Pharmaceuticals Ltd.] has been established with the farmers and their advisors, which is an excellent example of translational research. Garlic bulb aqueous extract (2% w/v) has also been adopted by farmers and the Government of Rajasthan in managing pests of Indian mustard. Use of quality strains of *Trichoderma, Pseudomonas fluorescens,* etc. in recommended quantity even as seed treatment has been found to be very successful in managing dreaded diseases of different field and horticultural crops, which could safeguard from high yield losses. When they are combined with soil application and/or foliar spray, they result in even better impact not only in reducing pests and increasing yields and economic benefits but also in safeguarding the environment from dangerous chemical pesticide load.

During the last 28 years, NCIPM has successfully launched IPM strategies for different crop sectors across the country; has been successful in promoting the IPM concept and in developing state-of-the-art technologies and strategies concepts like those in Astha village (Maharashtra), eight districts of Punjab and Jind (Haryana) models for cotton cultivation, IPM in rice (Bambawad, UP; Chhajpur, Haryana; Hooghly, WB), pulses (Gulbarga and Bidar, Karnataka; Jabalpur, MP; Mirzapur, UP), groundnut (Udaipur, Rajasthan; Kadiri, Andhra Pradesh), mustard (Alwar, Rajasthan; Mohindergarh, Haryana), vegetable (Anantpura, Rajasthan; onion at Singohi-Singoha-Rambha and Bitter Gourd at Padhana, Haryana), and fruit (mango at Navsari, Gujarat; citrus at Abohar, Punjab) crops. Other ICAR institutions have shown success with grapes at Nashik, Maharashtra, and pomegranate at Solapur, Maharashtra. NCIPM has also played a vital role in strengthening the capacity and capability in IPM in the country by organizing 55 programs to train 1582 participants, which include 476 KVK (Krishi Vigyan Kendra) personnel among existing 643 KVKs from all eight zones of India.

Recently, NCIPM in active collaboration with ICAR Complex for the North Eastern Hill (NEH) region and State Department of Agriculture,

Government of Tripura, launched a project to uplift the socioeconomic status of tribal farmers. The critical IPM input kits consisting *Trichoderma,* Trichocards, yellow sticky, and pheromone traps were distributed among the farmers. Demonstrations on the use of the inputs were also provided. Mode of operation and establishment of infrastructure for the launch of e-pest surveillance in Tripura has been planned. Potato grown in Tripura offer a great potential for organic cultivation as the application of chemical pesticides is minimal. The productivity of potato is low due to various biotic stresses of late blight, bacterial blight, common scab, and viral diseases. Under these circumstances, validation of IPM modules developed by NCIPM was undertaken at Barkhatal, Hazamara block, West Tripura. Similarly, tomato is also one of the important vegetable crops cultivated in Tripura as a postmonsoon crop. The crop is affected by several diseases like bacterial wilt, tomato leaf curl, etc. A field trial was conducted to evaluate the eco-friendly management practices, viz., staking, spray of neem oil, application of bleaching powder and combination thereof; it was observed that the combination provided significant reduction in incidence of diseases in tomato. Accordingly, these management strategies were discussed with the farmers, and efforts were undertaken to implement the same.

A few states have been more progressive in encouraging biological control of crop pests, viz., Gujarat, Tamil Nadu, West Bengal, etc. Safeguarding intellectual property on strains of bioagents is an important issue in the present era. Accordingly, there is a need to have DNA barcode data of all such strains in order to sustain IPR. There is a need to undertake a specific policy to encourage biopesticides, streamlining their label claim issues, simplification of process of registration for biopesticides with strict and adequate quality check from government departments (CIPMCs, SAUs (State Agricultural Universities), etc.), increased support to biopesticide industry for scaling up of production as a matter of government policy (viz., subsidies to biopesticides, higher taxes on chemical pesticide industries, etc.), which shall also enable generation of employment for small/micro-industries at the village level in line with concepts of model bio-village.

Precision pest management to reduce indiscriminate use of chemical pesticides could plan use of state-of-the-art technology through innovative and strategic research to enable devise integrated decision support system (IDSS) for crop protection services that suggests operational focus, research priorities, and evolution in a phased manner. Surveillance is the foundation of plant protection for early alert. But it is missing in most of the developing countries. In the recent past, the information communication technology (ICT)-based system of real time pest surveillance has played an important role in our country in collection and transfer of data from remote villages to main station through the Internet. The information is compiled and displayed on the website in tabulated and graphical form and that can be directly accessed by SAUs for issue of advisory through the State Agriculture Department by SMS to farmers and extension workers for implementation in farmers' fields. There is dramatic reduction in outbreak of any major pest on selected crops since the inception of ICT activity in different states. As the farmers are getting regular SMSs for IPM interventions, there is much awareness about IPM. Holistic planning provides farmers with the management tools they need to manage biological complex farming systems in a profitable manner. A successful IPM program requires time, money, patience, short- and long-term planning, flexibility, and commitment.

One of the most common approaches that have been adopted by many countries is pre-sowing treatment of seed. Seed treatment is defined as chemical or biological substances applied to seed or vegetatively propagated material to manage diseases organisms, insect pests, etc. Seed treatment pesticides include bactericides, fungicides, and insecticides. Most seed treatments are applied to true seeds such as corn, wheat, or soybean, which have a seed coat surrounding the embryo.

An IPM plan should identify important pests, determine pest management options, and blend them together to achieve the goals listed above. To use seed treatments effectively, it is important to understand the purposes of seed treatment, alternatives, or supplements to seed treatments, and the various advantages and disadvantages of seed treatments. Natural enemy cum beneficial fauna population such as coccinellids, spiders and *Chrysoperla*, pollinators, and honey bee remain unharmed due to seed treatment.

Bridging the gap between policy and practice is a herculean task in plant protection in the wake of extensive diversity (of crops, regions, pest problems, and practices) variegated resources (knowledge, experts, quality inputs, machinery, molecules, and financial support), require highly motivated and focused yet effective efforts (toward research, education, extension) to tackle the issues in plant protection. However, through well-coordinated efforts involving multistakeholders – policy makers, administrators, researchers, educationists, extension machinery, industry, media, etc. visible contributions can be made in this important area.

Although GoI has put IPM as part of its National Agricultural Policy, there is possibly a need to undertake a specific policy to push IPM by providing credits for greener pesticide molecules, streamlining label claim issues, simplification of process of registration for biopesticides with strict and adequate quality check from government departments (CIPMCs, SAUs, etc.), increased support to biopesticide industry for scaling up of production as a matter of government policy (viz., subsidies to biopesticides, higher taxes on chemical pesticide industries, etc.), which shall also enable generation of employment for small/micro-industries at the village level in line with concepts of model bio-village. This shall bring a shift in the chemical pesticide industries and transform them toward producing biopesticides.

KVKs and NGOs need to play a vital role in improving awareness levels of field functionaries and farmers in IPM, apart from fast-tracking of crop protection advisories. IPM happens to be a knowledge-intensive holistic approach wherein advanced planning, good agricultural practices (GAP) toward environment-safety linked to maximum residue limit (MRLs) of chemical pesticides and cost-benefit management, crop monitoring coupled with accurate diagnosis of the problem (pest), expert advice, timely decision-making, and quick action make the real difference to tackle unforeseen pest outbreaks.

Mankind has always been eager to know the unknown of the future to enable plan suitably for the same. With the generation of knowledge, the disease triangle and tetrahedron was considered involving the interaction among host and pathogen in a given environment over time. Plant disease forecasting is a management system used to predict the occurrence or change in the severity of plant diseases. At the field scale, these systems are used by growers to make economic decisions about disease management. The environment is usually the factor that controls whether disease develops or not, the vulnerability of the host, and the presence of the pathogen in a particular season through their effects on processes such as overseasoning or ability of the pathogen to cause disease. In these cases, a disease forecasting system attempts to define when the environment will be conducive to disease development.

The NEH region also has tremendous potential for IPM. Unfortunately, not a single IPM module has been developed for any crop for any part of the NEH region so far. As IPM is location-specific, the IPM modules developed for one region may not work in another. The NEH region is unique and different from the rest of the country in many ways. This is also true in the wider context of agriculture and more specifically, pests and natural enemies abundant therein. For example, the common blister beetle is a minor pest in the rest of the country, but in the NEH, it is a major pest. Certain pests are found here, which may not be found in the mainland. This is true for natural enemies too. The natural enemy diversity and abundance is more in the NEH region; therefore, natural control is very high here.

15.3 CONCLUSION

Today's challenge is to "produce more from less." The big questions are, do we have enough information on the fitness-cost of different pests of major crops, their injury profiles vis-á-vis attainable yields, and are we prepared enough to face the impacts of climate change in general and specifically on the pest scenario evolving as a result of the same? Now, the challenge is to bring continuous improvement in productivity, profitability, stability, and sustainability of major farming systems, wherein scientific management of plant pests holds a pivotal role. Crop loss models, representing a dynamic interaction between pests and host, are essential for forecasting losses thereof.

KEYWORDS

- biopesticide
- horticulture
- integrated
- intensive agriculture
- IPM
- management

REFERENCES

Ascherio, A., Chen, H., Weisskopf, M. G. O.,'Reilly, E., McCullough, M, L., Calle, E. E., Schwarzschild M. A., & Thun, M. J., (2006). Pesticide exposure and risk for Parkinson's disease. *Ann Neurol., 60*, 197–203.

Baldi, I., Lebailly, P., Mohammed-Brahim, B., Letenneur, L., Dartigues, J. F., & Brochard, P., (2010). Neurodegenerative diseases and exposure to pesticides in the elderly. *Am. J., Epidemiol. 157*, 409–414.

Beseler, C. L., Stallones, L., Hoppin, J. A., Alavanja, M. C., Blair, A., & Keefe, T., (2008). Depression and pesticide exposures among private pesticide applicators enrolled in the Agricultural Health Study. *Environ. Health Perspect, 116*, 1713–1719.

Gilden, R. C., Huffling, K., & Sattler, B., (2010). Pesticides and health risks. *Journal of Obstetric, Gynecologic, and Neonatal Nursing. 39*(1), 103–110.

Jurewicz, J., & Hanke, W., (2008). Prenatal and childhood exposure to pesticides and neurobehavioral development: review of epidemiological studies. *International Journal of Occupational Medicine and Environmental Health, 21*, 121–32.

McCauley, A., Linda Anger, K. W., Keifer, M., Langley, R., & Robson, G. M., (2006). Studying health outcomes in farm worker populations exposed to pesticide. *Environ Health Perspec, 114*, 6–8.

Mustafa, M., D, Garg Neha, Banerjee, B. D., Sharma, T., Tyagi, V., Ahmad, D. S., Guleria, K., Rafat, S., Ahmad, Vaid Neelam, & Tripathi, A. K., (2015). Inflammatory-mediated pathway in association with organochlorine pesticides levels in the etiology of idiopathic preterm birth, *Reproductive Toxicology, 15*, 345–49.

Sharma, T., Banerjee, B. D., Mazumdar, D., Tyagi, V., Thakur, G., Guleria, K., Rafat, S., Ahmed, & Tripathi, A. K., (2015). Association of organochlorine pesticides and risk of epithelial ovarian cancer: A case control study, *Journal of Reproductive Health and Medicine, 65*, 1–7.

Sheiner, E. K., Sheiner, E., Hammel, R., D, Potashnik, G., &Carel, R., (2003). Effect of occupational exposures on male fertility: literature review. *Ind Health, 41*(2), 55–62.

Tyagi, V., Garg, N., Mustafa, M. D., & Banerjee, B. D., Guleria, K., (2015). Organochlorine pesticide levels in maternal blood and placental tissue with reference to preterm birth: a recent trend in North Indian population, *Environ Monit Assess, 187*, 471–479.

Winchester, P. D., Huskins, J., & Ying, J., (2009). Agrichemicals in surface water and birth defects in the United States. *Acta Paediatr, 98*, 664–669.

CHAPTER 16

TECHNOLOGY FOR DIAGNOSIS OF PLANT VIRUSES IN HORTICULTURAL CROPS IN INDIA

BIKASH MANDAL

Advanced Centre of Plant Virology, Division of Plant Pathology, Indian Agricultural Research Institute, New Delhi–110012, India, E-mail: leafcurl@rediffmail.com

CONTENTS

ABSTRACT

Plant viruses are significant constraints in horticultural crops in India. Vegetables, fruits, and ornamental crops are known to be affected by several viruses in India. Identification of viruses based on visual observation of symptoms alone is not reliable. Molecular diagnostic kits based on the immunological and nucleic acids properties of viral protein and genome are commercially available for the detection of plant viruses in Europe and the USA. In India, the plant virus diagnostic kits are generally imported, and they are highly expensive. Of late, studies conducted in many Plant Virology laboratories in India on the viral genomics have facilitated developing molecular assays for several plant viruses occurring in horticultural crops in India. Engineered antigens and antibodies and the key reagents for immunodiagnosis of viruses have prepared, and diagnostic kit prototypes have been developed for several viruses infecting horticultural crops. For rapid and on-site detection, lateral flow assay (LFA) was developed for the selected viruses. Single, duplex, and multiplex polymerase chain reaction (PCR) methods have been developed for viruses affecting many important vegetables. The diagnostic reagents and procedures developed are suitable for up-scaling and commercial applications.

16.1   INTRODUCTION

Diagnosis of plant virus is a prerequisite for developing strategies for their management, ensuring biosecurity from the introduction of exotic viruses in a country, and providing credible phytosanitary certificate. Further, virus diagnostic methods are necessary for the identification and classification of viruses, generating virus-free planting materials, breeding for selective virus resistance, monitoring spread of viruses, and studying infection process. The diagnostic methods are broadly based on the properties of viral protein and nucleic acids, which are the essential constituents of viruses. There are several techniques available for the detection of plant viruses, among which enzyme-linked immunosorbent assay (ELISA)- and polymerase chain reaction (PCR)-based diagnoses are commonly utilized.

16.2   IMMUNODIAGNOSIS OF PLANT VIRUS

Immunological assay is the widely used method for plant virus diagnosis, which requires virus-specific antibody, the key reagent of diagnosis. Of the several immunoassays are available, ELISA is the most popularly used method for the detection of plant viruses. The immunoassays or nucleic acid-based assays are laboratory-based techniques that require technical manpower and expensive reagents and equipments. Currently lateral flow assay (LFA) or dip-stick assay has gained popularity in plant virus detection. Anyone without technical expertise can detect a specific virus within 10–15 min by using LFA.

16.2.1   RECOMBINANT ANTIGEN

To prepare antibody, traditionally plant virus is purified from the plant tissues and used as an antigen to immunize animals. Purification of adequate quantity of virus from plant tissues is tedious, and often the purified preparation contains plant proteins and other impurities, which influence the quality of antibody. Further, many plant viruses multiply in low rate, and some are difficult to propagate in suitable hosts. Therefore, adequate and renewable supply of high quality plant virus antigen is the major limitation in generating antibody to the specific virus. Both polyclonal and monoclonal antibodies are widely used for immunodiagnosis of plant viruses, and these are prepared traditionally through cumbersome methodology. In the alternative approach, instead of culturing viruses on plant and purifying from the plant tissues, the full-length or partial coat protein (CP) gene of virus is cloned and overexpressed as a recombinant protein in bacteria (*Escherichia coli*). The expressed viral protein is purified from bacterial cells and used as a recombinant antigen for raising antibody. The procedure of generation of recombinant antigen is suitable for up-scaling, and any quantity of antibody can be raised as and when is required.

The bacterial expressed recombinant antigen has been used for raising antibody against several individual plant viruses at the Advanced Center of Plant Virology (ACPV), Indian Agricultural Research Institute (IARI), New Delhi (Jain et al., 2005; Rani et al., 2010; ; Vijayanandraj et al., 2013; Phaneendra et al., 2014; Soumya et al., 2014). We have further demonstrated that the portion of CP of plant viruses from diverse genera can be combined into a fusion construct, and cocktail antibody can be generated, which can be used to detect mixed infections in plants (Mandal et al., 2011; Kapoor et al., 2013a,b). The dual (cucumber mosaic virus [CMV] (genus *Cucumovirus*) + papaya ringspot virus [PRSV] (genus *Potyvirus*) and Potato virus X + potato virus Y) and triple (CMV + PRSV + groundnut bud necrosis virus [GBNV] (genus *Tospovirus*) fusion proteins were successfully overexpressed in *E.*

coli. The cocktail antibodies raised from these fusion proteins were shown to detect the target viruses (Kapoor et al., 2013a,b).

16.2.2   RECOMBINANT ANTIBODY

Polyclonal antibody (PAb) and monoclonal antibody (MAb) are the key reagents for immunodiagnosis. Traditionally, both types of diagnostic antibodies are produced in an animal system, which requires elaborate arrangement and precaution of raring experimental animals that often invites ethical issues. Traditionally, PAb is produced by immunizing live animals with the viral antigen, whereas MAb is produced in an animal tissue culture. Immunoglobulin G (IgG) is a major antibody in the serum, which is composed of heavy chain (H) and light chain (L) peptides. The N termini of both the peptides are highly variable (V). The terminal regions contain paratopes that recognize the epitopes of an antigen. The V_H and V_L peptides can be directly overexpressed as recombinant antibody fragments (Fab, Fv, and scFv) in different hosts like bacteria, insect, yeast, plant, and mammalian cells. At the Advanced Center of Virology, IARI, New Delhi, we have generated recombinant antibodies to GBNV and PRSV (Yogita et al., 2015a,b). The V_H and V_L of 351 and 360 nucleotides, respectively, of PRSV were expressed individually as ~14 kDa proteins in *E. coli*. Both the antibody fragments individually or together detected PRSV efficiently in the crude sap (Maheshwari et al., 2015a). In another study, we combined 372 nucleotides of V_H and 363 nucleotides of V_L of GBNV into a single chain variable fragment (scFv). The *E. coli* cell-expressed recombinant ScFV antibody to GBNV detected the virus in the field samples of cowpea, groundnut, mungbean and tomato and differentiated watermelon bud necrosis virus, a close relative of GBNV in the serogroup-IV tospovirus (Maheshwari et al., 2015b).

16.3   NUCLEODIAGNOSIS OF PLANT VIRUSES

The partial or complete genome sequence information of alexivirus, badnavirus, begomovirus, carlavirus, cucumovirus, closterovirus, ilarvirus, macluravirus, mandrivirus, potexvirus, potyvirus, tobamovirus, and tospovirus has been generated at ACPV, IARI, New Delhi. The sequence information has been utilized for designing degenerate and specific primers for the detection of these viruses infecting several horticultural crops. Duplex and multiplex PCR methods were developed for important viruses affecting cucurbits and potato. The multiplex reverse transcription-polymerase chain reaction (RT-PCR) could simultaneously detect six RNA viruses infecting potato in India.

16.4 CONCLUDING REMARKS

The approaches to prepare the antigen and antibody have evolved with the application of molecular techniques, and recombinant antigen and antibodies are increasingly used in immunodiagnosis of plant viruses. At ACPV, IARI, New Delhi, the antigen and antibody engineering approach has been utilized for generating the key reagents for immunodiagnosis of plant viruses (Mandal and Jain, 2010, 2012). The technology for virus diagnostic kits based on ELISA and PCR have been developed and validated, which can be exploited for commercial applications.

KEYWORDS

- diagnosis
- horticulture
- immunological
- plant virus
- recombinant

REFERENCES

Jain, R. K., Pande, A., & Mandal, B., (2005). Immonodiagnosis of groundnut and water-melon bud necrosis virus using polyclonal antiserum to recombinant nucleocapsid protein of *Groundnut bud necrosis virus. J., Virol. Methods, 130*, 162–164.

Kapoor, R., Mandal, B., Paul, B. K., & Jain, R. K., (2013b). Simultaneous detection of potato viruses Y and X by DAC-ELISA using polyclonal antibodies raised against fused coat proteins expressed in *Escherichia coli. Journal of Plant Biochemistry and Biotechnology, 10.*1007/s13562-013-0251-5.

Kapoor, R., Mandal, B., Paul, P. K., Phaneendra, C., & Jain, R. K., (2013a). Production of cocktail of polyclonal antibodies using bacterial expressed recombinant protein for multiple virus detection. *Journal of Virological Methods, 196*, 7–14.

Maheshwari, Y., Verma, H. N., Jain, R. K., & Mandal, B., (2015a). Engineered antibody fragments for immunodiagnosis of papaya ringspot virus. *Molecular Biotechnology, 57*, 644–652. doi: 10.1007/s12033-015-9854-5.

Maheshwari, Y., Vijayanandraj, S., Jain, R. K., & Mandal, B., (2015b). Engineered single-chain variable fragment antibody for immunodiagnosis of groundnut bud necrosis virus infection. *Arch Virol. D.*, OI 10.1007/s00705-015-2345-y.

Mandal, B., & Jain, R. K., (2010). ELISA kit for tospoviruses detection. *ICAR News, 16*(3), 13.

Mandal, B., & Jain, R. K., (2011). *Molecular Diagnosis of Plant Viruses*: production of cocktail antibody to CMV and PRS. VNAIP Annual Report, pp.74–75.

Mandal, B., & Jain, R. K., (2012). Diagnostic kits for of some chronic and emerging plant viruses in India. *Virus Research News, 1*(1), 2–3.

Mandal, B., Kumar, A., Rani, P., & Jain, R. K., (2012). Complete genome sequence, phylogenetic relationships and molecular diagnosis of an Indian isolate of *Potato virus X. J., Phytopathol, 160*, 1–5.

Phaneendra, C., Sambasiva Rao, K. R. S., Kapoor, R., Jain, R. K., & Mandal, B., (2014). Fusion coat protein of Pumpkin yellow vein mosaic virus with maltose binding protein: applications in immunodiagnosis of begomoviruses. *Virus Disease, 25*(3), 390–393, D., OI 10.1007/s13337-013-0189-1.

Rani, P., Pant, R. P., & Jain, R. K., (2010). Serological detection of Cymbidium mosaic and Odontoglossum ring spot viruses on orchids with polyclonal antibodies produced against their recombinant coat proteins. *J., Phytopath., 158*, 542–545.

Soumya, K., Yogita, M., Prasanthi, Y., Anitha, K., Kavi-Kishor, P. B., Jain, R. K., & Mandal, B., (2014). Molecular characterization of Indian isolate of *Peanut mottle virus* and immunodiagnosis using bacterial expressed core capsid protein. *Virus Disease, 25*(3), 331–337.

Vijayanandraj, S., Yogita, M., Das, A., Ghosh, A., & Mandal, B., (2013). Highly efficient immunodiagnosis of large cardamom chirke virus using the polyclonal antiserum against *Escherichia coli.* D., OI 10.1007/s13337-013-0159-7.

CHAPTER 17

ACARICIDAL ACTIVITY OF PETROLEUM ETHER EXTRACT FROM SEED OF CUSTARD APPLE, *ANNONA SQUAMOSA* L. (ANNONACEAE) AGAINST RED SPIDER MITE, *OLIGONYCHUS COFFEAE* (NIETNER) INFESTING TEA

BIPLAB TUDU and SIBDAS BASKEY

Regional Research Station (Hill Zone), North Bengal Agriculture University, Kalimpong, Darjeeling, West Bengal – 734301, India, Tel. 03552255606, E-mail: btudu_bckv@rediffmail.com

CONTENTS

17.1　INTRODUCTION

Tea, *Camellia sinensis* (L) O. Kuntze, is one of the most popular beverages in the world. The crop suffers from the attack of a number of pests and pathogens, causing significant yield losses. The red spider mite (RSM) *Oligonychus coffeae* (Nietner) is one of the major pests of tea plantation, causing 5–15% crop loss in India. This mite is characterized by a high reproductive capacity, which leads to high population levels in a short period of time, thereby causing important economic damage (Das, 1959, 1960). Nymphs and adults of *O. coffeae* (Nietner) normally infest the upper surface of mature tea leaves and lacerate cells, producing minute characteristic reddish brown marks, and when severity of infestation increases, they move even to the lower surface of older leaves and tender tea shoots. Severe infestation ultimately leads to defoliation (Selvasundaram and Muraleedharan, 2003). The feeding activity of this mite on the leaves of tea induce drastic reduction in the level of the major photosynthetic pigment chlorophyll. Chlorophyll is an essential element of photosynthesis, and its content in plant leaves indicates their photosynthetic capacity (Jayakrishnan and Ramani, 2015). The modern usages of various synthetic chemicals for the control of pests have led to various environmental concerns. Botanical insecticides have long been touted as an attractive alternative strategy for pest management, because botanicals reputedly pose little threat to the environment or to human health. Annonaceae is the largest plant family in the order Magnoliales (Westra and Maas, 2012) and comprises around 2,500 species and 130 genera (Pirie et al., 2005). Except for two related North American genera (*Asimia* and *Deeringothamnus*), the family is entirely tropical (Thomas and Doyle, 1996). Among terrestrial plant families, Annonaceae has drawn considerable attention since the 1980s, owing to the presence of acetonins, a class of natural products with broad-spectrum insecticidal bioactivities. Crude extract from the plant species of Annonaceae have been extensively studied in recent years for bioactivity to insect pests and related arthropods worldwide. *Asimina triloba, Annona muricate,* and *Annona squamosa* are the species that have been most frequently tested for their insecticidal activities (Isman and Seffrin, 2014; Ocampo and Ocampo, 2006). The acetogenins are found in the leaves and branches and, predominantly, in the seeds of annonaceous plants. In this context, the present study was undertaken at Bidhan Chandra Krishi Viswavidyalaya (State Agriculture University), Mohanpur, Nadia, West Bengal, India, during 2010–2011 to evaluate acaricidal properties of petroleum ether extracts from the custard apple seed (CASE) *Annona squamosa* L. (Annonaceae) on red spider mite infesting tea plantations.

17.2 OBJECTIVES

1. To study the acaricidal activity of petroleum ether extract from custard apple seed on tea red spider mites.
2. To study the persistent toxicity of petroleum ether extract from custard apple seed on tea red spider mites.
3. To study the phytotoxicity on tea bushes, if any.

17.3 MATERIALS AND METHODS

17.3.1 COLLECTION AND EXTRACTION OF PLANT MATERIAL

The mature fruits of custard apple were collected from Bankura district under Red and Lateritic Zone of West Bengal. The seed materials were separated from fruits, dried in shade, and made into powder by using an electric grinder or blender. The seed extracts were prepared by the Soxhlet extraction method using petroleum ether as the solvent at its boiling range of 60 °C to 80 °C temperature. The powdered seed material and solvent were taken for the extraction, keeping the ratio of 1:5 (v/v). After 8 hours of extraction to obtain the desirable alkaloids present, if any in the seed material. The extract was filtered using a Whatman No. 1 filter paper. Later, they were evaporated to obtain concentrated slurry and kept in the refrigerator as stock solution. Further dilution was done with the distilled water to get the desired doses/concentrations during the spraying (Dwivedi and Venugopalan, 2001).

17.3.2 TEST SPECIES USED

The tea red spider mite *O. coffeae* Nietner (Acari: Tetrancychidae) was obtained from the nucleus culture maintained in the Acarology Laboratory, Department of Agricultural Entomology, Bidhan Chandra Krishi Viswavidyalaya (State Agriculture University), Kalyani, Nadia, West Bengal, India.

17.3.2.1 Maintenance of Mite Population

Standard modified leaf disc technique of mass rearing of mite was followed. With the help of binocular, the last instar female mite and one adult male were picked up with the help of camel brush and were transferred to a freshly washed tea leaf. The same step was repeated for a number of sets. The leaves were placed with the bottom side up on the wet cotton sponge in which the petiole was also cushioned. One of the healthy sets was selected and was successfully proliferated for conducting experiments.

17.3.2.2 Stage of the Specimens Used

Egg, one-day old larva, and adult stages of tea red spider mites were used for the experiment.

17.3.3 TEST CONDITION

The test was conducted in laboratory at 25–28°C temperature and 55–80% relative humidity (RH).

17.3.4 TREATMENT DOSAGE FOR BIOASSAY

A series of concentrations ranging from 0.1% to 0.5% of botanical extract with three replications were evaluated separately against egg, larval, and adult stages of the tea red spider mite and an untreated control (water spray) as a check to obtain a concentration Probit mortality curve. All purpose suspension agent (APSA) at the rate of 0.33 mL/L of water was added during treatment.

17.3.4.1   Bioassay Method

Fresh tea leaves of 2 cm × 2 cm area were cut into pieces. Three replications were taken for each treatment, and every replication was composed of five leaf discs. For different experiments, in disc, 20 numbers of adult, nymph, and egg stages were treated separately. For eggs, firstly adults were released and given 24 hours to lay sufficient eggs. On the next day, the adults were removed, and spraying was done on eggs. In all the replications, Petri dishes were turn upside down, where absorbent cotton pieces were placed and soaked in water. The leaf discs were placed facing the bottom of the disc on the surface of water soaked cotton with dorsal surface of the leaf disc facing upward and sprayed with the desired concentration of the botanical solution by using a glass atomizer at a distance of 1.5 ft.

17.3.4.2   Post Treatment Observation

Observation with respect to adult and larval stages was recorded at the 3rd, 5th, and 7th days after treatment (DAT) and in case of egg stages at the 8th DAT. Moribund mites (adult and larvae) and eggs turned black, collapsed, or shriveled were considered as dead.

17.3.5   TREATMENT DOSE FOR THE STUDY OF PERSISTENT TOXICITY

For the study of persistent toxicity, tea bushes were sprayed with 0.58% concentration of botanical extract with a high volume sprayer @ 400 l spray fluids/ha.

17.3.5.1   Site of Experiment

The field experiments were conducted at the eco-friendly tea garden, Gayeshpur, Nadia, West Bengal, and the laboratory experiments were carried out at the Acarology Laboratory, Bidhan Chandra Krishi Viswavidya-laya (State Agriculture University), Kalyani, Nadia, West Bengal.

17.3.5.2   Methods of Persistent Toxicity Study

Tea bushes were sprayed separately using a high volume Knapsack sprayer, taking care to avoid drip. Only one spraying with the high volume sprayer @ 400 L/ha was done as high volume spraying. The leaves were collected from the treated plants at the 0, 1st, 3rd, 5th, 7th, and 10th DAT starting from 2 hours from treatment and brought into the laboratory.

Three replications were taken for each treatment, and every replication comprised five leaf discs. For different treatments, in each dish, 20 numbers of adult and larval stages of mites were treated separately.

17.3.5.3   Post Treatment Observation

The mortality count was taken after the 7th DAT for adult and larval stage of mites. In case of egg stage, if the egg was laid, the reading was taken on the 8th DAT.

17.3.6   EXPERIMENT ON PHYTOTOXICITY

17.3.6.1   Site of Experiment

The phytotoxicity study was conducted at the eco-friendly tea garden, Gayeshpur, Nadia, West Bengal.

17.3.6.2   Treatment Dosages for Study of Phytotoxicity

The phytotoxicity study of the botanicals was carried out at the recommended dose and two and three times higher than the recommended dose of botanical extract, i.e., 0.58%, 1.16%, and 1.74% with three replications.

17.3.6.3 Methods of Phytotoxicity Study

The total quantity of each botanical extract for particular treatment was measured for spraying, and plot size was then marked. The calculated amount of botanical extract for each replicated plot was diluted with water, and they were sprayed separately with the help of the high volume Knapsack sprayer, taking care to avoid drip. Only one spraying with the Knapsack sprayer @ 400 L/ha was done as high volume spraying.

17.3.6.4 Posttreatment Observation

Phytotoxicity symptoms were recorded continuously for the first 15 days by the observing plots in each treated plot for any adverse effect on plants, like leaf chlorosis, leaf tip burning, leaf necrosis, leaf epinasty, leaf hyponasty, vein clearing, wilting and resetting, etc., according to Central Insecticides Board, Government of India guidelines. Phytotoxicity Scale: (No phytotoxicity = 0 scale; 1% to 10% phytotoxicity = 1 scale, 11 to 20% phytotoxicity = 2 scale ....91 to 100% phytotoxicity = 10 scale).

17.3.7 Statistical Methods Utilized

The data obtained in percentage were first subjected to angular transformation (Fisher and Yate, 1938). The percent mortality observed in different treatments was corrected using modified Abbott's formula (Abbott, 1925).

$$P(\%) = \frac{(P/-C)}{(100-C)} \times 100$$

where P = percentage corrected mortality, P^1 = percentage observed mortality; C = percentage control mortality.

To calculate the LC_{50} values, data obtained on mortality were subjected to Probit analysis (Finney, 1971). Relative toxicity of the botanicals was calculated on the basis of LC_{50} value of one of the higher LC_{50} values as unity.

For persistent toxicity, the rate of mortality (percent) of the red spider mites to different doses at various time intervals after treatment was recorded. The average residual toxicity, index of persistent toxicity, and the order of relative efficacy were calculated by the criterion elaborated by Pradhan and Venkataraman (1962). To calculate the PT_{50} values, the data were subjected to Probit analysis as in case of LC_{50} values.

17.4 RESULTS AND DISCUSSION

17.4.1 ACARICIDAL ACTIVITY

It is evident from studies that the LC_{50} value of CASE was the lowest, i.e., 0.10% on one-day old larval stage and highest against the egg stage with LC_{50} value of 0.29% (e.g., see Table 17.1). This indicates that CASE was most toxic to the one-day old larval stage followed by adult and egg stage of RSM. The LC_{50} values for crude extracts from the leaves of *A. senegalensis* and *A. squamosa* were 0.67 and 0.64 µg/mL, respectively, for *Culex quinquefaciatus* (Magadula et al., 2009). Begum et al. (2010) investigated the toxic effects of ethanol extracts of seeds of *A. squamosa* and *Calotropis procera* (Asclepiadaceae) against different developmental stages of the housefly *Musca domestica* L. (Diptera: Muscidae). LC_{50} values for the extracts of *C. procera* and *A. squamosa* seeds were 870 and 345 mg/L, respectively. An ethanolic extract of *A. squamosa* leaves showed potent activity against the rice weevil *S. oryzae*. The extract produced significant knockdown (KDT_{50}) at 1% (23.1 min) and 5% w/v (11.4 min). Complete mortality was achieved at 39.6 ± 1.4 and 14.5 ± 1.1 min for 1% and 5% w/v, respectively (Kumar et al., 2010). Kamaraj et al. (2011) assessed the larvicidal activities of hexane, chloroform, ethyl acetate, acetone, and methanol dried leaf and bark extracts of *A. squamosa*, *Chrysanthemum indicum*, and *Tridax procumbens* against 4th instar larvae of the malaria

TABLE 17.1 Acaricidal Activity of CASE on Tea Red Spider Mites after 72 h of Treatment

Stage of mites	d.f.	χ^2	Regression equation	Fiducial limit	$LC_{50}(\%)$
Egg	7	21.78	Y=55+1.02X	-0.54 ± 0.04	0.29
Larva	7	1.03	Y=6.40+1.38X	-1.02 ± 0.04	0.10
Adult	7	2.83	Y=5.59+0.64X	-0.92 ± 0.04	0.12

vector, *Anopheles subpictus* Grassi, and the Japanese encephalitis vector *Culex tritaeniorhynchus* Giles (Diptera: Culicidae). All plant extracts showed moderate effects after 24 h of exposure; however, the most toxic were the methanolic bark extract of *A. squamosa*, leaf ethyl acetate extract of *C. indicum*, and leaf acetone extract of *T. procumbens* against the larvae of *A. subpictus* (LC_{50} = 93.80, 39.98, and 51.57 mg/L, respectively) and methanolic bark extract of *A. squamosa*, leaf methanol extract of *C. indicum*, and leaf ethyl acetate extract of *T. procumbens* against the larvae of *Cx. tritaeniorhynchus* (LC_{50} = 104.94, 42.29, and 69.16 mg/L, respectively). Attia et al. (2011) reported that garlic juice at a concentration of 7.49% showed LD_{50} value against *Tetranychus urticae* Koch. Grzybowski et al. (2013) tested crude ethanolic extracts of *A. muricata* L. seeds and *Piper nigrum* L. fruits against *A. aegypti* larvae. The LC_{50} value for *A. muricata* was 93.48 μg/mL and for *P. nigrum* 1.84 μg/mL. The index of persistent toxicity (PT) of CASE when applied @ 0.58% on the egg stage was 230.19, followed by one-day old larva and adult at 99.34 and 90.27, respectively. Further, the average residual toxicity (T), i.e., 39.04, was highest in the egg stage than in other stages and persisted for period (P) of 5 days (e.g., see Table 17.2). PT_{50} values of this botanical extract on egg, one-day old larva, and adults of tea red spider mite were 1.01, 0.27, and 0.22 day, respectively (e.g., see Table 17.3).

17.4.2 BIOEFFICACY OF CASE ON THREE DEVELOPMENTAL STAGES OF RSM

The performance of petroleum ether fractions of custard apple seed @ 0.10%, 0.15%, and 0.20% caused 80.60%, 89.14%, and 97.09% larval mortality at the 7th DAT, respectively, and all the higher dosages, i.e., 0.25%, 0.30%, 0.35%, 0.40%, 0.45%, and 0.50% at the 7th DAT showed 100% larval mortality. Lowest dose, i.e., 0.10% at the 3rd DAT caused 54.30% larval mortality. Eggs of *Oligonychus coffeae* were successfully killed up to 77.67% at concentration of 0.50% of CASE. Results also revealed that petroleum ether extract of custard apple seed at concentration of 0.25% at the 7th DAT recorded 84.65% mortality and at concentration of 0.10%, 0.15%, and 0.20% caused mortality of 77.37%, 79.13%, and 80.52%, respectively, of adult stage of *O. coffeae*. At the highest dose,

TABLE 17.2 Persistent Toxicity of CASE at a Concentration of 0.58 % on Tea Red Spider Mites

Stage of mites	% mortality at various days interval (days)						P	T	PT
	0.083	1	3	5	7	10			
Egg	91.64	83.64	5.64	3.23	N/A	N/A	5	46.04	230.19
Larva	84.61	11.47	3.26	N/A	N/A	N/A	3	33.11	99.34
Adult	78.86	9.14	2.26	N/A	N/A	N/A	3	30.09	90.27

P = Period in days; T = Average residual toxicity; PT = Index of persistent toxicity.

TABLE 17.3 PT_{50} Values of CASE on Tea Red Spider Mites

Stage of mites	d.f.	χ^2	Regression equation	Fiducial limit	PT_{50} (days)
Egg	4	128.63	Y=5.01–2.23X	0.01±0.03	1.01
Larva	4	0.99	Y=3.88–1.97X	–0.57±0.03	0.27
Adult	4	0.37	Y=3.72–1.92X	–0.67±0.03	0.22

i.e., 0.5%, 93.47% adult mortality was observed at the 7th DAT (e.g., see Table 17.4). Similarly, Pavela (2009) reported 100% mortality of *Tetranychus urticae* Koch by spraying pongam oil at 1% and 3% concentration. Lin et al. (2009) tested the cold pressed oil from the seeds of *A. squamosa*. The oil was effective in controlling the Kanzawa spider mite *Tetraanychus kanzawai* Kishida (Acari: Tetranychidae) on soybean leaves. The high level of mortality of head louse, *Pediculus humanus capitis*, with petroleum ether extract from Indian neem, *A. indica,* and a seed of *A. squamosa* at concentration of 0.1%, 1%, and 10% have been reported by Kosalge and Fursule (2009). The acetone, chloroform, hexane, petroleum ether, and ethanol extracts of *A. squamosa* foliage were studied against the early 4th instar larvae of *A. aegypti*, *Anopheles stephensi*, and *Culex quinquefasciatus*. Larval mortality was observed after 24 h exposure. All extracts showed moderate larvicidal effects; however, the greatest larval mortality was obtained with a petroleum ether extract (Kumar et al., 2011). Gonzalez Esquinca et al. (2012) used water and ethanolic extracts to determine the activity of stem and leaf extracts of *A. muricata* L., *A. diversifolia* Saff., and *A. lutescens* Saff. against larvae of *Anastrepha ludens* (Mexican fruit fly). Extracts of the three *Annona* species showed time-dependent larvicidal activity against *A. ludens*, with variable mortality rates at 72 h

TABLE 17.4 Effect of CASE on Tea Red Spider Mites (Mean of Two Applications and Three Replications)

Sl. No.	Concen-tration	% mortality of mites at various days after treatment						
		Larval			Adult			Egg
		3rd DAT	5th DAT	7th DAT	3rd DAT	5th DAT	7th DAT	8th DAT
1.	0.1%	54.30 (47.47)	56.84 (48.93)	80.60 (63.87)	51.06 (45.61)	56.60 (48.79)	77.37 (61.59)	38.28 (38.22)
2.	0.15%	58.60 (49.95)	58.91 (50.13)	89.14 (70.76)	53.21 (46.84)	61.49 (51.64)	79.13 (62.82)	39.46 (38.91)
3.	0.20%	64.80 (53.61)	69.53 (56.50)	97.09 (80.25)	54.97 (47.85)	69.19 (56.28)	80.52 (63.81)	42.11 (40.46)
4.	0.25%	71.16 (57.52)	75.87 (60.58)	100.00 (90.00)	55.13 (47.94)	69.95 (56.76)	84.65 (66.94)	45.07 (42.17)
5.	0.30%	74.47 (59.65)	79.21 (62.88)	100.00 (90.00)	56.57 (48.77)	73.58 (59.07)	86.21 (68.20)	45.77 (42.57)
6.	0.35%	79.66 (63.20)	81.65 (64.63)	100.00 (90.00)	58.01 (49.61)	75.86 (60.57)	86.53 (68.47)	47.12 (43.35)
7.	0.40%	81.20 (64.31)	82.71 (65.43)	100.00 (90.00)	64.02 (53.15)	80.21 (63.58)	88.34 (70.04)	47.44 (43.53)
8.	0.45%	83.36 (65.93)	84.20 (6.58)	100.00 (90.00)	65.75 (54.18)	82.78 (65.49)	90.96 (72.51)	58.10 (49.66)
9.	0.50%	84.13 (66.52)	86.00 (68.02)	100.00 (90.00)	69.74 (56.63)	84.56 (66.86)	93.47 (75.20)	77.67 (61.80)
10.	Untreated	8.29 (16.73)	11.97 (20.24)	19.31 (26.07)	6.56 (14.83)	8.33 (16.77)	11.91 (20.19)	10.07 (18.50)
C. D. at 5% level		0.87	0.55	0.83	0.62	0.47	0.52	0.46

N.B.: Figure in the parentheses is angular transformed values.

of exposure as follows: *A. lutescens* 87–94%, *A. diversifolia* 70–90%, *and A. muricata* 63–74%. Radhakrishnan and Prabhakaran (2014) reported that aqueous extracts of *Allamanda catharitica* and *Conyza bonariensis* showed 100% and 80% adult mortality, respectively, at 5% concentration after 96 h of observation under laboratory condition. Mech et al. (2015) found that methanolic extract from *Parthenium hysterophorus* leads to highest mortality against the adults of *O. coffeae* followed by petroleum ether and chloroform extracts. The LC_{50} value of methanol extract against the adult mite was 0.12% (48 h).

17.4.3   PHYTOTOXIC EFFECT OF BOTANICAL EXTRACT

None of the treated concentrations, i.e., 0.58%, 1.16%, and 1.74% of CASE showed phytotoxicity on tea leaves (e.g., see Table 17.5). Selvasundaram (2004) reported that Exodus, a new herbal product derived from *Sophora flavescens*, was tested against the red spider mite *Oligonychus coffeae* under laboratory and field conditions. Field studies conducted from January to May 2004 in Anamallais, Tamil Nadu, India, indicated that spraying the herbal product at 250 and 500 mL/ha reduced the number of red spider mites. Application of this product did not result in phytotoxicity to tea at the tested concentrations. Acaricidal and ovicidal activities of *Clerodendrum viscosum Ventenat* (Verbenaceae), a common weed of India, were investigated on tea red spider mite, *Oligonychus coffeae* Nietner (Acarina: Tetranychidae). Different solvent extracts (water, methanol, acetone, and petroleum ether) of *C. viscosum* at different concentrations (1, 2, 4, and 8%) were used. No phytotoxic effect (score, 0–5% and grade 1) was observed in the field from tea bushes sprayed with different doses of extracts of *C. viscosum* (Roy et al., 2011). Aqueous seed extract of *Melia azedarach* (L.) was evaluated against the tea red spider mite *O. coffeae* (Nietner), in relation to mortality of adult mites, viability of eggs and subsequent adult emergence and oviposition deterrence in the laboratory, and the extract underwent field evaluation in terms of percent reduction of the mite population. No phytotoxic effect (score 0–5% and grade 1) was observed in the field when tea bushes were sprayed with different doses of aqueous seed extract of *M. azedarach* (Roy and Mukhopadhyay, 2012).

17.5   SUMMARY AND CONCLUSION

Finally, it may be concluded that the LC_{50} value of CASE was the lowest (0.10%) on one-day old larval stage and highest against the egg stage (0.29%). This indicates that CASE was most toxic to the one-day old larval stage, followed by adult and egg stage of RSM. The index of persistent toxicity (PT) of CASE when applied @ 0.58% on egg stage was 230.19, followed by one-day old larva and adult at 99.34 and 90.27, respectively.

TABLE 17.5 Evaluation of Different Treatment Schedules of CASE for Phytotoxicity on Tea Bushes during March–April 2010 at Gayeshpur Tea Garden, West Bengal (Based on One Application and Ten Replications)

Sl. No.	Treat-ment	Dose (%)	Visual rating (phytotoxicity) in 0 – 10 scale of grading										
			0	1	2	3	4	5	6	7	8	9	10
			0.0%	1– 10%	11– 20%	21– 30%	31– 40%	41– 50%	51– 60%	61– 70%	71– 80%	81– 90%	91– 100%
1.	CASE	0.58	0	0	0	0	0	0	0	0	0	0	0
2.	CASE	1.16	0	0	0	0	0	0	0	0	0	0	0
3.	CASE	1.74	0	0	0	0	0	0	0	0	0	0	0
4.	Untreated	N/A	0	0	0	0	0	0	0	0	0	0	0

N.B.: Observations were based on 1, 3, 7 and 15 days after spraying on necrosis, epinasty, hyponasty, leaf tip injury, leaf surface injury, wilting, vein clearing etc.

Further, the average residual toxicity (T), i.e., 39.04, was highest in the egg stage than in other stages and persisted for period (P) of 5 days. PT_{50} values of this botanical extract on egg, one-day old larva and adults of tea red spider mite were 1.01, 0.27, and 0.22 day, respectively. The present investigation revealed that petroleum ether extract from the custard apple seed *Annona squamosa* L. (Annonaceae) has good acaricidal activity. The results of this study indicate the plant-based compounds may be an effective alternative to conventional acaricides for the management of RSM. Plant allelochemicals may be quite useful in increasing the efficacy of biological control agents because plant produces a large variety of compounds that increase their resistance to insect attack (Senthil Nathan et al., 2005). Hence, by using extracts of custard apple seed in their field, tea planters may reduce the incidence of RSM in tea plantations and may find scope in integrated pest management system of *Oligonychus coffeae* (Nietner). However, further investigation is necessary for optimization of the bioactive compound.

ACKNOWLEDGMENT

We are in great pleasure to express our deepest sense of gratitude to late Dr. A. K. Somchoudhury, Professor and former Dean, Post Graduate Studies, Bidhan Chandra Krishi Viswavidyalaya, Mohanpur, West Bengal, for his learned and affectionate guidance, helpful criticism, and encouragement during the entire course of investigation.

KEYWORDS

- acaricidal
- bioassay
- custard apple
- red spider mite
- tea
- toxicity

REFERENCES

Abbott, W. S., (1925). A method of computing the effectiveness of an insecticide. *J. Eco. Ent., 18*, pp. 265–267.

Attia, S., Grissa, K. L., Mailleux, A. C., Lognay, G., Heuskin, S., Mayoufi, S., & Hance, T., (2011). Effective concentrations of garlic distillate (*Allium sativum*) for the control of *Tetranychus urticae* (Tetranychidae). *Journal of Applied Entomology, 135*(4), pp. 300–312.

Begum, N., Sharma, B., & Pandey, R. S., (2010). Toxicity potential and anti ache activity of some plant extracts in *Musca domestica. J. Biofert. Biopestic, 2*, pp. 108–114.

Das, G. M., (1959). Bionomics of tea red spider *Oligonychus coffeae* (Nietner). *Bulletin of Entomological Research, 50*(2), pp. 265–275.

Das, G. M., (1960). Occurrence of red spider *Oligonychus coffeae* (Nietner) on tea in North East India in relation to pruning and defoliation. *Bulletin of Entomological Research, 51*(3), pp. 415–426.

Dwivedi, S. C., & Venugopalan, S., (2001). Evaluation of leaf extracts for their ovicidal action against callosobruchus chinensis (L.). *Asian J. Exp. Sci., 16*(1–2), pp. 29–34.

Finney, D. J., (1971). *Probit Analysis*. Cambridge University Press, London, pp. 333.

Fisher, R. A., & Yates, F. A., (1938). Statistical tables for biological, agricultural and medical research. *Oliver and Boyd*. London, pp. 146.

Gonzalez, E. A. R., Luna, C. L. M., Schlie, G. M. A., De la, C. C. I., Hernandez, G. L., Breceda, S. F., & Gerardo, P. M., (2012). In *vitro* larvicidal evaluation of *Annona muricata* L. *A., diversifolia* Saff. and *A., lutescens* Saff. extracts against *Anastrepha ludens* larvae (Diptera, Tephritidae). *Interciencia, 37*, pp. 264–289.

Grzybowski, A., Tiboni, M., Silva, M. A., Chitolina, R. F., Passos, M., & Fontana, J. D., (2013). Synergistic larvicidal effect and morphological alterations induced by ethanolic extracts of *Annona muricata* and *Piper nigrum* against the dengue fever vector *Aedes aegypti. Pest. Manag. Sci., 69*, pp. 589–601.

Isman, M. B., & Seffrin, R., (2014). Natural insecticides from Annonaceae: a unique example for developing biopesticides. In: *Advances in Plant Biopesticides*, Singh, D., (edn.), Springer India, pp. 21–33.

Jayakrishnan, T. V., & Ramani, N., (2015). Reduction of major photosynthetc pigments induced by *oligonychus coffeae* (Nietner) (Acari: Tetranychidae) infesting *camellia sinensis* (L) Kuntze, O. *International Journal of Recent Scientific Research., 6*(5), pp. 3947–3950.

Kamaraj, C., Bagavan, A., Elango, G., Zahir, A. A., Rajakumar, G., Marimuthu, S., Santhoshkumar, T., & Rahuman, A. A., (2011). Larvicidal activity of medicinal plant extracts against *Anopheles subpictus* and *Culex tritaeniorhynchus. Indian J. Med. Res., 134*, pp. 101–106.

Kosalge, S. B., & Fursule, R. A., (2009). Investigation of licicidal activity of some plants from Satpura Hills. *Int. J. Pharm Tech. Res., 1*, pp. 564–567.

Kumar, J. A., Rekha, T., Shyamala, D. S., Kannan, M., Jaswanth, A., & Gopal, V., (2010). Insecticidal activity of ethanolic extract of leaves of *Annona. J. Chem. Pharm Res., 2*, pp. 177–180.

Kumar, S. V., Mani, P., John, B. T. M. M., Arun, K. R., & Ravikumar, G., (2011). Larvicidal, oviposition deterrent and repellent activity of *Annona squamosa* extracts against hazardous mosquito vector. *Int. J. Pharm Tech., 3*, pp. 3143–3155.

Lin, C. Y., Wu, D. C., & Ko, W. H., (2009). Control of silverleaf whitefly, cotton aphid and kanzawa spider mite with oil and extracts from seeds of sugar apple. *Neotrop. Entomol.*, *38*, pp. 531–536.

Magadula, J. J., Innocent, E., & Otieno, J. N., (2009). Mosquito larvicidal and cytotoxic activities of three *Annona* species and isolation of active principles. *J. Med. Plant. Res.*, *3*, pp. 674–680.

Mech, J., Bhuyan, P. D., & Bhattacharyya, P. R., (2015). Acaricidal activities of *Parthenium hysterophorus* L., against red spider mite, *Oligonychus coffeae* Nietner (Acarina: Tetranychidae) of tea. *Int. J. Sci. and Res.*, *6*(14), pp. 901–904.

Ocampo, D., & Ocampo, R., (2006). Bioactividad de la familia Annonaceae. *Rev. Univ. Caldas.*, *26*, pp. 135–155.

Pavela, R., (2009). Effectiveness of some botanical insecticides against *Spodoptera littoralis* Boisduvala (Lepidoptera: Noctuidae), *Myzus persicae* Sulzer (Hemiptera :Aphidae) and *Tetranychus urticae* Koch (Acari: Tetranychidae). *Plant Protection Science*, *45*(4), pp. 161–167.

Pirie, M. D., Chatrou, L. W., Erkens, R. H. J., Maas, J. W., Van der Niet, T., Mols, J. B., & Richardson, J. E., (2005). Phylogeny reconstruction and molecular dating in four neotropical genera of annonaceae: the effect of taxon sampling in age estimations. *Plant Species Level Syst. New Perspect Pattern Process*, *143*, pp. 149–174.

Pradhan, S., & Venkataraman, T. V., (1962). Integration of chemical and biological control of *Chilo zonellus* (Swinch.), the stalk borer of maize and jower. *Proceedings of the Symposium on Advancing Frontiers of Life Sciences. Bull. Natn. Inst. Sci.*, India., *19*, pp. 119–125.

Radhakrishnan, B., & Prabhakaran, P., (2014). Biocidal activity of certain indigenous plant extracts against red spider mite, *Oligonychus coffeae* (Nietner) infesting tea. *J. Biopest.*, *7*(1), pp. 29–34.

Roy, S., & Mukhopadhyay, A., (2012). Bioefficacy assessment of *Melia azedarach* (L.) seed extract on tea red spider mite, *Oligonychus coffeae* (Nietner) (Acari: tetranychidae). *Int. J. Acarol.*, *38*(1), pp. 79–86.

Roy, S., Mukhopadhyay, A., & Gurusubramanian, G., (2011). Anti-mite activities of *clerodendrum viscosum ventenat* (Verbenaceae) Extracts on tea red spider mite, *Oligonychus coffeae* nietner (Acarina: Tetranychidae). *Archives of Phytopathology and Plant Protection*, *44*(16), pp. 1550–1559.

Selvasundaram, R., & Muraleedharan, N., (2003). Red spider mite - biology and control. *Hand Book of Tea Culture, Sect.*, *18*, UPASI Tea Research Foundation, Valparai, pp. 4.

Selvasundaram, R., Muraleedharan, N., James, S. P., & Sudarmani, D. N. P., (2004). A herbal product for red spider mite control. *Newsletter UPASI Tea Research Foundation*, *14*(1), pp. 4.

Senthil Nathan, S., Kalaivani, K., Murugan, K., & Chung, P. G., (2005). The toxicity and physiological effect of neem limonoids on *Cnaphalocrocis medinalis* (Guene), the rice leaf folder. *Pesticide Biochemistry and Physiology*, *81*, pp. 113–122.

Thomas, A., & Doyle, J. A., (1996). Geographic relationships of malagasy annonaceae. *Biogéogr Madag*, pp. 85–94.

Westra, L. Y. T., & Maas, P. J. M., (2012). *Tetrameranthus* (Annonaceae) revisited including a new species. *PhytoKeys*, *12*, pp. 1–21.

CHAPTER 18

EVALUATION OF SOME PLANT EXTRACTS AGAINST EARLY BLIGHT OF TOMATO UNDER NET HOUSE CONDITION

MANISHA DUBEY,[1] T. S. THIND,[1] S. K. JINDAL,[2] and R. K. DUBEY[3]

[1]Department of Plant Pathology, [2]Department of Vegetable Science and [3]Department of Floriculture and Landscape, Punjab Agricultural University, Ludhiana–141004, India, E-mail: manishalandscape@rediffmail.com

CONTENTS

The antifungal effect of extracts of three plants, viz., *Allium sativum, Eucalyptus chamadulonsis*, and *Azadirachta indica* against *Alternaria solani*, the cause of early blight of tomato, was evaluated under in vitro and field conditions. Effect of plant extracts and the chemical fungicide Indofil M-45 (as standard check) at various concentrations on mycelial growth of *A. solani* was determined by the poisoned food method. The leaf extract of *E. chamadulonsis* and *A. indica* at 20% concentration caused highest reduction of mycelia growth of *A. solani* (73.3% and 63.2%, respectively), while *A. sativum* at 20% concentration caused the lowest inhibition of fungal growth. In net house experiments, the highest reduction of disease severity was achieved by the fungicide (Indofil M-45) at 79.3%, followed by extracts of *E. chamadulonsis* and *A. sativum* at 20% concentration. Fungicide and plant extracts increased the fruit yield by 80.0%, 71.4%, 66.7%, and 60.0% compared to untreated control. All treatments of plant extracts and fungicide (Indofil M-45) significantly reduced the early blight disease as well as increased the yield of tomato under field condition.

Tomato (*Solanum lycopersicon* L.) is one of the most important "protective food" because of its special nutritive value and its wide production. Among the fungal diseases affecting tomato crop, early blight caused by *Alternaria solani* (Ell. & Martin) causes considerable yield losses. Most tomato cultivars are susceptible to this disease and depending upon age of the plant, season, and environmental factors at the time of infection, and yield losses range between 50% and 86% (Mathur and Shekhawat, 1986). The fungus *Alternaria* can cause the disease on all parts of the plant (leaf blight, stem lesions, and fruit lesion at the pedicel end) that results in severe damage during all stages of plant development (Abada et al., 2008). Control of early blight disease has been accomplished primarily by the application of chemical fungicides (Jones et al., 1991). Several effective fungicides have been recommended for use against *A. solani*, but these are not considered to be long-term solutions due to concerns of expense, exposure risks, fungicide residues and other health and environmental hazards. In an attempt to ameliorate this condition, some alternative methods of control have been adopted. Natural products isolated from various plants appear to carry minimal environmental impact and danger to consumers in contrast to synthetic pesticides (Varma and Dubey, 1999).

The use of plant extracts has been shown to be ecofriendly and effective against many plant pathogens (Bowers and Lock, 2004; Saadabi, 2006). A number of plant species have been reported to possess natural substances that are effective against several plant pathogenic fungi (Goussous et al., 2010). Sallam (2011) evaluated the effect of different plant extracts on mycelial growth of *A. solani* and found that leaf extracts of some plants, i.e., *Tamarix aphylla* and *Salsola bayosma*, totally inhibited the growth of the pathogen. Obagwu (1997)[9] reported that garlic extracts significantly reduced the early blight disease of tomato. The present study was conducted to determine the efficacy of leaf extracts of *E. chamadulonsis* and *A. indica* and bulb extract of *Allium sativum* for the control of early blight of tomato under net house and field conditions. The treatments were compared with the commonly used fungicide Indofil M-45.

18.1　MATERIALS AND METHODS

The causal fungus was isolated from naturally infected tomato leaves and fruits showing blight symptoms. Pathogenicity tests of *Alternaria solani* isolates were carried on tomato plants grown under net house of the Vegetable Department, PAU, Ludhiana. The inoculum was prepared by scraping *A. solani* colony surface in sterilized water. The resulting conidial and mycelia fragment suspension was adjusted to 5×10^6 cfu/mL using a hemocytometer and used for inoculation of 15 tomato plants in pots in net house. After inoculation, the net house was covered with a polythene sheet for 48 h to maintain high humidity conditions. After 48 h, the polythene sheet was removed. The pots were arranged in a completely randomized design under net house conditions. Disease severity was recorded 2 weeks after inoculation. Ten plants were randomly selected and scored individually using 0–5 rating scale based on leaf area, stem and fruit covered by blight symptoms following the rating scale described by Pandey et al. (2003).

Percentage disease index (PDI) was calculated as follows:

$$PDI = \frac{\text{Sum of all rating}}{\text{Total no.of observation} \times \text{Maximum rating growth}} \times 100$$

The mean value of the PDI from 10 individual plants was calculated for each of the observations at 90 and 120 days after transplanting and averaged.

18.1.1　PREPARATION OF EXTRACTS

Leaves of two plants namely *Eucalyptus chamadulonsis* and *Azadirachta indica* and bulbs of *Allium sativum* were collected from the field of Punjab

Agricultural University, Ludhiana, Punjab and used for preparing extracts. Plant leaf extract was prepared according to Sallam (2011). Ten grams of fresh leaf material of each plant species was collected, washed with water, and crushed in a grinder by adding sterile distilled water at the rate of 10 mL/gm of plant tissue, and the homogenates were centrifuged at 10,000 × g for 15 min and the supernatant solutions were collected. The plant extract was diluted further to obtain 10% and 20% concentration (v/v). These fractions were sterilized using 0.22-μm Millipore filters and used for assay of antimicrobial activity as described below.

18.1.2 IN VITRO SCREENING OF PLANT EXTRACTS

The filtrate of each plant extract was mixed with autoclaved potato dextrose agar (PDA) medium @ 10% and 20% concentration. The plant extract supplemented medium was poured in sterilized Petri plates and allowed to solidify. These Petri plates were inoculated at the center with a 5-mm agar disc obtained from the 7-days-old fungal culture of *Alternaria solani*. In the control, a Petri plates containing PDA medium with the requisite amount of sterilized water instead of a plant extract was inoculated with the test pathogen. Indofil M-45 (mancozeb), the standard fungicide recommended for control of early blight of tomato, @ 0.2% was used as a positive control for comparison. The inoculated Petri plates were then incubated at 25±2°C for 7 days. The diameter of the fungal colony was measured using a meter rule along two diagonal lines drawn on the reverse side of each petri plate 24 h after inoculation. Each treatment was replicated three times with three plates per replication. Percentage inhibition of mycelial growth was calculated, using the formula:

$$\% \text{ growth inhibition} = \frac{dc - dt}{dc} \times 100 \times 100$$

where dc= average diameter of fungal colony in control plates; dt= average diameter of fungal colony in treated plates.

Three plates per each treatment were used as replications. The diameter of the fungal colony was measured using a meter rule along two diagonal lines drawn on the reverse side of each Petri plate 7 days after inoculation. Each treatment was replicated three times with three plates per replication.

18.1.3 EFFICACY OF THE TESTED PLANT EXTRACTS AGAINST A. SOLANI IN A NET HOUSE

Plant extract treatments at 20% concentrations were applied as foliar application on 7-weeks-old tomato plants and every 15 days up to 60 days. The fungicide Indofil M-45 (mancozeb, 75%) @ 0.2% was applied as the standard check. Tomato plants were inoculated with 20 mL of *A. solani* suspension containing 5×10^6 cfu/mL. The first spray was carried out as soon as the first symptom of early blight was seen in the net house. Two weeks after inoculation, disease severity was recorded. Ten plants were randomly selected and scored individually using the 0–5 rating scale (Table 18.1, Figure 18.1) based on leaf area, stem, and fruit covered by blight systems following the rating scale described by Pandey et al. (2003).

Observations on disease severity were recorded 2 weeks after the last application. Percent disease severity (PDI) was calculated by the standard formula as follows

$$PDI = \frac{\text{Sum of all rating}}{\text{Total no.of observation} \times \text{Maximum rating growth}} \times 100$$

Statistical analysis of the data was done using completely randomized block design (CRD).

TABLE 18.1 In Vitro Effect of Plant Extracts on the Mycelial Growth of *Alternaria solani*

Treatments	Concentration (%)	Percent of mycelial growth reduction (%)
Eucalyptus chamadulonsis	10	59.13
	20	73.30
Allium sativum	10	53.30
	20	63.00
Azadirachta indica	10	56.10
	20	63.20
Indofil M-45 (mancozeb 75%)	0.2	80.00
Control (water)	-	0.00
CD	-	7.20

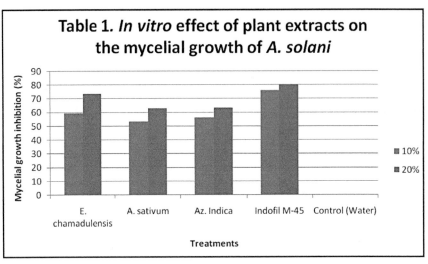

FIGURE 18.1 **(See color insert.)** In vitro effect of plant extracts on the mycelial growth of *Alternaria solani.*

18.2 RESULTS AND DISCUSSION

18.2.1 EFFECT OF PLANT EXTRACTS ON MYCELIAL GROWTH OF A. SOLANI

Three plant species and the fungicide Indofil M-45 were evaluated for the antifungal activity against *A. solani*. Leaf extracts of *Eucalyptus chamadulonsis and Azadirachta indica* and bulb extract of *Allium sativum* at 10% and 20% concentration were effective in inhibiting the radial growth of *A. solani* as compared to untreated control. The leaf extract of *E. chamadulonsis* and *A. indica* at 20% concentration caused highest reduction of mycelial growth of *A. solani* (73.3% and 63.2%, respectively). Bulb extract of *Allium sativum* at 20% concentration caused the lowest inhibition of mycelial growth of the pathogen. Overall, Indofil M-45 at 2.0 g/L caused the highest reduction of the pathogen by 80.0% (Table 18.2).

18.3 EFFECT OF PLANT EXTRACTS ON EARLY BLIGHT IN A NET HOUSE

The protective action of the plant extracts relative to the recommended fungicide Indofil M-45 against early blight in tomato plant that evaluated

TABLE 18.2   Effect of Plant Extracts on Early Blight of Tomato in a Net House

Treatments	Phytotoxicity	Disease severity (%)	Disease reduction (%)	Fruit yield (% increase)
Eucalyptus chamadulonsis (20%)	NO	19.50	68.30	71.40
Allium sativum (20%)	NO	22.50	63.40	66.70
Azadirachta indica (20%)	NO	24.90	59.40	60.00
Indofil M-45 (0.2%)	NO	12.70	79.30	80.00
Control (water)	-	61.40	0.00	-
CD		-	-	5.20

NO = Not observed.

under net house conditions is shown in Table 18.2. The results showed that Indofil M-45 was the most effective treatment against *A. solani* followed by *Eucalyptus chamadulonsis, Allium sativum* and *A. indica,* respectively, at 20% concentration. There was no phytotoxicity of the tested plant extracts observed on tomato.

The plant extracts of *E. chamadulonsis, A. sativum,* and *A. indica* caused significant reduction in the linear growth of *A. solani.* This reduction was gradually increased by increasing concentration of extracts in the growth medium. Indofil M-45 was more effective than the plant extracts. Similar effects of various other plant products effective against *Alternaria* spp have been reported by several workers (Dushyant and Bohra, 1997; Bowers and Lock, 2004; Latha et al., 2009). The bulb extracts of *A. sativum,* leaf extract of *Aegle marmelos,* and flower extract of *Catharanthus roseus* inhibited the spore germination and mycelia growth of *A. solani* (Vijayan, 1989). Garlic and neem products have also shown some antimicrobial properties and have been used in the control of fungal pathogens (Stoll, 1998). Methanolic extracts of peppermint (15%) and eucalyptus (15%) were reported as the best in preventing the spore germination of *Alternaria sesame* (Zaker, 2013).

ACKNOWLEDGMENT

The authors are thankful to the Head, Department of Vegetable Science, for providing necessary facilities to conduct the experiment, and Department of Science and Technology New Delhi, for sponsoring the study under the WOS B scheme.

KEYWORDS

- *Alternaria solani*
- early blight
- fungicide
- plant extracts
- tomato

REFERENCES

Abada, K. A., Mostafa, S. H., & Mervat, R., (2008). Effect of some chemical salts on suppressing the infection by early blight disease of tomato. *Egyptian Journal of Applied Science, 23,* 47–58.

Bowers, J. H., & Locke, J. C., (2004). Effect of formulated plant extracts and oils on population density of *Phytophthora nicotianae* in soil and control of *Phytophthora blight* in the green house. *Plant Dis., 88,* 11–16.

Dushyent, G., & Bohra, A., (1997). Effect of extracts of some halophytes on the growth of *A., solani. Journal of Mycological Plant Pathology, 27,* 233.

Goussous, S. J., Abu-El-Samen, F. M., & Tahhan, R. A., (2010). Antifungal activity of several medicinal plants extracts against the early blight pathogen (*Alternaria solani*). *Archives of Phytopathology and Plant Protection, 43,* 1746–1758.

Jones, J. B., Jones, J. P., Stall, R. E., & Zitter, J. A., (1991). Infectious diseases : Diseases caused by fungi. In: Compendium of Tomato diseases. *Am. Phytopathol, Soc. St.* Paul, M. N., pp. 9–25.

Latha, P., Anand, T., Raghupati, N., Prakasam, V., & Samiyappan, R., (2009). Antimicrobial activity of plant extracts and induction of systemic resistance in tomato plants by mixtures of PGPR strains and Zimmu leaf extract against *Alternaria solani. Biol. Control, 50,* 85–93.

Mathur, K., & Shekhawat, K. S., (1986). Chemical control of early blight in kharif sown tomato. *Indian J. Mycol. Pl. Pathol., 16*, 236–238.

Obagwu, J., Emechebe, A. M., & Adeoti, A. A., (1997). Effects of extracts of garlic (*Allium sativum*) bulb and neem (*Azadirachta indica*) seed on the mycelial growth and sporulation of *Collectotrichum capsici. J. Agric. Technology, 5*, 51–55.

Pandey, K. K., Pandey, P. K., Kalloo, G., & Banerjee, M. K., (2003). Resistance to early blight of tomato with respect to various parameters of disease epidemics. *J. Gen. Pl. Pathol., 69*, 364–71.

Saadabi, A. M. A., (2006). Antifungal activity of some Saudi plants used in traditional medicine. *Asian J. Plant Sci., 5*, 907–909.

Sallam, M. A., (2011). Control of tomato early blight disease by certain aqueous plant extracts. *Plant Pathology Journal, 10*(4), 187–191.

Stoll, G., (1998). *Natural Crop Protection in Tropics*. AGRECOL. Margraf Verlag, Weikersheim, Germany, pp. 188.

Varma, J., & Dubey, N. K., (1999). Prospective of botanical and microbial products as pesticides of tomorrow. *Curr. Sci., 76*.

Vijayan, M., (1989). Studies on early blight of tomato caused by *Alternaria solani* (Ellis and Martin) jones and grout. *M., Sc. Thesis,* Tamil Nadu, Agricultural University, Coimbatore, India.

Zaker, M., (2013). Screening some medicinal plant extracts against *Alternaria sesame*, the causal agent of Alternaria leaf spot of sesame. *J. Ornam. Hortic. Plants, 3*(10), 1–8.

CHAPTER 19

BIO-EFFICACY OF ECO-FRIENDLY PESTICIDES ON THE MANAGEMENT OF *SPODOPTERA LITURA* FAB. ON CABBAGE

R. MANDI and A. PRAMANIK

Department of Agricultural Entomology, Bidhan Chandra Krishi Viswavidyalaya, Mohanpur, Nadia, 741252, West Bengal, India, E-mail: rbnmandi@gmail.com

CONTENTS

ABSTRACT

The bio-efficacy of eco-friendly pesticides (botanicals: Azadirachtin, 10000 ppm, neem seed kernel extract (NSKE), Mittimax; microbials: *Bacillus thuringiensis* 5% WP, *Beauveria bassiana,* Spinosad 45% SC; and IGR (insect growth regulator): diflubenzuron + deltamethrin 22% SC) was evaluated for use in managing cabbage head borer (*Spodoptera litura*

Fab.) population. An experiment was conducted at the Central Research Farm, Gayeshpur, Bidhan Chandra Krishi Viswavidyalaya, Nadia, West Bengal, during the rabi season of 2007–2010. The experiment was carried out in randomized block design (RBD) with three replications. Cabbage (cv. Green express) seedlings were transplanted in the plot of 4 m × 3 m area with 50 cm × 50 cm spacing. The counting of *S. litura* population was made one day before as pretreatment and 1, 3, 7, and 10 days after treatment. All the insecticides were superior in controlling the *Spodoptera* population in comparison to untreated control. Among the different treatments, diflubenzuron + deltamethrin 22% SC recorded the highest 75.81% population reduction and proved to be the most effective treatment followed by *B. bassiana* @ 0.5% and NSKE, with population reduction of 60.72% and 60.35%, respectively. The percent increase of yield over control ranged from 11.41% to 50.13% in pesticide treatments with the highest yield in Spinosad 45% SC as against 32.79 t/ha in untreated control. The research indicated that all the eco-friendly pesticides were effective against cabbage head borer.

19.1 INTRODUCTION

India is mostly an agro-based country, and agriculture is a major component of the Indian economy; more than 75% of Indian people have their livelihood as agriculture and agriculture-oriented works (Thenmozhi and Thilagavathi, 2014). Among the cruciferous vegetables, Cabbage (*Brassica oleracea* L. var. *capitata*) is the most popular and grown throughout India. It is used as salad, boiled, and dehydrated vegetable as well as in cooked curries and pickles. The main edible part of cabbage is head/card, i.e., leaf is a good source of protein 1.6%; vitamins A, B_1, B_2, and C; sulfur; amino acids; minerals (calcium, iron, magnesium, phosphorus, and potassium); low amount of calories 2.4%; fat 0.2%; carbohydrate 4.8%; and substantial amount of β-carotene (Hanif et al., 2006). Insect pests, diseases, and weeds are the major constraints limiting agricultural productivity growth. It is estimated that herbivorous insects eat about 26% of the potential food production (Sing and Sharma, 2004). The cabbage borer *Spodoptera litura* Fabricius (Noctuidae: Lepidoptera) is an economically important, regular, multivoltine, gregarious, polyphagous pest that seriously harms cabbage, cauliflower, other vegetables, soybeans, cotton, and cash crops. It is widely distributed throughout tropical and temperate Asia, Australia, and Pacific islands (Kranthi et al., 2002). Indiscriminate use of chemical pesticides to manage this pest has resulted in resistance to chemical pesticides, resurgence, harmful effects on nontarget organisms, environmental pollution, and human health hazards (Vastrad et al., 2003, 2004; Haseeb et al., 2004). With the increasing concern on the health hazards, environmental pollution, and pest resistance due to over use of synthetic pesticides, the usage of eco-friendly pesticides (botanicals, microbials, IGRs) in the pest control is gaining importance, which can minimize the use of these synthetic chemicals. Botanical pesticides (Nathan and Kalaivani, 2005), microbial pesticides (Asi et al., 2013), and IGRs (Deb Prasad et al., 2012) are highly effective, safe, and ecologically acceptable. In the present investigation, the bio-efficacy of eco-friendly pesticides was evaluated against an economically important agricultural devastating pest *S. litura.*

19.2 MATERIALS AND METHODS

The field experiment was conducted with cabbage *var.* "Green express" in the experimental field of Central Research Farm, Bidhan Chandra Krishi Viswavidyalaya, Gayeshpur, Nadia, West Bengal, during the *rabi* season of 2007–2008, 2008–2009, and 2009–2010. The experiment was carried out in randomized block design (RBD) with three replications and nine treatments including untreated control. The seedlings were transplanted in the plot of 4 m × 3 m area with 50 cm × 50 cm spacing during the last week of December for the 3 seasons.

The pesticides evaluated were one bacterial pesticide T_1= *Bacillus thuringiensis* - 5% WP @ 0.2% *a.i.*, one fungal pesticide T_2= *Beauvaria bassiana* – 2 × 10^9 spore/g @ 0.5% *a.i.* and T_3= *B. bassiana* – 2 × 10^7 spore/g @ 1.0% *a.i.*, two neem-based pesticides T_4= azadirachtin 10,000 ppm @ 0.002% *a.i.* and T_5= neem seed kernel extract (NSKE) @ 5% *a.i.*, one IGR T_6= diflubenzuron + deltamethrin 22% SC @ 0.022% *a.i.*; one soil actinomycetes T_7= spinosad 45% SC @ 0.01% *a.i.*, one organic pesticide T_8= mittimax @ 0.15% *a.i.*, and T_9= untreated control.

When the cabbage head borer population was evenly distributed, the test pesticides were applied as foliar spray by the back pack hydraulic sprayer

(Aspee, Mumbai) with a hollow cone nozzle, with the spray fluid of 500–600 L/ha depending on the stage of the crop growth twice at 15 days interval except in the untreated control plots. The counting on the larval population of *S. litura* was made one day before as pretreatment and 1, 3, 7, and 10 days after treatment on 10 randomly selected plants in each plot. The percent efficacy was calculated on the basis of mortality of larvae at the above intervals after treatments. Reduction percent of different pests was calculated by applying a correction factor given by Henderson and Tilton (1995).

$$\text{Percent reduction over control} = 100 \left[1 - \{ (T_a \times C_b)/(T_a \times C_a) \} \right]$$

where, T_a = number of insects after treatment; T_b = number of insects before treatment.; C_a = number of insects in untreated plot after treatment; C_b = number of insects in untreated plot before treatment. The cabbage card harvested from each plot was recorded and calculated as tons/ha.

The critical difference (CD) at 0.05% level of significance were worked out from the data of percent reduction population of replication before treatment and various days interval after treatment of two consecutive sprays per year. The data analyzed in RBD were subjected to Duncan's multiple range test (DMRT) at 5% level after making necessary transformation wherever needed.

19.3   RESULTS AND DISCUSSION

Field experiment of eco-friendly pesticides on *S. litura* Fab. and its effect on yield parameters was observed during three consecutive years, 2008, 2009, and 2010, and are described as follows:

19.3.1   EFFECT OF ECO-FRIENDLY PESTICIDES ON LARVAL POPULATION OF CABBAGE HEAD BORER DURING 2008–2010

19.3.1.1   Effect of Eco-Friendly Pesticides on *S. litura* Infesting Cabbage during 2008

The results achieved on the bio-efficacy of different botanical, microbial, and chemicals against *S. litura* (Fab.) on cabbage during 2008 are presented in

Table 19.1. On the first day of first and second spraying, it was found that the highest (67.67% and 70.50%, respectively) reduction in larval population was recorded in NSKE @ 5.0%. The highest population reduction on the third day after first and second spraying (91.67% and 90.50%, respectively) was obtained in diflubenzuron + deltamethrin 22% SC @ 0.022%. After the seventh day of first spraying, the maximum population reduction (85.17%) was recorded in Mittimax @ 0.15%, which was significantly at par with 82.33% in diflubenzuron + deltamethrin − 22% SC @ 0.022%. On the seventh day after the second treatment, maximum (85.17%) population reduction was recorded in diflubenzuron + deltamethrin 22% SC @ 0.022% and was significantly at par with that of (85.00%) *B. bassiana* @ 1.0%. On the 10th day after first and second spraying, maximum larval mortality of 90.00% and 88.67%, respectively, was recorded in *B. bassiana* @ 1.0%. Considering the overall population reduction during the year 2008, the highest population reduction of 77.46% was recorded in diflubenzuron + deltamethrin 22% SC @ 0.022% and the lowest population reduction of 35.60% was obtained in Mittimax @ 0.15%. Considering the overall mean population reduction during the year 2008 the highest population reduction of 77.46% was recorded in diflubenzuron + deltamethrin 22% SC @ 0.022% and the lowest population reduction of 35.60% was obtained in Mittimax @ 0.15%.

19.3.1.2   Effect of Eco-Friendly Pesticides on *S. litura* Infesting Cabbage during 2009

The results of 2009 are presented in Table 19.2. A more or less similar trend was also observed during *rabi* season of 2009; from the overall mean population reduction after 1st to 10th day of two spraying, highest larval population reduction (76.63%) was recorded in diflubenzuron + deltamethrin 22% SC @ 0.022% and lowest population reduction (35.25%) was obtained in Mittimax @ 0.15%.

19.3.1.3   Effect of Eco-Friendly Pesticides on *S. litura* Infesting Cabbage during 2010

Filed experiment of eco-friendly pesticides on the cabbage head borer in cabbage during 2010 is depicted in Table 19.3. On the first and third day

TABLE 19.1 Effect of Ecofriendly Pesticides on *Spodoptera litura* (Fab.) Infesting Cabbage During 2008

| Treatments | Dosage | Pre treatment larval population/plant | Mean percent efficacy (% Reduction Over Control) at different days after spraying | | | | | | | | Overall mean of post spray count |
| | | | After 1st Spray | | | | After 2nd Spray | | | | |
			1st day	3rd day	7th day	10th day	1st day	3rd day	7th day	10th day	
T$_1$	0.2%	4.55	27.67 (31.99)bc	54.00 (47.58)c	65.67 (54.45)c	82.67 (65.81)b	35.00 (36.55)cd	49.17 (44.81)c	66.50 (54.95)c	81.33 (64.78)b	57.75
T$_2$	0.5%	4.50	30.00 (33.40)bc	56.00 (48.73)cd	63.33 (53.08)cd	84.00 (66.82)b	42.50 (40.98)b	61.83 (52.17)b	74.00 (59.75)b	80.00 (63.80)b	61.46
T$_3$	1.0%	4.61	17.33 (24.97)e	40.67 (39.91)e	71.67 (58.17)b	90.00 (72.06)a	20.33 (27.09)e	39.50 (39.22)d	85.00 (67.63)a	88.67 (70.81)a	56.65
T$_4$	0.002%	5.78	33.67 (35.77)b	51.00 (45.86)d	61.00 (51.66)d	66.33 (54.85)d	38.67 (38.74)bc	50.83 (45.77)c	59.33 (50.67)d	60.33 (51.26)d	52.65
T$_5$	5.0%	5.44	67.67 (55.74)a	79.00 (63.08)b	63.00 (52.83)cd	43.00 (41.26)e	70.50 (57.42)a	60.00 (51.06)b	38.00 (38.34)e	45.50 (42.71)e	58.33
T$_6$	0.022%	5.17	65.00 (54.03)a	91.67 (73.80)a	82.33 (65.55)a	74.00 (59.70)c	67.00 (55.26)a	90.50 (72.55)a	85.17 (67.76)a	64.00 (53.44)c	77.46

TABLE 19.1 (continued)

Treat-ments	Dosage	Pre treat-ment larval population/plant	Mean percent efficacy (% Reduction Over Control) at different days after spraying								Overall mean of post spray count
			After 1st Spray				After 2nd Spray				
			1st day	3rd day	7th day	10th day	1st day	3rd day	7th day	10th day	
T₇	0.01%	4.05	19.33 (26.43)ᵈᵉ	31.00 (34.14)ᵍ	43.17 (41.36)ᵉ	64.00 (53.43)ᵈ	19.00 (26.19)ᵉ	40.33 (39.72)ᵈ	60.00 (51.06)ᵈ	63.67 (53.24)ᶜ	42.56
T₈	0.15%	5.67	23.67 (29.44)ᶜᵈ	37.33 (37.95)ᶠ	85.17 (67.76)ᵃ	24.67 (30.09)ᶠ	33.33 (35.57)ᵈ	39.00 (38.94)ᵈ	19.67 (26.66)ᶠ	22.00 (28.30)ᶠ	35.60
T₉	0.00	7.00	0.00 (4.05)ᶠ	0.00 (4.05)ʰ	0.00 (4.05)ᶠ	0.00 (4.05)ᵍ	0.00 (4.05)ᶠ	0.00 (4.05)ᵉ	0.00 (4.05)ᵍ	0.00 (4.05)ᵍ	
SEm±			1.32	0.48	0.76	0.74	0.81	0.94	0.96	0.53	
CD at 0.05%			3.94	1.43	2.26	2.23	2.43	2.83	2.87	1.58	

* Figures in parentheses are angular transformed values.
* Means followed by common letter are not significantly different by DMRT ($p = 0.05$).

TABLE 19.2   Effect of Eco-Friendly Pesticides on *Spodoptera litura* (Fab.) Infesting Cabbage During 2009

Treat-ments	Dosage	Pre treat-ment larval popula-tion/plant	Mean percent efficacy (% Reduction Over Control) at different days after spraying								Overall mean of post spray count
			After 1st Spray				After 2nd Spray				
			1st day	3rd day	7th day	10th day	1st day	3rd day	7th day	10th day	
T_1	0.2%	5.55	42.67 (41.07)c	57.17 (49.41)c	69.83 (57.01)bc	82.83 (65.93)b	34.83 (36.47)d	55.17 (48.26)cd	67.00 (55.24)c	79.00 (63.08)b	61.06
T_2	0.5%	4.78	46.33 (43.18)c	57.00 (49.32)c	71.33 (57.95)b	84.00 (66.86)b	37.17 (37.86)c	59.50 (50.79)bc	66.00 (54.63)c	77.50 (62.04)b	62.35
T_3	1.0%	4.56	25.00 (30.31)f	43.67 (41.65)d	83.33 (66.33)a	91.33 (73.48)a	19.50 (26.56)g	45.00 (42.42)de	72.00 (58.37)b	82.00 (65.30)a	57.73
T_4	0.002%	6.67	36.33 (37.36)d	52.67 (46.81)c	63.83 (53.34)cd	69.83 (57.05)c	37.83 (38.25)c	51.83 (46.34)cd	57.00 (49.31)e	66.17 (54.75)cd	54.44
T_5	5.0%	5.61	73.50 (59.37)a	82.00 (65.28)b	50.00 (45.29)e	45.67 (42.80)d	69.67 (56.90)a	68.50 (56.54)b	61.33 (51.85)d	44.17 (41.94)e	61.85
T_6	0.022%	4.28	66.00 (54.64)b	90.67 (73.02)a	82.00 (65.46)a	70.17 (57.21)c	67.17 (55.35)b	88.00 (70.25)a	80.67 (64.30)a	68.33 (56.07)c	76.63
T_7	0.01%	6.39	23.17 (29.05)f	37.83 (38.24)e	59.00 (50.49)d	67.67 (55.66)c	22.67 (28.76)f	32.67 (35.15)f	42.67 (41.07)f	64.00 (53.43)d	43.71

TABLE 19.2 (continued)

Treatments	Dosage	Pre treatment larval population/plant	Mean percent efficacy (% Reduction Over Control) at different days after spraying								Overall mean of post spray count
			After 1st Spray				After 2nd Spray				
			1st day	3rd day	7th day	10th day	1st day	3rd day	7th day	10th day	
T_8	0.15%	5.33	29.67	40.18	18.33	24.67	28.00	38.33	81.00	21.83	35.25
			(33.27)e	(39.63)de	(25.64)f	(30.02)e	(32.27)e	(38.54)ef	(64.55)a	(28.19)f	
T_9	00	6.39	0.00	0.00	0.00	0.00	0.00	0.00	0.00	0.00	
			(4.05)g	(4.05)f	(4.05)g	(4.05)f	(4.05)h	(4.05)g	(4.05)g	(4.05)g	
SEm±			0.94	0.96	1.44	1.06	0.36	2.19	0.43	0.55	
CD at 0.05%			2.82	2.89	4.31	3.18	1.09	6.55	1.27	1.66	

* Figures in parentheses are angular transformed values.
* Means followed by common letter are not significantly different by DMRT ($p = 0.05$).

TABLE 19.3 Effect of Eco-Friendly Pesticides on Spodoptera Litura (Fab.) Infesting Cabbage During 2010

Treat-ments	Dosage	Pre treat-ment larval popula-tion/plant	Mean percent efficacy (% Reduction Over Control) at different days after spraying								Overall mean of post spray count
			After 1st Spray				After 2nd Spray				
			1st day	3rd day	7th day	10th day	1st day	3rd day	7th day	10th day	
T_1	0.2%	3.67	29.00 (32.86)d	38.33 (38.54)d	62.75 (52.69)c	83.08 (66.12)c	36.50 (37.45)d	54.50 (47.87)c	66.67 (55.05)c	83.17 (66.18)b	56.75
T_2	0.5%	4.39	32.67 (35.16)c	42.92 (41.21)c	61.58 (51.99)cd	85.33 (67.89)b	40.83 (39.99)c	52.50 (46.72)c	72.17 (58.49)b	78.83 (62.96)c	58.35
T_3	1.0%	4.78	16.17 (24.03)f	18.25 (25.66)f	67.25 (55.41)b	90.48 (72.54)a	19.00 (26.19)e	41.83 (40.59)e	84.50 (67.22)a	90.50 (72.55)a	53.50
T_4	0.002%	4.45	33.67 (35.77)c	36.67 (37.56)d	52.25 (46.58)e	69.28 (56.66)d	36.33 (37.36)d	48.50 (44.43)d	63.67 (53.27)c	65.83 (54.54)d	50.78
T_5	5.0%	4.17	69.17 (56.59)a	73.00 (59.03)a	59.48 (50.76)d	42.33 (40.88)f	72.50 (58.71)a	81.17 (64.65)b	44.33 (42.03)d	45.00 (42.42)e	60.87
T_6	0.022%	4.44	62.33 (52.44)b	67.92 (55.81)b	79.38 (63.36)a	69.42 (56.74)d	67.83 (55.76)b	89.50 (71.57)a	84.50 (67.22)a	65.83 (54.53)d	73.34

TABLE 19.3 (continued)

Treat-ments	Dosage	Pre treat-ment larval popula-tion/plant	Mean percent efficacy(% Reduction Over Control) at different days after spraying								Overall mean of post spray count
			After 1st Spray				After 2nd Spray				
			1st day	3rd day	7th day	10th day	1st day	3rd day	7th day	10th day	
T_7	0.01%	3.00	18.17	20.50	37.62	63.65	20.83	32.83	40.50	64.17	37.28
			(25.59)f	(27.27)f	(38.12)f	(53.22)e	(27.50)e	(35.27)f	(39.82)d	(53.53)d	
T_8	0.15%	3.39	21.17	33.58	79.48	21.00	33.83	80.50	85.83	22.00	47.17
			(27.74)e	(35.72)e	(63.43)a	(27.56)g	(35.87)d	(64.16)b	(68.31)a	(28.29)f	
T_9	0.00	2.67	0.00	0.00	0.00	0.00	0.00	0.00	0.00	0.00	0.00
			(4.05)g	(4.05)g	(4.05)g	(4.05)h	(4.05)f	(4.05)g	(4.05)e	(4.05)g	
SEm±			0.69	0.55	0.52	0.57	0.72	0.57	0.79	0.45	
CD at 0.05%			2.06	1.65	1.57	1.70	2.17	1.72	2.27	1.35	

* Figures in parentheses are angular transformed values.

* Means followed by common letter are not significantly different by DMRT ($p = 0.05$).

after first spraying, the highest population reduction (69.17% and 73.00%, respectively) was obtained in NSKE @ 5.0%, and on the third day after second spraying, the highest (89.50%) population reduction was recorded in diflubenzuron + deltamethrin- 22% SC @ 0.022%.

Considering the overall population reduction after 1st to 10th day of two spraying in 2010, the highest population reduction of 73.34% was recorded in diflubenzuron + deltamethrin 22%SC @ 0.022% and the lowest population reduction (37.28%) was recorded in spinosad 45%SC @ 0.01%.

19.3.1.4   Comparative Efficacy of Eco-Friendly Pesticides on *S. litura* Infesting Cabbage (2008, 2009, and 2010)

The pooled data (Table 19.4) indicated that all the insecticidal treatments were significantly superior over control in reducing the larval population of *S. litura* at 1, 3, 7, and 10 days after pesticide application. The treatment with diflubenzuron + deltamethrin 22% SC was found significantly superior as compared to other pesticides. The highest population reduction of 83.42 and 81.24% in first spraying and 89.33 and 83.44% in second spraying was recorded at 3 and 7 days after application, respectively.

The results of the present studies are in conformity with the earlier findings of Hussain et al. (2002), Zaz (1989), and Venkadasubramanian and David (1999) who also observed that the microbial pesticide *B. thuringiensis* at 500 mL/ha and 1000 mL/ha was effective against *S. litura* on cabbage. *B. thuringiensis* at 500 mL/ha and 1000 mL/ha was superior to untreated control in causing larval decline of *S. litura* with a mean percent reduction of 28.46% to 70.40% after spraying (Hussain et al., 2002). According to Zaz (1989), *B. thuringiensis* caused 5.0–72.5% larval mortality in *S. litura*. Venkadasubramanian and David (1999) reported that *B.t.* products and other botanicals recorded 20.37–65.55% mortality of *S. litura* on cabbage crop. The efficacy of spinosad (Tracer 45 SC) and *B. bassiana* against *S. litura* was studied in Bapatla, Andhra Pradesh, India, during the rabi of 2002–2003. Spinosad and *B. bassiana* gave 61.8%, and 62.3% larval mortality, respectively (Srinivas and Nagalingam, 2005), and it supports the present investigation.

TABLE 19.4 Comparative Efficacy of Eco-Friendly Pesticides on *Spodoptera* Infesting Cabbage (2008, 2009 and 2010)

Treatments	Dosage	Pre treatment larvae population/plant	Mean percent efficacy (% Reduction Over Control) at different days after spraying								Overall mean of post spray count
			After 1st Spray				After 2nd Spray				
			1st day	3rd day	7th day	10th day	1st day	3rd day	7th day	10th day	
T_1	0.2%	4.59	33.11 (35.31)d	49.83 (45.18)d	66.08 (54.72)c	82.86 (65.95)b	35.44 (36.82)e	52.94 (46.98)d	66.72 (55.08)d	81.17 (64.68)b	58.52
T_2	0.5%	4.55	36.33 (37.25)c	51.97 (46.42)c	65.42 (54.34)c	84.44 (67.19)b	40.17 (39.61)c	57.94 (49.89)c	70.72 (57.63)c	78.78 (62.93)c	60.72
T_3	1.0%	4.65	19.50 (26.44)f	34.19 (35.74)g	74.08 (59.97)b	90.61 (72.69)a	19.61 (26.61)g	42.11 (40.74)e	80.50 (64.41)b	87.06 (69.55)a	55.96
T_4	0.002%	5.63	34.56 (36.30)cd	46.78 (43.41)e	59.03 (50.53)e	68.48 (56.19)d	37.61 (38.12)d	50.39 (45.51)d	60.00 (51.08)b	64.11 (53.52)e	52.62
T_5	5.0%	5.07	70.11 (57.23)a	78.00 (62.46)b	57.49 (49.63)e	43.67 (41.65)b	70.89 (57.67)a	69.89 (57.42)b	47.89 (44.07)g	44.89 (42.35)b	60.35
T_6	0.022%	4.63	64.44 (53.71)b	83.42 (67.54)a	81.24 (64.79)a	71.19 (57.88)c	67.33 (55.45)b	89.33 (71.46)a	83.44 (66.43)a	66.06 (54.68)d	75.81
T_7	0.01%	4.48	20.22 (27.03)f	29.78 (33.21)h	46.59 (43.32)f	65.11 (54.10)e	20.83 (27.48)g	35.28 (36.71)f	47.72 (43.98)g	63.94 (53.40)e	41.18
T_8	0.15%	4.80	24.83 (30.15)e	37.03 (37.77)f	60.99 (52.28)d	23.44 (29.22)g	31.72 (34.57)f	52.61 (47.21)d	62.17 (53.17)c	21.94 (28.26)g	39.34
T_9	00	5.35	0.00 (4.05)g	0.00 (4.05)i	0.00 (4.05)g	0.00 (4.05)h	0.00 (4.05)hi	0.00 (4.05)g	0.00 (4.05)h	0.00 (4.05)h	
SEm±			0.59	0.40	0.57	0.47	0.38	0.82	0.44	0.30	
CD at 0.05%			1.67	1.15	1.62	1.34	1.09	2.32	1.24	0.84	

* Figures in parentheses are angular transformed values.

* Means followed by common letter are not significantly different by DMRT ($p = 0.05$).

19.3.2 CABBAGE FRESH YIELD AND PERCENT INCREASE IN YIELD OVER CONTROL DURING 2008, 2009, AND 2010

The results of different treatments were reflected in the yield of cabbage. Treatment-wise average saleable yield (tons/ha) and percent yield increase over control in 2008, 2009, and 2010 and pooled mean value are presented in Table 19.5.

All the treatments provided higher yields than that of control. In 2008, the maximum yield of cabbage (48.03 tons/ha) was found in diflubenzuron + deltamethrin 22% SC followed by (47.78 tons/ha) in Spinosad 45% SC. A similar trend was observed in the next 2 years, 2009 and 2010. In 2009, the highest cabbage yield (49.05 tons/ha) was obtained in diflubenzuron+ deltamethrin 22% SC which was at par with that of (47.60 tons/ha) in Spinosad 45% SC. In 2010, the maximum cabbage yield (50.61 tons/ha) was obtained in diflubenzuron + deltamethrin 22% SC, which was statistically at par with that of (50.18 tons/ha) Spinosad 45% SC. From the 3-year pooled mean value, the highest cabbage yield (49.23 tons/ha) was obtained in diflubenzuron+ deltamethrin 22% SC followed by (48.52 tons/ha) Spinosad 45% SC and the lowest yield (32.79 tons/ha) was recorded in untreated control.

The treatment diflubenzuron + deltamethrin 22% SC exhibited 39.92% increase in the yield of cabbage over control followed by 39.18% in Spinosad 45% SC. A similar trend was observed in the 2nd and 3rd year (2009 and 2010). In 2009, the highest increase in cabbage yield of 55.43% was noticed in diflubenzuron + deltamethrin 22% SC, which was statistically at par with that of 50.85% yield increase in Spinosad 45% SC. In 2010, diflubenzuron+ deltamethrin 22% SC showed 55.77% increase in yield of cabbage over control followed by 54.43% in Spinosad 45% SC. From the pooled mean value, highest percent increase in yield (50.13%) was obtained in diflubenzuron + deltamethrin 22% SC followed by 47.96% in Spinosad 45% SC and 34.06% in NSKE. Higher yield (42.4 tons/ha) of cabbage from *B.t.* (Halt)-treated plot was earlier reported by Ghosh et al. (2001), which is in conformity with the present studies.

TABLE 19.5 Cabbage Fresh Yield and % Increase in Yield Over Control During 2008, 2009 and 2010 and Pooled

Treat-ments	Dosage	Cabbage yield (tons/ha)							
		2008		2009		2010		Pooled	
		Fresh Yield	% increase over control	Fresh Yield	% increase over control	Fresh Yield	% increase over control	Fresh Yield	% increase over control
T_1	0.2%	$(41.93)^{bc}$	22.14	$(42.05)^b$	33.25	$(43.80)^{bc}$	34.81	$(42.59)^c$	29.89
T_2	0.5%	$(47.78)^a$	39.18	$(47.60)^a$	50.85	$(50.18)^a$	54.43	$(48.52)^a$	47.96
T_3	1.0%	$(40.38c)^d$	17.63	$(40.16)^c$	27.25	$(42.84)^{cd}$	31.83	$(41.13)^d$	25.41
T_4	0.002%	$(39.26)^d$	14.35	$(39.45)^c$	25.01	$(42.22)^{cd}$	29.94	$(40.31)^d$	22.92
T_5	5.0%	$(43.10)^b$	25.54	$(43.17)^b$	36.79	$(45.63)^b$	40.42	$(43.96)^b$	34.06
T_6	0.022%	$(48.03)^a$	39.92	$(49.05)^a$	55.43	$(50.61)^a$	55.77	$(49.23)^a$	50.13
T_7	0.01%	$(35.97e)^f$	4.79	$(35.30)^d$	11.85	$(38.34)^e$	17.99	$(36.54)^f$	11.41
T_8	0.15%	$(37.96)^{de}$	10.57	$(38.47)^e$	21.90	$(40.71)^d$	25.28	$(39.04)^e$	19.06
T_9	00	$(34.33)^f$		$(31.56)^e$		$(32.49)^f$		$(32.79)^g$	
SEm±		0.81		0.63		0.74		0.24	
CD at 0.05%		2.43		1.88		2.22		1.20	

* Figures in parentheses are angular transformed values.

* Means followed by common letter are not significantly different by DMRT ($p = 0.05$).

KEYWORDS

- cabbage head borer
- eco-friendly pesticides
- *Spodoptera litura*

REFERENCES

Asi, M. R., Bashir, M. H., Afzal, M., Zia, K., & Akram, M., (2013). Potential of entomopathogenic fungi for biocontrol of *Spodoptera litura* fabricius (Lepidoptera: Noctuidae). *J. Anim. Plant Sci., 23*(3), 913–918.

Ghosh, S. K., Choudhuri, N., Ghosh, J., Chattarjee, H., & Senapati, S. K., (2001). Field evaluation of pesticides against the pest complex of cabbage under terai region of West Bengal. *Pestology, 25*(2), 95–97.

Hanif, R., Iqbal, Z., Iqbal, M., Hanif, S., & Rasheed, M., (2006). Use of vegetables as nutritional food: role in human health. *Journal of Agricultural and Biological Science, 1*, 18–22.

Haseeb, M., Liu, T. X., & Jones, W. A., (2004). Effects of selected insecticides on *Cotesia plutellae* (Hymenoptera: Braconidae), an endolarval parasitoid of *Plutella xylostella* (Lepidoptera:Plutellidae). *Biocontrol, 49*, 33–46.

Henderson, L. F., & Tilton, E. W., (1995). Tests with acaricides against brown wheat mite. *Journal of Economic Entomology, 48*(2), 152–154.

Hussain, M. A., Pachori, R., & Choudhary, B. S., (2002). Population dynamics of *Spodoptera litura* Fab. on cabbage in Jabalpur, Madhya Pradesh. *JNKVV-Research Journal, 36*(1/2), 106–107.

Kranthi, K. R., Jadhav, D. R., Kranthi, S., Wanjari, R. R., Ali, R. R., & Russell, D. A., (2002) Insecticide resistance in five major insect pests of cotton in India. *Crop Protection, 21*, 449–460.

Nathan, S., & Kalaivani, K., (2005). Efficacy of nucleopolyhedrovirus (NPV) and azadirachtin on *Spodoptera litura* Fabricius (Lepidoptera :Noctuidae). *Biological Control, 34*, 93–98.

Ray, D. P., Dutta, D., Srivastava, S., Kumar, B., & Saha, S., (2012). Insect growth regulatory activity of *Thevetia nerifolia* Juss. against *Spodoptera litura* (Fab.). *J. Appl. Bot. Food Qual., 85*, 212–215.

Singh, A., & Sharma, O. P., (2004). Integrated pest management for sustainable agriculture. Pratap, S., Birthal, O. P., Sharma, (Eds) proceedings 11 In: *Integrated Pest Management in Indian Agriculture*. NCAP and NCIPM. New Delhi, India, pp. 11–24.

Srinivas, M. S., & Nagalingam, B., (2005). Influence of certain microbial agents on *Spodoptera litura* (Hubner) and coccinellids in groundnut. *Journal of Entomological Research, 29*(1), 31–34.

Thenmozhi, S., & Thilagavathi, P., (2014). Impact of agriculture on Indian economy. *IRJARD, 3*(1), 96–105.

Vastrad, A. S., Lingappa, S., & Basavana Goud, K., (2004). Ovicides for managing resistant populations of diamondback moth, *Plutella xylostella, L. Resistant Pest Management Newsletter, 14*, 16–17.

Vastrad, A. S., Lingappa, S., & Basavanagoud, K., (2003). Management of insecticide resistant populations of diamondback moth, *Plutella xylostella* (L.) (Yponomeutidae: Lepidoptera). *Pest Management in Horticultural Ecosystem, 9*(1), 33–40.

Venkadasubramanian, V., & David, P. M. M., (1999). Insecticidal toxicity of commercial *Bacillus thuringiensis* (Berliner) products in combination with botanicals *to Spodoptera litura* (Fabricius) and *Heliciverpa armigera* (Hubner). *Journal of Biological Control, 13*, 85–92.

Zaz, G. M., (1989). Relative effectiveness of *Bacillus cereus* Frankland and frankland, *Bacillus thuringiensis* Benliner and endosulfan against *Spodoptera litura* (Fab.) on cauliflower. *Indian Journal of Plant Protection, 18*(1), 85–88.

CHAPTER 20

FOLIOSE LICHEN SPECIES: A POTENTIAL SOURCE FOR BIO-CONTROL AGENT AGAINST *COLLETOTRICHUM CAPSICI*

M. CHINLAMPIANGA,[1] A. C. SHUKLA,[1] A. R. LOGESH,[1] and D. K. UPRETI[2]

[1]*Department of Horticulture, Aromatic and Medicinal Plants, Mizoram University, Aizwal – 796004, India*

[2]*Lichenology Laboratory, Plant Biodiversity, Systematics and Herbarium Division, CSIR-National Botanical Research Institute, Lucknow – 226001, Uttar Pradesh, India*

CONTENTS

ABSTRACT

Colletotrichum capsici Butler & Bisby is an important phytopathogenic fungi causing many diseases (anthracnose or die back in chili, capsicum, betel vine, turmeric, ashwagandha, black pepper, cucurbits, etc.) on various horticultural crops; it not only reduces the quality and yield of the infected crops but also enhances the cost of production. The objective of the present study was to determine the antifungal efficacy of foliose macrolichen species, viz., *Parmotrema reticulatum* and *Everniastrum cirrhatum* against *C. capsici*, so that an alternative to the synthetic antifungals can be explored. Keeping these views in mind, an in vitro antifungal study of the acetone extracts of *P. reticulatum* and *E. cirrhatum* (at 5% concentration) was investigated against *C. capsici*. The observations recorded that efficacy of both the lichens at low doses were not much effective; however, as the doses increased, efficacy also increased. Further, it was recorded that the acetone extracts of *P. reticulatum* and *E. cirrhatum* (5% at 50 mL/L concentration) showed cidal efficacy and having similar efficacy to that of the synthetic antifungal "Mancozeb." This was determined by using the modified spore germination inhibition technique (MSGIT).

Findings of the present investigation show that after detailed in vitro investigations, the active constituents of lichens can be used as a potential substitute of synthetic fungicides.

20.1 INTRODUCTION

Phytopathogens, particularly, fungi are responsible for poor establishment and stand loss in a variety of commercial horticultural crops. *Colletotrichum capsici* diseases account for marked reduction in productivity and economic return in various horticultural crops including chilies. Anthracnose (both pre- and postharvest) is one of the most important diseases in chili. It results in yield loss (up to 50%) and deterioration of fruit quality. The disease is known to be caused by various species of *Colletotrichum* such as *C. capsici*, *C. acutatum*, *C. gloeosporiodes*, *C. coccodes*, and *C. dematium*. Among these, *C. capsici* is the one of the important phytopathogens, causing enormous losses in horticultural crops (Than et al., 2008; Narasimhan et al., 2012; Susheela, 2012; Masoodi et al., 2013).

Although there are number of chemicals that are widely used for the management of plant diseases, they are not beneficial in many respects such as high cost, breakdown of resistance, residual problem, and deleterious effect on non-target organisms including humans (Cowan, 1999; Poorniammal and Sarathambal, 2009). Therefore, search for alternatives to control plant diseases has been a challenging task for the plant scientist/ researchers, since the last few years.

Plants and their associated epiphytic plants, including lichens, serve mankind as an important source of food, nutraceuticals, and medicine from ancient time. They are also used for the management of diseases caused by biotic agent/pathogens such as bacteria, fungi, mycoplasma, actinomycetes, and nematodes, as they possess some defense mechanisms against enemies, due to some unique substances synthesized through the metabolic pathway.

Natural products such as plant-based formulations, solvent, and cow urine extracts of plants, lichens and their metabolites, etc., have been reported to possess potential antifungal effect against various types of phytopathogenic fungi. Moreover, these agents are nontoxic and are easily decomposed (Halama and Haluwin, 2004; Goel et al., 2011; Dileep et al., 2013; Kambar et al., 2013; Rakesh et al., 2013; Vivek et al., 2013).

Lichens are symbiotic associations of a fungus (mycobiont) and a photosynthetic partner (photobiont) that can be either green algae or cyanobacteria (Ahmadjian, 1995). Lichens occurs in different growth forms, viz., crustose, foliose, fruticose, leprose, and squamulose and are well known to grow on rocks and soil as well as epiphytes on the trees, leaves, and twigs. Lichens synthesize a wide variety of secondary metabolites. Lichens are well known for their prolific sources of biologically active natural products, as they produce a diverse range of secondary metabolites that are not reported in any other plant groups. Slow growth and often-harsh living make production of protective metabolites a necessity to lichens, and these metabolites are believed to have antigrowth, anti-herbivore, and antimicrobial activity (Hale, 1983; Rankovic et al., 2008). The Eastern Himalayan region in northeast India and Western Ghats in south India represents luxuriant and diverse growth of a number of lichen species. So far, only a few lichen species from India have been screened for their antifungal activities (Shahi et al., 2001, 2011; Tiwari et al., 2011). Owing to the rich diversity of lichens in the country, there is a lot of scope for such studies. Therefore, in the present study, an attempt has been made to evaluate the antifungal efficacy of lichen *Parmotrema reticulatum* especially against phytopathogenic fungi, which are responsible for the huge loss of agricultural and horticultural yield throughout the world.

20.2   MATERIALS AND METHODS

20.2.1   COLLECTION AND IDENTIFICATION OF LICHENS

Lichens species were collected in September 2014 from two areas of the Murlen National Park, Champhai district, Mizoram, India. *P. reticulatum* (Taylor) Choisy (LWG Acc. No. 14-019172), corticolous habit, was collected from the western part of the study area, while *Everniastrum cirrhatum* (Fr.) Hale ex Sipman (LWG Acc. No. 14-031427), corticolous habit, was collected from the eastern part of the study area. The lichens were identified on the basis of morphology, anatomy, and chemical tests. Color tests were carried out on cortex and medulla of lichens by using 25% potassium hydroxide (K), Steiner"s stable paraphenylene-diamine solution (P), and calcium hypochlorite solution (C). Secondary metabolites were detected by thin layer chromatography (TLC) using solvent system A (toluene 180 mL: 1-4 dioxane 60 mL: acetic acid 8 mL) following Walker and James (1980). The lichen specimens were authenticated and deposited at the Lichenology Laboratory, Plant Biodiversity, Systematics and Herbarium Division, CSIR-National Botanical Research Institute (NBRI), Lucknow, Uttar Pradesh.

20.2.2   EXTRACTION

The lichen samples/materials (whole thallus) used for in vitro antifungal activity collected from the study area were sorted, cleaned/washed, and then air dried at room temperature. Before extraction, the dried samples were pulverized to powder form. Each powdered form of lichen sample was stored in a sterile glass bottle in the refrigerator. A stock solution of acetone extract was prepared by macerating 10 g of lichen material in 20 mL of acetone by using a pestle and mortal, followed by filtering through a muslin cloth and a Millipore filter (pore size 0.22 mm). The solvent extraction was carried out at the specific boiling temperature at 56°C for 48 h for complete extraction of secondary compounds. The 10 g portion of sieved powder was added to 100 mL of each solvent and left for 3 days at room temperature. The crude extract was prepared by decanting, followed by filtration through a muslin cloth, and further filtered with a Whatman No. 1 filter paper to obtain a clear filtrate. The filtrates were sterilized by membrane filtration using 0.45-µm pore size filters. The extracts were then evaporated to dryness under reduced pressure and again dissolved in respective solvents to attain the required concentrations of 5% for antifungal bioassay. The lichen extracts were kept at 4°C till used.

20.2.3 ANTIFUNGAL ACTIVITY OF LICHEN EXTRACTS

The test fungus *Colletotrichum capsici* Butler & Bisby (MTCC 8473) was procured from IMTECH, Chandigarh, India. The antifungal activity of lichen extracts was determined using the modified spore germination inhibition technique (MSGIT) of Shahi et al. (1997) with slight modification of Shukla (2011). Potato dextrose broth was prepared and amended with penicillin G (5 mg/L) and streptomycin sulfate (5 mg/L) in the medium at 40°C in order to prevent bacterial growth, as suggested by Gupta and Banerjee (1997). Culture discs containing spores (5 mm diameter) cut out from the 7-day-old cultures grown in petri dishes were transferred aseptically in flasks (100 mL) containing the broth and shaken thoroughly for homogenous distribution of spores. The numbers of spores were counted per microscopic field using "Modified Cytometer Technique" (MCT) (Shahi et al., 1997). The diameter of microscopic field was measured by micrometer, and the area and volume of the microscopic field was calculated using the following formula:

$$AMF = \pi r^2$$

$$VMF = (AMF)\,h$$

where AMF = area of microscopic field; VMF = volume of microscopic field; h = thickness of medium (in between slide and cover glass) 0.1 mm.

The number of spore (average count value of five microscopic fields) was counted by eliminating the overlapped spores. The number of spores in the volume of microscopic fields (NSV) was calculated using the formula:

$$NSV = ANS/VMF$$

where ANS = average number of spores in the microscopic field.

The volume of liquid medium (VLM) per microscopic field was calculated by the formula:

$$VLM = 2rh$$

The total inoculums density (TID) was calculated in the initial volume of medium by using the formula:

$$TID = (NSV/VLM)\ IVM$$

where IVM = initial volume of medium.

The effective concentration of acetone extract was determined by dissolving requisite quantity of extracts in acetone (2% of the required quantity of the seed medium) and mixing it with the standardized inoculum suspension in the culture tubes. In control, distilled water was used in place of the lichen extracts. Besides, a comparison of the lichen extracts with a synthetic fungicide "Mancozeb" (75%) was also made. Culture tubes thus prepared were incubated at 27°C ± 1°C, and the observations were recorded at the interval of 24 h up to 96 h by counting the number of germinated spores. Percentage of zone of inhibition/spore germination (SGI) by the extracts against test fungal culture was calculated using the formula:

$$SGI\ (\%) = (Gc - Gt)/Gc \times 100$$

where Gc = number of germinated spore in control; Gt = number of germinated spore in treatment.

The experiments were repeated twice and each of the tests was made in triplicates.

20.2.4 STATISTICAL ANALYSIS

The results recorded in Table 20.2 are the mean values.

20.3 RESULTS AND DISCUSSION

The details of color tests and TLC of *P. reticulatum* and *E. cirrhatum* are recorded in Table 20.1. The antifungal efficacy of the 5% acetone extracts of *Parmotrema* and *Everniastrum* lichen species against *C. capsici* are recorded in Table 20.2 and Figure 20.1.

The observations show that an inhibition of >50% was recorded in both the extracts at 5% concentration of 10 mL/L, and it was cidal at 50 mL/L, respectively (Table 20.2 and Figure 20.1). Further, it was also observed

TABLE 20.1 Observations of Color Tests and TLC of *Parmotrema* and *Everniastrum* spp.

S. No.	Lichen	Colour test	TLC
1	*E. cirrhatum* (Parmeli-aceae)	Cortex K + yellow; medulla K+ yellow turning red, C-, KC-, P+ orange.	Atranorin, salazinic and protoliches-terinic acid
2	*P. reticulatum* (Parmeli-aceae)	Medulla K$^+$ yellow then red,C-, KC-, PD$^+$ orange – red	Salazinic acid and consala-zinic acids

TABLE 20.2 Antifungal Activity of Extract of *P. reticulatum* and *E. cirrhatum* against *C. Capsici* and Their Comparison with Synthetics

5% Conc (ml/L)	Control	Spore germination inhibition (%)		Synthetic Antifungal 'Mancozeb' (2.6 mg/L)
		P. reticulatum	*E. cirrhatum*	
00.05	00	9.00	6.00	13.00
00.10	00	12.03	13.03	26.25
02.00	00	24.05	25.05	39.50
05.00	00	43.11	46.10	57.40
10.00	00	55.02	52.08	78.15
20.00	00	87.00	87.00	100.00^c
50.00	00	100.00^c	100.00^c	100.00^c

^{s}indicates static, ^{c} indicates cidal in nature; Conc. indicates concentration.

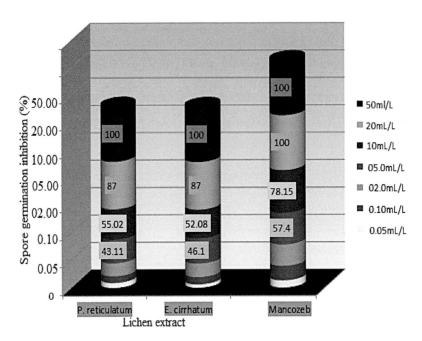

FIGURE 20.1 **(See color insert.)** Antifungal efficacy (%) of the test samples against *C. capsici* – A graphical representation.

that the activity of the two extracts and the synthetic fungicide "Mancozeb" (75%) exhibited more or less similar inhibitory efficacy against *C. capsici* (Figure 20.1).

Further, diagrammatic representation of the findings, as recorded in Table 20.2, is shown in Figure 20.1.

Literature revealed that the antifungal efficacy of 50% ethanolic extract from some macrolichens, *Parmelia tintorum, Ramalina* sp., *Teloschistes flavicans,* and *Usnea undulate* had been tested by Dikshit (1991) against the pathogenic fungi *Aspergillus flavus.* Acetone extract of *Stereocaulon* sp. was found to be more effective than *Ramalina* sp. against plant pathogenic fungi, viz., *Alternaria alternata, Aspergillus flavus*, and *Penicillium italicum* (Shukla et al., 2011). Three *Parmotrema* species, viz., *P. tinctorum, P. grayanum,* and *P. praesorediosum* were reported as bio-control agents against *C. capsici* (Kekuda et al., 2014). Further, in another study, Preeti et al. (2014) reported the activities of solvent extracts of *Parmotrema reticulatum* against some phytopathogenic fungi.

Similarly, aqueous extract of *Heterodermia leucomela* was found to exhibit significant inhibition of germination of spores of some phyto-pathogenic fungi (Shahi et al., 2001). A lichen-forming fungus isolated from *Heterodermia* sp. was found to exhibit strong inhibitory activity against *Pythium* spp. (Hur et al., 2003). Similarly, acetone extracts of *Evernia prunastri* and *Hypogymnia physodes* were found effective in inhibiting *Pythium ultimum, Ustilago maydis*, and *Phytophthora infes-tans*. Lichenic acids, viz., evernic acid, usnic acid (±), protolchesterinic acid, and atranorin were reported to have strong inhibitory activity against two phytopathogenic fungi (Kekuda et al., 2011, 2012). Metha-nol extract of *Parmelia sulcata, Flavoparmelia caperata*, and *Evernia prunastri* were shown to possess antifungal activity against a panel of human and phytopathogenic fungi (Mitrovic et al., 2011). However, in the present study, antifungal efficacy of acetone extracts of *P. reticula-tum* and *E. cirrhatum* against *C. capsici* was recorded cidal in nature (5% concentration at 50 mL/L), respectively (Table 20.2 and Figure 20.1).

20.4 CONCLUSION

The acetone extracts of *P. reticulatum* and *E. cirrhatum* showed fungicidal efficacy against *C. capsici* (5% concentration at 50 mL/L). To the best of our knowledge, this is the first report on inhibitory effect of *Evernias-trum* species and *Parmotrema* species from Mizoram, India, against the test fungi. Therefore, based on these findings as well as detailed in vivo investigations, lichen-based formulations can be effective agents for the management of the disease caused by the fungal pathogen *C. capsici* in various commercial horticultural crops and can be an alternative source for developing natural fungicides.

ACKNOWLEDGMENTS

The authors are thankful to the Director, CSIR-National Botanical Research Institute, Lucknow, and the authorities of Mizoram University, Aizawl, for the necessary facilities as well as to the Council of Scientific and Industrial Research, New Delhi, for financial support [CSIR project No. 38(1376)/ 13/ EMR-II; dated. 1.10.2013)].

KEYWORDS

- **anthracnose**
- **antifungal**
- *Colletotrichum capsici*
- *everniastrum*
- **modified spore germination inhibition technique**
- *parmotrema*

REFERENCES

Ahmadjian, V., (1995). Lichens are more important than you think. *Bioscience, 45*, 124.

Cowan, M. M., (1999). Plant products as antimicrobial agents. *Clini. Microbiol. Reviews, 12*(4), 564–582.

Dikshit, A., (1991). Antifungal activity of some micro-lichens. *Proc. Int. Conf. on Global Envir. Diver.(abs.),* pp. 53.

Dileep, N., Junaid, S., Rakesh, K. N., Kekuda, P. T. R., & Nawaz, A. S. N., (2013). Antifungal activity of leaf and pericarp extract of *Polyalthialongifolia* against pathogens causing rhizome rot of ginger. *Sci. Technol. and Arts Res. J., 2*(1), 56–59.

Goel, M., Sharma, P. K., Dureja, P., Rani, A., & Unilal, P. L., (2011). Antifungal activity of extracts of the lichens *Parmelia reticulate, Ramalina roesleri, Usnea longissima* and *Stereocaulon himalayense. Arch. Phytopathol. and Pl. Protec., 44*(13), 1300–1311.

Halama, P., & Van Haluwin, C., (2004). Antifungal activity of lichen extracts and lichenic acids. *Bio Control, 49*(1), 95–107.

Hale, M. E., (1983). *The Biology of Lichens.* 3rd Ed. Edward Arnold Ltd. London.

Hur, J., Kim, H. J., Lim, K., & Koh, Y. J., (2003). Isolation, cultivation and antifungal activity of a lichen-forming fungus. *Pls. Pathol. J., 19*(2), 75–78.

Kambar, Y., Vivek, M. N., Manasa, M., Kekuda, P. T. R., & Nawaz, N. A. S., (2013). "Inhibitory effect of cow urine against *Colletotrichumcapsici* isolated from anthracnose of Chilli (*Capsicum annuum* L.)" *Sci. Techno. and Arts Res. J., 2*(4), 91–93.

Kekuda, P. T. R., Raghavendra, H. L., Swathi, D., Venugopal, T. M., & Vinayaka, K. S., (2012). Anitfungal and cytotoxic activity of *Everniacirrhatum* (Fr.) Hale. *Chiang Mai. J. Sci., 39*(1), 76–83.

Kekuda, P. T. R., Vinayaka, K. S., Swathi, D., Suchitha, Y., Venugopal, T. M., & Mallikarjun, N., (2011). Mineral composition, total phenol content and antioxidant activity of a macro lichen *Everniacirrhatum* (Fr.) Hale (Parmeliaceae). *E. J, Chem., 8*(4), 1886–1894.

Kekuda, P. T. R., Vivek, M. N., Yashoda, K., & Manasa, M., (2014). Biocontrol potential of *Parmotrema* species against *Colletotrichumcapsici* isolated from anthranose of chilli. *J, Biolog. and Sci. Opin, 2*(2), 116–169.

Masoodi, L., Anwar, A., Ahmed, S., & Sofi, T. A., (2013). Cultural, morphological and pathogenic variability in *Colletotrichumcapsici* causing die back and fruit rot in chilli. *Asian J. Pt. Path., 7*(1), 29–41.

Mitrovic, T., Stamenkovic, S., Cvetkovic, V., Tosic, S., Stankovic, M., Radojevicć, I., Stefanovic, O., Čomic, L., Dacic, D., Ćurcic, M., & Markovic, M., (2011). Antioxidant, antimicrobial and antiproliferative activities of five lichen species. *Int. J. Mol. Sci., 12*, 5428–5448.

Narasimhan, A., & Shivakumar, S., (2012). Study of mycolytic enzymes of *Bacillus* sp. against *Colletotrichumgloeosporiodes* causing anthracnose in Chilli. *Acta. Biologica. Indica, 1*(1), 81–89.

Poorniammal, R., & Sarathambal, C., (2009). Comparative performance of plant extracts bio control agents and fungicides on the diseases of sunflower. *Indian J. Weed Sci., 41*(3/4), 207–209.

Prashith Kekuda, P. T. R., Vivek, M. N., Yashoda, K., & Manasa, M., (2014). Boicontrol potential of *Parmotrema* species against *Colletotrichumcapsici* isolated from Anthracnose of chilli. *J. Biol Sci. Opin, 2*(2), 166–169. http:/dx. doi. org/10.7897/2321–6328.02238.

Preeti S., Babiah, D. K., Upreti, & John, S. A., (2014). An *in vitro* analysis of antifungal potential of lichen species *Parmotrema reticulatum* against phytopathogenic fungi. *Int. J. Curr. Microbiol. App. Sci., 3*(12), 511–518.

Preeti, S., Babiah, Upreti, D. K., & John, S. A., (2014). An *in vitro* analysis of antifungal potentialof lichen species *Parmotrema reticulatum* against phytopathogenic fungi. *Int. J. Curr. Microbiol. App. Sci., 3*(12), 511–518.

Rakesh, K. N., Dileep, N., Junaid, S., Kekuda, P. T. R., Vinayaka, K. S., & Nawaz, N. A. S., (2013). Inhibitory effect of cow urine extracts of selected plants against pathogens causing rhizome rot of ginger. *Sci. Technol. and Arts Res. J., 2*(2), 92–96.

Rankovic, B., Misic, M., & Sukdolak, S., (2008). The antimicrobial activity of substances derived from the lichens *Physciaaipolia, Umbilicariapolyphylla, Parmeliacaperata* and *Hypogymniaphysodes*. *World J. Microbiol. and Biotech., 24*(7), 1239–1242.

Shahi, S. K., Shukla, A. C., Dikshit, A., & Upreti, D. K, (2001). Broad spectrum antifungal properties of the lichen *Heterodermialeucomela*. *Lichenologist, 33*, 177–179.

Shahi, S. K., Shukla, A. C., Dikshit, S., & Dikshit, A.,(1997). Modified spore germination inhibition technique for the evaluation of candidate fungi toxicants (*Eucalyptus* sp.): *Diagnosis and Identification of Plant Pathog*ens.(eds. Dhene, H. W., Adam, G., Diekmann, M., Frahm, J., Maular Machnik, A., & Halteren, P., Van), 257–263, Dordrecht, Kluwer.

Shukla, A. C. M., Chinlampianga, Archana,V., Anupam, D., & Upreti, D. K., (2011). Efficacy and potency of lichens of Mizoram as antimycotic agents. *Indianphytopath, 64*(4), 367–370.

Shukla, A. C., (2010). Bioactivities of the major active constituents isolated from the essential oil of *Cymbopogonflexuosus* (Steud) Wats and *Trachyspermumammi* L., Sprague as a herbal grain protectants. D. Sc. Thesis, Allahabad University, Allahabad.

Shukla, A. C., Pandey, K. P., Mishra, R. K., Dikshit, A., & Shukla, N. 2011. Broad spectru-mantimycotic plant as a potential source of therapeutic agent. *J. Natural Products, 4*, 42–50.

Susheela, K., (2012). Evaluation of screening methods for anthracnose disease in chilli. *Pest Management in Horticultural Ecosystems, 7*(1), 29–41. http://dx.doi.org/10.3923/ajppaj.2013.29.41.

Than, P. P., Prihastuti, H., Phoulivong, S., Taylor, P. W. J., & Hyde, K. D., (2008). Chilli-anthranose disease caused by *Colletotrichumcapsici. Pls. Pathol. J., 26*(3), 273–279.

Tiwari, P., Rai, H., Upreti, D. K., Trivedi, S., & Shukla, P., (2011). Assessment of antifun-gal activity of some Himalayan foliose lichen against plant pathogenic fungi. *Am. J. Plant Sci., 2*, 841–846.

Vivek, M. N., Kambar, Y., Manasa, M., Pallavi, S., & Kekuda, P. T. R., (2013). Bio control potential of *Pimentadioca* and *Anacardium occidentale* against *Fusariumoxyspo-rumf.* sp. *zingiberi. J. Biolo. and Sci. Opin., 1*(3), 193–195.

CHAPTER 21

FUNGAL DISEASES OF MEDICINAL AND AROMATIC PLANTS AND THEIR BIOLOGICAL MANAGEMENT

RAMESH S. YADAV,[1] D. MANDAL,[2] and AMRITESH C. SHUKLA[2]

[1]Department of Plant Pathology, Sardar Vallabhbhai Patel University of Agriculture & Technology, Meerut–250110, India

[2]Department of Horticulture, Aromatic and Medicinal Plants, Mizoram University, Aizawl–796004, India, E-mail: kritikajune2004@gmail.com

CONTENTS

ABSTRACT

Biological control of fungal diseases constitutes a very broad category of control, consisting of a wide variety of different organisms, mechanisms, interactions, and processes. It has enormous potential to supplement and complement existing disease control strategies. However, biological control also has different properties, requirements, and constraints than previous conventional controls and needs to be properly implemented and integrated with current production strategies. Biological control depends on the effective functioning of the appropriate antagonist strains within each particular plant-pathogen ecosystem. Identifying the appropriate antagonist strains is generally the first step in this process. Understanding how, where, when, and why the biocontrol works may also be crucial to successful development of the biological control system. Because of complexity of soil microbial communities and the role of the biocontrol organism within these communities, an ecological approach to the development of the biological control system is recommended. Evaluating the ecological interactions of the biocontrol organism with the pathogen, host plant, surrounding microbial community, and the environment will be useful in developing the best strategies for the implementation and management of the biological control system. For maximum effectiveness, biocontrol organisms that are locally adapted to the particular environments and patho-systems where they are suited may need to be developed.

21.1   INTRODUCTION

Over the centuries, the use of medicinal plants and their products have been an important part of daily life despite the progress in modern medicine and pharmaceutical research. Approximately 3000 plant species are reported in India to have medicinal properties (Prakash, 1998). The Rigveda (3700 BC) mentions the use of medicinal plants. Our traditional system of medicine, viz., Ayurveda, Yunani, Siddha, Homeopathy, etc., uses herbs or herb products for the treatment of various diseases of human beings and animals. It is estimated that 80% of the world population depends directly on plant-based medicine for their healthcare (WHO, 2003). During the last 20 years, there has been a complete transformation of the medicinal system in the world and especially in those countries of the West that were totally dependent on modern medicine. Because of the realization of toxicity associated with the use of antibiotics and synthetic drugs, the Western society is obsessed with the idea that drugs from medicinal herbs are safer than synthetics. The importance of plants as a major source of clinical agents is much more in developing countries of Asia, Africa, and South America, where traditional medicine is practiced widely. Considerable emphasis has also been laid in the research programs of academic institutions/scientific organizations and drug companies during the last 10 years to trap the efficacy of the plant kingdom for treatment of such diseases for which the modern medicine does not have any effective treatment, including virus diseases like AIDS and herpes, cancer, arthritic disorder, liver ailments, among others (Bhakuni and Jain, 1995). In India, medicinal plants offer low cost and safe healthcare solution.

Medicinal plants also suffer from various kinds of diseases that are similar to those affecting animals and man. Plant growth and yield depend on protecting the plants from diseases. Anything that affects the health of medicinal plants is likely to affect their growth and yield and also seriously reduce their usefulness to mankind. Plant diseases are thus important to man because they not only damage plants but also the plant products. Plant diseases are caused by fungi, bacteria, viruses, phytoplasma, spiroplasma, viroids, nematodes, algae, protozoa, and parasitic higher plants apart from environmental and physiological factors.

Fungi are small, generally microscopic, eukaryotic, usually filamentous, branched, spore-bearing organisms that lack chlorophyll. Fungi have cell wall that contains chitin and glucans (but no cellulose) as the skeleton components. These are embedded in a matrix of polysaccharides and glycoprotein. Fungal cells have distinct nucleus that contain genetic material and is surrounded by a special envelope called nuclear membrane. Fungi are unicellular or multicellular and can reproduce sexually or asexually. The unicellular forms of fungi are larger than bacteria. They obtain nourishment from soil, water, animal, or plant hosts. Vast number of fungi inhabit on the Earth in which more than 10,000 species of fungi can cause disease in plants (Agrios, 2005).

There are several kinds of fungal diseases affecting the medicinal plants, and each plant can be affected by a number of fungal plant pathogens. Disease conditions in the medicinal plants are recognized according to the symptoms produced by the pathogens. The usual disease symptoms are root rots, wilts, collar rot, stem rot, leaf spots, blights, canker, anthracnose, rust, mildews, smuts, damping-off seedling, etc. Fungal diseases of medicinal plants, thus, can be identified by the symptoms and presence of the pathogens especially on the site of infection.

Although there are number of synthetic fungicides available in the market for the control of fungal diseases of medicinal plants, concerns about the health, safety, and environmental effects of chemicals in our water, soil and food require that these inputs be minimized. In addition, biological control may especially important for use in patho-systems in which chemical controls are not economical or ineffective. Biological control may also lessen other problems associated with certain chemical controls, such as the development of pathogen resistance to chemicals, reduction in beneficial organisms' populations, and the creation of biological vacuums (Cook and Baker, 1983). Biological control generally has more specific effects, with only the target pathogen organism(s) being adversely affected, leaving other beneficial organisms and a diverse soil microbial community intact to provide for healthier plants and roots. Thus, biological control can be safer for humans, the plants, and the environment. Biological control has the potential to be more stable and longer lasting than some other controls and is compatible with the concept and goals of integrated plant disease management and sustainable agriculture.

The foundation for the development of contemporary biological control was laid by the two landmark books by Cook and Baker (1983) and Baker and Cook (1974). In addition, there have been numerous books, reviews, and research papers on more recent development in biological control concepts, product formulations, and implementations (Cambell, 1994; Chet, 1994; Wilson and Wisniewski, 1994; Lumsden et al., 1995; Mukherjee and Mukhopadhyay, 1995; Fravel and Larkin, 1996; Mukhopadhyay, 1996, 2003; Fravel et al., 1997; Mishra et al., 1998, 2000; Butt et al., 2001; Gnanamanickam, 2002; Singh et al., 2004; Khilari et al., 2008, 2009; Yadav et al., 2009, 2010). The objective of this chapters is to high-

light the current status of biological control work including product formulations and delivery systems and to optimize biocontrol products with
organic substrates (farm yard manure (FYM), vermicompost, and press
mud) at the grower field for the control of fungal pathogens of medicinal
plants particularly in Indian context.

21.2   BIOCONTROL PRODUCTS IN MEDICINAL PLANTS DISEASE MANAGEMENT

Most of the efforts on biocontrol of plant diseases with antagonistic fungi
and bacteria were diverted against pre- and postemergence seedling,
root rot and wilt diseases, and only a few reports deals with the foliar
pathogens. Major pathogens targeted are *Rhizoctonia solani, Sclerortium
rolfsi, Fusarium* (particularly *F. oxysporum*), *Sclerotinia, Macrophomina,
Pythium, Phytophthora,* etc. Almost all present generation fungicides are
static in their action. Therefore, they are unable to kill completely rest-
ing structures like sclerotia, chlamydospores, rhizomorph, etc. formed by
many soil-borne pathogens. Higher doses and more frequent application
of fungicides to soil disturb ecological balance. Therefore, biological control alone or in combination with cultural and other practices seems to be
the only practical answer for these types of pathogens. A critical review of
the literature reveals that soil-borne diseases such as pre- and post-emer-
gence damping off, root rots, wilts, collar rots, stem rots, etc., under field
condition were successfully suppressed by applying biocontrol products
(Table 21.1).

21.3   MICROBIAL AND BIOCONTROL PRODUCT FORMULATIONS

A wide variety of microorganisms including fungi, bacteria, and actinomycetes have been shown to have biocontrol activity against various fungal
pathogens or the disease they cause and have been studied as biocontrol
agents in several patho-systems. Because of high reproductive ability,
short generation time, targeted host specificity, and capability of surviving
in adverse conditions as either saprobes or resistant structures, scientist

TABLE 21.1 Microbial species with Demonstrated Success as Biocontrol Products

Plants	Diseases	Target pathogen(s)	Biocontrol product with (source)	Mode of application of products
Chlorophytum borivilianum (Safed musli)	Anthracnose	*Colletotricum sp.*	**BioJect Spot-Less** (*Psedomonas aureofaciens*) Trichodex (*Trichoderma harzianum*)	Root dip treatment, Soil treatment
Planttago ovata (Isabgol-Indian Psyllium)	Damping-off	*Pythium ultimum, Rhizoctonia solani*	**Companion** (*Bacillus subtilis* str. GB03) **Gliogard** (*Gliocladium virens* GL-21)	Seed treatment Soil treatment
	Wilt	*Fusarium oxysporum*	**HiStick N/T** (*Bacillus subtilis* str. MB1600) **Trichodex** (*Trichoderma harzianum*)	Soil treatment
	Powdery mildew	*Erysiphae cichoracearum*	**AQ10 Biofungicide** (*Ampelomyces quisqualis* M-10)	Spray
Papaver somniferum (Opium Poppy)	Powdery mildew	*Erysiphae polygoni*	**AQ10 Biofungicide** (*Ampelomyces quisqualis* M-10)	Spray
	Root rot	*Macrophomina phaseolina, Rhizoctonia bataticola, Sclerotinia sclerotiorum*	**Ecofit** (*Trichoderma viride*) **Kodiak** (*Bacillus subtilis* str. GB03); **Contans WG** (*Coniothyrium minitans*)	Soil treatment
	Wilt	*Fusarium oxysporum*	**Fusaclean**-Nonpathogenic *Fusarium oxysporum*	Soil treatment
Rauolfia serpentine (Surpgandha)	Wilt	*F. oxysporum f sp. rauvolfii*	**Biotox C**- Nonpathogenic *Fusarium oxysporum* **Root Shield** (*Trichoderma harzianum*)	Soil treatment
Cantharanthus roseus (Periwinkle)	Foot rot	*Sclerotium rolfsii*	**Trichoderma 2000** (*Trichoderma harzianum*)	Soil treatment
Cassia angustifolia (Senna)	Damping-off at seedling stage	*Macrophomina phaseolina, Rhizoctonia bataticola*	**Promote** (*Trichoderma harzianum* and *T. viride*)	Soil treatment

TABLE 21.1 (Continued)

Plants	Diseases	Target pathogen(s)	Biocontrol product with (source)	Mode of application of products
Atropa belladonna (Belladona)	Root rot	*Pythium butleri*	**T-22** (*Trichoderma harzianum*)	Soil treatment
	Collar rot	*Sclerotium rolfsii*	**Trichoderma 2000** (*Trichoderma harzianum*)	Soil treatment
Cinchona ledgeriana (Cinchona)	Damping off	*Pythium vexans*	*Chaetomium globosum*	Soil treatment
	Collar rot	*Sclerotium rolfsii*	**Trichoderma 2000** (*Trichderma harzianum*)	Soil treatment
Coleus forskohlli (Coleus-Pathar Chur)	Leaf blight	*Rhizoctonia solani*	**Conquer** (*Pseudomonas fluorescens*) **Deny** (*Burkholderia cepacia*)	Spray
Datura stramonium (Datura)	Root rot	*Sclerotium rolfsii*	**Trichoderma 2000** (*Trichoderma harzianum*)	Soil treatment
Digitalis purpurea (Digitalis)	Anthracnose	*Colletotrichum fuscum*	**Trichoject** (*T. harzianum* and *T.viride*)	Soil treatment
Dioscoria dettoidea (Dioscoria)	Tuber rot	*Aspergillus niger*	**HiStick N/T** (*Bacillus subtilis* str. MB1600)	Tuber treatment Soil treatment
Panax ginseng (Ginseng)	Anthracnose	*Colletotrichum dematium*	**BioJect Spot-Less** (*P. aureofacens*)	Soil treatment
	Powdery mildew	*Erysiphae panax*	**AQ10 Biofungicide** (*Ampelomyces quisqualis* M-10)	Spray
	Damping-off	*Pythium sp.,* *Rhizoctonia solani*	**Deny** (*Burkholderia cepacia*)	Soil treatment
Gycyrrhiza glabra (Liquorice-Mulathi)	Root rot	*Rhizoctonia bataticola*	**Gliogard** (*Gliocladium virens* GL-21)	Rhizome treatment
	Wilt	*Fusarium sp.*	*Chaetomium globosum*	Soil treatment
Withania somnifera (Ashwagandha)	Damping off	*Pythium sp.*	**Root Shield** (*T. harzianum*)	Soil treatment
	Seedling rot	*Alternaria alternata*	**Kodiak** (*Bacillus subtilis* str. GB03)	Soil treatment

TABLE 21.1　(Continued)

Plants	Diseases	Target pathogen(s)	Biocontrol product with (source)	Mode of application of products
Mentha spp. (Mint)	Stolon rot	*Macrophomina phaseolina*	**Ecofit** (*Trichoderma viride*)	Stolen treatment Soil treatment
	Verticillium wilt	*Verticillium albo-atrum*	*Verticillium biguttatum*	-
	Powdery mildew	*Erysiphae cichocearum*	**AQ10 Biofungicide** (*Ampelomyces quisqualis* M-10)	Spray
Chrysanthemum cinerariifolium (Pyrethrum)	Damping-off	Pythium sp.	**Polygandron** (*Pythium oligandrum*)	Soil treatment
	Verticillium wilt	*Verticillium albo-atrum*	*Talaromyces flavus*	-
Curcuma longa (Turmeric)	Rhizome and root rot	*Pythium aphanidermatum P. graminicolum*	**Intercept** (*Burkholderia cepacia*) **BINAB-T WP** (*Trichoderma. harzianum* and *T.polysporum*)	Rhizome treatment Soil treatment
Carica papaya (Papaya)	Stem or foot rot	*Pythium aphanidermatum*	**Deny** (*Burkholderia cepacia*) **BINAB-T WP** (*Trichoderma. harzianum* and *T. polysporum*)	Seed treatment Soil treatment
Musa sapientum (Banana)	Panama wilt	*Fusarium oxysporum* f. sp. *cubense*	**Biotox C**- Nonpathogenic (*Fusarium oxysporum*) **Root Shield** (*T. harzianum*)	Sucker treatment Soil treatment
Lycopersicon esculentum (Tomato)	Damping-off	Pythium sp.	**Trichopel** (*Trichoderma. harzianum* and *T. viride*)	Soil treatment Seed treatment
	Fusarial wilt	*Fusarium oxysporum* f. sp. *lycopersici*	**Ecofit** (*Trichoderma viride*) **HiStick N/T** (*Bacillus subtilis* str. MB1600	Soil treatment Seed treatment

manipulated the ability of microorganisms for the control of diseases of crop plants and developed several formulations.

Formulations are an important component in the development of an effective biological control system (Lewis and Papavizas, 1991). Their development and application can have a major impact on the success of a biocontrol agent (Vidhyasekaran and Muthamilan, 1995). Formulations must deliver the antagonist to the appropriate location in sufficient quantities and in the proper form. The activity and growth of the biocontrol agent depend greatly on the endogenous nutritional status of the formulated propagule and the energy source provided in the formulation. An effective formulation provides nutrients for germination and sporulation of the biocontrol organism, withstands harsh environmental conditions, interacts with other organisms to the advantage of the biocontrol agent, can be integrated with cultural practices, and can be applied with existing machinery or methods. The formulation must also be inexpensive, easy to produce, and have an appropriate shelf-life (Lumsden et al., 1995). Currently, there are over 40 different formulations of biocontrol agents commercially available around the world for the control of various plant diseases (Fravel et al., 1997). The composition of formulation varies greatly with the intended use (Fravel et al., 1997). Fungi and actinomycetes are currently on the market as pellets, wettable powders, granules, micro-granules, water dispersible granules, and dusts. Fungi are also impregnated in sticks and produced on grain. Actinomycetes have been specifically produced as powders for drenches or sprays or added through irrigation systems. Bacteria have been formulated as aqueous suspensions of fermenter biomass to spray, liquid suspension for drench or for drip irrigation, dry powders and wettable powders. Bacterial and fungal products can also be applied to seed with a sticker and a peat and clay carrier (Fravel et al., 1997).

In recent years, many small and medium entrepreneurs have entered into commercial production of biocontrol agents, resulting in several products into Indian market, and to date, approximately 80 biocontrol products/biopesticides have been developed so far (Singh et al., 2004).

At present, the market share of biopesticides is around 2.5% of total pesticide market compared to 1% in 2001 (Singh et al., 2004). This increase is due to utilization of biopesticides mainly in medicinal plants and vegetable crops, sugarcane, cotton and paddy. The projected share of these biopesticides by the end of 2015 is 15% of the total pesticide market. They are now well accepted in "organic farming" and are major part of private market mainly in Punjab, Haryana, western Uttar Pradesh, Uttarakhand,

Madhya Pradesh, Maharashtra, and Tamil Nadu. The Indian biopesticide industry is the umbrella of biopesticides products available in the country today (Table 21.2). Many products are being exported from India now. India has over 400 biopesticide production units with 8–10 major players (Singh et al., 2004). Technologies for joint ventures in foreign countries are also available. Commercialization of biopesticides is a multistep process involving a wide range of activities (Singh et al., 2004).

21.4   SHELF-LIFE OF FORMULATION

Shelf-life of a biocontrol agent plays a crucial role in storing a formulation. In general, the antagonist multiplied in an organic food base has greater shelf life than the inert or inorganic food bases (Vidhyasekaran and Muthamilan, 1995; Fravel et al., 1997). Shelf-life of *Trichoderma harzianum* in coffee husk was more than 18 months. Talc, peat, lignite, and kaolin-based formulation of *Trichoderma viride* had a shelf-life of 4 months. Shelf-life of this species in gypsum-based formulation was 4 months (Fravel et al., 1997).

21.5   MECHANISMS OF ACTION

Biocontrol mechanism involving antagonistic microbial species/strains operates by way of competition, antibiosis, parasitism, and induced resistance.

21.5.1   COMPETITION

Competition has been defined as the active demand in excess of the immediate supply of material on the part of two or more organisms (Clarke, 1965). The result is a restriction on population size or microbial activity of one or more of the competitors (Paulitz and Baker, 1987). Competition between microorganisms generally refers to competition for nutrients such as available carbon, nitrogen, iron, or trace elements or competition for space such for colonization or infection sites on the root or seed sur-

TABLE 21.2 Microbial Agents Registered and Commercially Marketed as Biocontrol Products

Biocontrol Products	Microbial	Developing agency
Antagon- TV	*Trichoderma. viride*	Green Tech Agro Products, Coimbatore
Biocon	*T. viride*	Tocklai Experimental Station, Tea Research Association, Jorhat, Assam
Bioderma	*T. viride + T. harzianum*	Biotech International Limited, New Delhi
Bioguard	*T. viride*	Krishi Rasayan Export Pvt. Ltd., Solan (H.P.)
Ecoderma	*T. viride + T. harzianum*	Margo Biocontrol Pvt. Ltd., Bangalore
Ecofit	*Trichoderma viride*	Hoechst and Schering AgroEvo Ltd., Mumbai
Funginil	*T. viride*	Crop Health Bioproduct Research Centre, Ghaziabad
Kalisena SD		
Kalisena SL	*Aspergillus niger* AN-27	Cadilla Pharmaceuticals Ltd., Ahmedabad
Pant Biocontrol Agent-1	*T. harzianum*	G.B. Pant University of Agriculture Technology, Pantnagar
Sun-Derma	*T. viride*	Sun agro Chemicals, Chennai
Trichoguard	*T. viride*	Anu Biotech International Ltd., Faridabad
Tricho	*T. viride*	Excel Industries Ltd., Mumbai
Antifungus	*Trichoderma spp.*	Grondortsmettingen De Cuester, Belgium
AQ10	Ampelomyces quisqualis isolate M-10	Ecogen, Inc. Israel
Aspire	*Candida oleophila I-182*	Ecogen, Inc. Israel
Bas-derma	*T. viride Basarass*	Biocontrol Res. Lab., India
Biact/Paecil	*Paecilomyces lilacinus*	Technological Innovation Corporation Pty. Ltd., Australia
BINAB T	*Trichoderma harzianum* (ATCC 20476) and *Trichoderma polysporum* (ATCC 20475)	Bio-Innovation Eftr AB UK Bredholmen, Sweden
Bioderma	*Trichoderma viride/ T. harzianum*	Biotech International Ltd., India

TABLE 21.2   (Continued)

Biocontrol Products	Microbial	Developing agency
Biofungus	*Trichoderma* spp.	Grondortsmettingen deCuester n. v., Belgium
Bio-Trek 22G	*Trichoderma harzianum*	Bio works, Inc. of Geneva, NY
Blue Circle	*Burkholderia cepacia*	ecoScience Corp., USA
Coniothyrin	*Coniotyrium minitans*	Russia
Contans	*Coniotyrium minitans*	Prophyta Biologischer Pflanzenschutz GmbH', Germany
BioJect Spot-Less	*Peudomonas aureofaciens*	Eco soil systems, USA
Fusaclean	*Fusarium oxysporum (Nonpathogenic)*	Natural Plant Protection, France
Kali sena	*Aspergillus niger*	Cadilla Phrma., India
Polyversum	*Pythium oligandrum*	Biopreparaty Ltd., Czech Republic
Polyversum/ Polygandron	*Pythium oligandrum*	Biopreparaty Ltd., Czech Republic
Prestop, Prima-stop	*Gliocladium catenulatum*	Kemria Agro. Oy, Finland
Root Pro.	*Trichoderma harzianum*	Bioworks Inc., USA
Root shield, Plant Shield, T-22 Planter Box	*Trichoderma harzianum Rifai strain KRL-AG (T-22)*	Bioworks Inc., USA
Rootstop	*Phlebia gigantean*	Kemira Agro. Oy, Finland
Gliogard	*Gliocladium Virens strain GL-21*	Thermo Trilogy, USA
Supresivit	*Trichoderma harzianum*	Borregaard and Reitzel, Czech Republic
Trichoderma 2000	*Trichoderma sp.*	Myocontrol Ltd., Israel
Trichodex, Trichopel	*Trichoderma harzianum*	Makhteshim Chemical Works Ltd., USA
Trichopel, Trichoject, Trichodowels, Trichoseal	*Trichoderma harzianum* and *Trichoderma viride*	Agrimm Technologies Ltd., New Zealand
Tri-control	*Trichoderma spp.*	Jeypee Biotechs, India

TABLE 21.2 (Continued)

Biocontrol Products	Microbial	Developing agency
Trieco	*Trichoderma viride*	Ecosense Labs Pvt. Ltd., Mumbai, India
TY	*Trichoderma sp.*	Myocontrol, Israel
Vaminoc	*Mycorrhizal Fungi*	Biological Crop Protection Ltd., USA
Biofox C	*Fusarium oxysporum*	SIAPA, Italy
Companion	*Bacillus subtilis* GB03	Growth Products, USA
Deny/Intercept	*Burkholderia cepacia*	Stine Microbial Products, Shwance, KS
HitStick N/T	*Bacillus subtilis* str. MB1600	MicroBio Group, UK
Kodiak	*Bacillus subtilis str.* GB03	Gustaffson Inc., USA
Conquer	*Peudomonas fluorescens*	Mauri Foods, Australia
Koni	*Coniothyrium minitans*	BIOVED Ltd, Hungary
Victus	*Peudomonas fluorescens*	Sylvan Spawn Laboratory, USA

face. Nutrients from roots and seeds support microbial growth and other activities in the spermosphere and rhizosphere (Paulitz, 1990). Nutrients from roots and seeds are derived from several sources including exudates, secretion lysates, and mucilages. These nutrients are chemically diverse and include carbohydrates, amino acids, peptides organic acids, and other plant metabolites (Curl and Truelove, 1986).

Competition can be an effective biocontrol mechanism when the antagonist organism is present in sufficient quantities at the correct time and location and can utilize limited nutrients or other resources more efficiently than the pathogen. An example of this interaction was observed in Fusarium wilt suppressive soils in France (Alabouvette et al., 1993). High populations of saprophytic strains of *Fusarium oxysporum* effectively competed with the pathogenic strains of *F. oxysporum* for reduced carbon sources in these soils, resulting in an inhibition of pathogen propagule germination, reduced saprophytic growth of the pathogen, and low levels of disease. Competition for iron between biocontrol bacteria and plant pathogens has been well documented (Leong, 1986). Iron is required for growth by microbes, but is typically limited in availability in soil. Strains of certain biocontrol bacteria, such as *Pseudomonas fluorescens*, produce siderophore, which have a high affinity for soluble ferric iron (Fe^{+3}). Siderophores bind iron, allowing these biocontrol bacteria to effectively compete with pathogens for iron. Production of siderophores and the limited availability of iron in soil have been associated with disease suppression by several bacterial antagonists (Buysens et al., 1996).

21.5.2 ANTIBIOSIS

Antibiosis refers to the inhibition or destruction of the pathogen by a metabolite product of the antagonist, such as the production of specific toxin, antibiotics, or enzymes. This interaction can result in suppression of activity of pathogen or destruction of pathogen propagules. To be effective, antibiotics must be produced in situ in sufficient quantities at the precise time and place where they will interact with the pathogen. The production of antibiotics by various strains of *P. fluorescens,* which include phenazine compounds and 2,4-diacetylphloroglucinol has been shown to be important in biocontrol through mutational analysis. Mutant strains of *P. fluorescens* that do not produce these antibiotics have reduced efficacy as biocontrol agents (Keel et al., 1992).

Antibiosis has also been shown to be important in biocontrol interactions between fungal biocontrol agents and fungal plant pathogens. *Gliocladium virens*, an important biocontrol fungus, has activity against several soil-borne plant pathogens, including *Pythium ultimum* and *Rhizoctonia solani* (Papavizas and Lewis, 1989). Different strains of *G. virens* produce a variety of metabolites, including gliotoxin and glioviridin, which are toxic or inhibitory to several fungal pathogens (Wilhite et al., 1994). In addition, gliotoxin has been detected in a number of soils colonized by *G. virens*, and quantities of gliotoxin in these soils have been correlated with disease suppression (Lumsden et al., 1992). Gliotoxin is not found in any commercial formulations, but when the spores of *G. virens* begin to grow in the soil, they produce the antibiotic. Furthermore, gliotoxin is sensitive to oxidation and probably poses no health risk because of rapid degradation.

Trichoderma spp. secrete diverse secondary metabolites with antibiotic properties, including polyketides, terpenoids, polypeptides, and metabolites derived from alpha-amino acids (Taylor, 1986). *T. harzianum* produces trichorzianins, trichkindins, trichozins, and harzianic acids that exhibit antifungal activity (Corley et al., 1994). Furthermore, a novel antifungal protein from *T. viride* "tricholin" causes cessation of growth and uptake of amino acids and is effective against *Rhizoctonia solani* (Lin et al., 1994).

21.5.3   PARASITISM

Parasitism occurs when the antagonist feeds on or within the pathogen, resulting in the direct destruction or lysis of propagules and structures. When one fungus parasitizes another fungus, it is called mycoparasitism (Lumsden, 1992). *Trichoderma, Chaetomium,* and *Gliocladium* spp. are well-known parasites of a wide variety of pathogenic fungi. Biocontrol fungi coil around and parasitize mycelia of the fungal pathogen. This complex process involves the activities of cell wall hydrolytic enzymes such as glucanases, chitinases, proteases, and lipase produced by biocontrol fungi (Schirmbock et al., 1994; Haran et al., 1995). Glucanse and chitinase activities have been detected in sterile soil containing mycelium of *Sclerotium rolfsii, Rhizoctonia solani*, and *Pythium aphanidermatum* inoculated with *T. harzianum* (Elad et al., 1982). Cell wall-degrading enzymes produced by *T. harzianum* and *G. virens* inhibit conidia germination and germ tube elongation of *Botrytis cinerea* in vitro (Lorito et al., 1993).

21.5.4   INDUCED RESISTANCE

Induced resistance, when an antagonist induces defense responses within the host plant; resulting in resistance to diseases through reducing, restricting, or blocking the ability of the pathogen to produce disease. This can happen by prevention of infection, restricting pathogen growth within the plant, or some other mechanism of defense activation within the plant. Induction of systemic resistance may involve activation of multiple potential defense mechanisms, including increased activity of chitinases, beta 1,3-glucanases, peroxidases, and other pathogenesis related (PR) proteins, and accumulation of antimicrobial compounds such as phyto-alexins and formation of protective biopolymers such as lignin, callose, and hydroxyproline-rich glycoproteins (Kloepper et al., 1996). Examples of this interaction are presented by certain rhizobacteria (such as strains of *P. fluorescens*) that can induce systemic resistance to a number of different pathogens (Zhou and Paulitz, 1994; Liu et al., 1995). In cucumber, induction of resistance by a single rhizobacterial strain provided protection against several different pathogens, including fungi, bacteria, and viruses as well as reduction in insect feeding (Kloepper et al., 1996). Protection of crops by induced resistance has also been demonstrated in the field for some crops (Tuzum et al., 1992; Wei et al., 1996). Another example of induced systemic resistance (ISR) is that incited by certain nonpathogenic or avirulent strains of *F. oxysporum*, which are able to parasitically colonize roots without causing disease and induce resistance to Fusarial wilt in several crops (Leeman et al., 1995; Larkin et al., 1996). Because there is generally an induction period of one to several days required after exposure of the plant to the antagonist before ISR occurs, this mechanism is most effective when the antagonist is applied prior to exposure to the pathogen, such as with a seed or seedling treatment operations. Although the mechanism of induction is unknown, a role for salicylic acid has been confirmed in some ISR reactions.

Treatment of plants with exogenous salicylic acid induced PR protein synthesis and enhanced resistance to a number of pathogens. In addition, endogenous salicylic acid levels rise specifically during resistance response in plants (Chen et al., 1995).

21.6 DELIVERY OF BIOCONTROL PRODUCTS UNDER FIELD CONDITIONS

One of the areas of biocontrol products research has the delivery system. It is rather necessary to have an efficient economically viable mode of application of biocontrol products in seed, soil, and seedling under field condition.

21.6.1 SEED TREATMENT

Seed coating with biocontrol products has emerged as a feasible way of delivering the antagonist for the management of plant diseases. For sowing purpose, a powdered product of antagonist is used @ 5–8 gm powder/ kg seed, based on seed size and formulations of antagonist (Mukhopadhyay et al., 1992; Mukherjee and Mukhopadhyay, 1995; Vidhyasekaran and Muthamilan, 1995). *Trichoderma hamatum, T. harzianum, T. virens,* and *T. viride* are effective seed protectants against *Pythium spp.* and *Rhizoctonia solani* (Mukherjee and Mukhopadhyay, 1995). A large number of seed, seedling, root, stem, foliar, and panicle diseases have been suppressed by seed treatment with biocontrol products.

21.6.2 SEEDLING TREATMENT

Treatment of planting materials with beneficial microorganisms is becoming increasingly important. Seedling roots can be treated with spore or cell suspension of antagonists. This method is generally used for vegetable crops, rice, etc., where transplanting is practiced. Seedlings of root should be dipped in water suspension of the powdered product of antagonist @ 10 g/L of water for 30 minutes in shady conditions. Root dipping in antagonist suspension not only reduces disease severity but also enhances seedling growth in rice, tomato, brinjal, chili, and capsicum (Mishra and Sinha, 2000).

21.6.3 SOIL TREATMENT

Numerous attempts have been made to control several soil-borne diseases by incorporating natural substrates colonized by a powdered product of antagonists into field (Vidhyasekaran and Muthamilan, 1995; Sen, 2000). For the treatment of one hectare area, 2.5 to 3 kg powdered products of antagonist can be mixed in 70–100 kg vermicompost or well-rotted farm yard manure (FYM) or well-rotted press-mud in about 7–10 days, and when the substrate is fully colonized, it should be incorporated into the soil before sowing.

21.7 CONCLUSION

Biocontrol may function through competition, antibiosis, parasitism, induced résistance, or a combination of these mechanism. These various mechanisms may control disease by the reduction or inhibition of the pathogen inoculums, protection of the infection court (prevention of pathogen infection), or by limiting disease development after pathogen infection. Each organism, mechanism, and activity has different traits, conditions, and requirements associated with it that are essential to biocontrol activity. Other areas of research that may be critical for further development and improvement of biological control systems are the development and use of multiple antagonists having different mechanism of action, incorporating the influence of host plant on microbial communities to enhance biocontrol organism's activity, better integration of biocontrol with other disease control strategies, and improved formulations and delivery systems to provide the optimal starting point.

ACKNOWLEDGMENT

The authors are thankful to the authorities of Sardar Vallabhbhai Patel University of Agriculture and Technology, Meerut, as well as to Mizoram University, Aizawl, for providing various facilities during the research.

KEYWORDS

- **bio-control**
- **biological management**
- **formulation**
- **fungal**
- **medicinal plants**
- **microbial**

REFERENCES

Agrios, G. N., (2005). *Plant Pathology*, Elsevier Academic Press, San Diago, California, USA.

Alabouvette, C., Lemanceau, P., & Steinberg, C., (1993). Recent advances in the biological control of *fusarium* wilts, *Pest. Sci., 37*, 365–373.

Baker, K. F., & Cook, R. J., (1974). *Biological Control of Plant Pathogens,* Freeman, San Francisco (Reprinted 1982, American Phytopathological Society, St. Pau, MN).

Bhakuni, D. S., & Jain, S., (1995). Chemistry of cultivated medicinal plants. In: *Advances in Horticulture* (eds. Chadha, K. L., and Gupta, R.), vol. 11, pp. 45–119, Malhotra Publishing House, New Delhi.

Butt, T. M., Jackson, C. W., & Magan, N., (2001). *Fungi as Biocontrol Agents-Progress, Problems and Potential*, CBI Publishing, Wallingford, Oxon OX10 SDE. UK.

Buysens, S., Heungens, K., Poppe, J., & Hofte, M., (1996). Involvement of pyochelin and pyoverdin in suppression of *Pythium* induced damping-off of tomato by *Pseudomonas aerguginosa* 7NSK2, *Appl. Environ. Microbiol., 62*, 865–871.

Campbell, R., (1994). Biological control of soil-borne diseases: some present problems and different approaches, *Crop Prot., 13*, 4–13.

Chen, Z. X., Malamy, J., Henming, J., Conrath, U., Sanchezcasas, P., Silva, H., Ricingliano, J., & Klessig, D. F., (1995). Induction, modification and transduction of the Salicylic acid signal in plant defense responses, *Proc. Natl. Acad. Sci.* USA, *92*, 4134–4137.

Chet, I., (1994). Biological control of fungal pathogens, *Appl. Biochem. Biotechnol., 48*, 37–43.

Clarke, F. E., (1965). The concept of competition in microbial ecology, in *Ecology of Soil-Borne Plant Pathogens* (eds. Baker, K. F., & Snyder, W. C.), pp. 339–345, University of California Press, Berkeley, CA.

Cook, R. J., & Baker, K. F., (1983). *The Nature and Practice of Biological Control of Plant Pathogens,* American Phytopathological Society, St Paul, MN.

Corley, D. G., Miller-Wideman, M., & Durley, R. C., (1994). Isolation and structure of harzianum A: a new trichothecene from *Trichoderma harzianum, Journal of Natural Products, 57*, 422–425.

Curl, E. A., & Truelove, B., (1986). *The Rhizosphere,* Springer-Verlag, New York.

Elad, Y., Chet, I., & Henis, Y., (1982). Degradation of plant pathogenic fungi by *Trchoderma harzianum, Phytopathology, 28*, 719–725.

Fravel, D. R., & Larkin, R. P., (1996). Availability and application of biocontrol products, *Biol. Cult. Tests, 11*, 1–7.

Fravel, D. R., Connick, W. J., Jr. & Lewis, J. A., (1997). Formulation of microorganisms to control plant diseases, *Formulation of Microbial Biopesticides, Beneficial Microorganisms and Nemadotes* (ed. H. D. Burges), Chapman and Hall, London.

Gnanamanickam, S. S., (2002). *Biological Control of Crop Diseases*, Global Research, Valent BioSciences Corporation, Long Grave, Illinois, USA.

Haran, S., Schickler, H., Oppenheim, A., & Chet, I., (1995). New components of the chitinolytic system of *Trichoderma Harzianum, Mycol. Res., 99*, 441–446.

Keel, C., Schnider, U., Maurhofer, M., Voisard, C., Laville, J., Burger, U., Wirthner, P., Hass, D., & Defago, G., (1992). Suppression of root diseases by *Pseudomonas fluo-*

rescens CHAO: Importance of the bacterial secondary metabolite 2,4-diacetylphoro-glucinol, *Molec. Plant-Microbe Interact., 5,* 4–13.

Khilari, K., Mukhopadhyay, A. N., & Yadav, R. S., (2008). Effect of volatile and non volatile compounds of *T., koningii* and *G., virens* on growth of Rolfsii, S., Solani, R., & Oxysporum, F., f. sp. ciceri, *Environment & Ecology, 26*(4B), 1961–1964.

Khilari, K., Mukhopadhyay, A. N., & Yadav, R. S., (2009). Seed treatment with air dried spore powder of *T. koningii* and *G. virens* on germination, nodulation and growth promotion of chickpea (*Cicer aeritinum*), *Progressive Agriculture, 9*(1), 35–38.

Kloepper, J. W., Zehnder, G. W., Tuzun, S., Murphy, J. F., Wei, G., Yao, C., & Raupach, G., (1996). Toward agricultural implementation of PGPR-mediated induced systemic resistance against crop pests, *in Advances in Biological Control of Plant Diseases* (eds T., Wenhua, R. J., Cook and A., Rovira), pp.165–174, China Agricultural University press, Beijing, China.

Larkin, R. P., Hopkins, D. H., & Martin, F. N., (1996). Suppression of fusarium wilt of watermelon by nonpathogenic *Fusarium oxysporum* and other microorganism recovered from a disease suppressive soil, *Phytopathology, 86,* 812–819.

Leeman, M., Van Pelt, J. A., Den Ouden, F. M., Heinsbrook, M., Bakker, P. A., & Schippers, B., (1995). Induction of systemic resistance by *Pseudomonas fluorescens* in radish cultivars differing in susceptibility to fusarium wilt using a novel bioassay, *Eur. J. Plant Pathol., 101,* 655–664.

Leong, J., (1986). Siderophores: their chemistry and possible role in the control of plant pathogens, *Annu. Rev. Phytopathol., 24,* 187–209.

Lewis, J. A., & Papavizas, G. C., (1991). Biocontrol of cotton damping-off caused by Rhizoctonia solani in the field with formulation of *Trichoderma* and *Gliocladium virens, Crop Prot., 10,* 396–402.

Lin, A., Lee, T. M., & Rern, J. C., (1994). Tricholin, a new antifungal agent from *Trichderma viride* and its action in biological control *of Rhizoctonia solani, Journal of Antibiotics* (Tokyo), *47,* 799–805.

Liu, L., Kloepper, J. W., & Tuzun, S., (1995). Induction of systemic resistance in cucumber against Fusarium wilts by plant growth promoting rhizobacteria, *Phytopathology, 85,* 695–698.

Lorito, M., Harman, G. E., Hayes, C. K., Broadway, R. M., Tronsmo, A., Woo, S. L., & Di Pietro, A., (1993). Chitinolytic enzymes produced by *Trichoderma harzianum*: antifungal activity of purified endochitinase and chitobiosidase, *Phytopathology, 83,* 302–307.

Lumsden, R. D., (1992). Mycoparasitisms of soil borne plant pathogens, in the fungal community: *Its Organization and Role in the Eecosystem* (eds. Carroll, G. C., & Wicklow, D. T.), Marcel Dekker, New York, pp. 275–293.

Lumsden, R. D., Lewis, J. A., & Fravel, D. R., (1995). Formulation and delivery of biocontrol agents for used against soil borne plant pathogens, *in Biorational Pest Control Agents: Formulation and Delivery* (eds. Hall, F. R., & Barry, J. W.), *Amer. Chem. Soc.,* Washington DC, pp. 166–182.

Lumsden, R. D., Locke, J., Adkins, S. T., & Ridout, C. J., (1992). Isolation and localization of the antibiotic gliotoxin produced by *Gliocadium virens* from alginate prill in soil and soilless media, *Phytopathology, 89,* 230–235.

Mishra, D. S., & Sinha, A. P., (2000). Plant growth-promoting activity of some fungal and bacterial agents on rice seed germination and seedling growth, *Trop. Agri.*, *77*, 188–191.

Mishra, R. C., Yadav, R. S., Chaturvedi, C., & Dikshit, A., (1998). Ecofriendly biopesticide for the management of collar rot & wilt of patchouli. In: *New Trends in Microbial Ecology* (eds. Bharat Rai & Dkhar, M. S.), pp. 255–261, Publisher International Society for Conservation of Natural Resources, Banaras Hindu University, Varanasi, India.

Mishra, R. C., Yadav, R. S., Chaturvedi, C., & Dikshit, A., (2000). Biological management of *Rhizoctonia solani*, *Current Science, 78*(3), 230–232.

Mishra, R. C., Yadav, R. S., Singh, H. B., & Dikshit, A., (2000). *In situ* efficacy of *Trichoderma harzianum* as mycoparasite on *Sclerotium rolfsii* and *Rhizoctonia solani*, *Tropical Agriculture* (Trinidad), *77*, 205–206.

Mukherjee, P. K., & Mukhopadhyay, A. N., (1995). Evaluation of *Trichoderma harzianum* for biocontrol of Pythium damping-off of cauliflower, *Indian Phytopathol.*, *48*, 101–102.

Mukhpadhyay, A. N., (1996). *Recent Innovations in Plant Disease Control by Ecofriendly Biopesticides*, Presidential address, Agriculture Sciences Section, 83rd Indian Science Congress, Patiala, India.

Mukhpadhyay, A. N., Shrestha, S. M., & Mukherjee, P. K., (1992). Biological seed treatment for the control of soil borne plant pathogens, *FAO Plant Prot. Bull.*, *40*, 21–30.

Mukohpadhyay, A. N., (2003). *Pesticides Usage Scenario in India and Viable Alternatives*, Voluntary Health Association of India, New Delhi, pp. 137.

Papavizas, G. C., & Lewis, J. A., (1989). Effect of *Gliocladium* and *Trichoderma* on damping-off of snap been caused by *Sclerotium rolfsii* in the green house, *Plant Pathol.*, *38*, 277–286.

Paulitz, T. C., (1990). Biochemical and ecological aspects of competition in biological control, *in New Direction in Biological Control: Alternatives for Agricultural Pests and Diseases* (eds. Baker, R., Dunn, P. E.), Alan, R., Liss, New York, pp. 713–724.

Schirmbock, M., Lorito, M., Wang, Y. L., Hayes, C. K., Arisan-Atac, I., Scala, F., Harman, G. E., & Kubicek, C. P., (1994). Parallel formation and synergism of hydrolytic enzymes and peptaibol antibiotics, molecular mechanisms involved in the antagonistic action of *Trichoderma harzianum* against phytopathogenic fungi, *Applied Environmental Microbiology, 60*, 4364–4370.

Sen, B., (2000). Biological control: A success story, *Indian Phytopathol, 53*, 243–249.

Singh, U. S., Zaidi, N. W., & Sharma, M. C., (2004), Integrated pest management in India: Availability of inputs and associated problem. *In: Proceeding of Brain Storming Session on Chickpea Productivity* (NCIPM), New Delhi.

Taylor, A., (1986). Some aspects of the chemistry and biology of the genus Hyppocrea and its anamorph, *Trichoderma* and *Gliocladium*. *Proceedings of the Nova Scotia Institute of Science, 36*, 27–58.

Tuzun, S., Juarez, J., Nesmith, W. C., & Cuc, J., (1992). Induction of systemic resistance in tobacco against metalaxyl-tolerant strains of *Peronospora tabacina* and the natural occurrence of the phenomenon in Mexico, *Phytopathology, 82*, 425–429.

Vidhyasekaran, P., & Muthamilan, M., (1995). Development of formulations of *Pseudomonas fluorescence* for control of chickpea wilt, *Plant Dis. 79*, 782–786.

Wei, G., Kleopper, J. W., & Tuzun, S., (1996). Induced systemic resistance to cucumber diseases and increased plant growth by plant growth promoting rhizobacteria under field conditions, *Phytopathology, 86*, 221–224.

Wilhite, S. E., Lumsden, R. D., & Straney, D. C., (1994). Mutational analysis of gliotoxin production by the biocontrol fungus *Gliocladium virens* in relation to suppression of *Pythium* damping-off, *Phytopathology, 84*, 816–821.

Wilson, C. L., & Wisniewski, M. E., eds. (1994). *Biological Control of Postharvest Diseases: Theory and Practice*, CRC Press, Boca Raton, FL.

Yadav, R. S., Khilari, K., & Kumar, R., (2010). Eco-Friendly Management of Alternaria Leaf Spot Disease of Cabbage, *Environment and Ecology, 28*(4), 2200–2202.

Yadav, R. S., Khilari, K., Yadav, M. K., & Shukla, A. C., (2009). Fungal Products: An Eco-friendly Input for Integrated Crop Disease Management, *Vegetos, 22*(2), 1–15.

Zhou, T., & Paulitz, T. C., (1994). Induced resistance in the biocontrol of *Pythium aphanidermatum* by *Pseudomonas* spp. on cucumber, *J. Phytopathol., 142*, 51–63.

PART III

HORTICULTURE FOR
HEALTH AND NUTRITION

CHAPTER 22

BROAD SPECTRUM OF INDIAN PEPPERMINT OIL AGAINST DISEASE-CAUSING HUMAN BACTERIAL PATHOGENS

AWADHESH KUMAR,[1] AMRITESH C. SHUKLA,[1] and ANUPAM DIKSHIT[2]

[1]Department of Horticulture, Aromatic and Medicinal Plants, Mizoram University, Aizawl–796004, India, E-mail: kumarawadhesh9@gmail.com

[2]Biological Product Lab, Department of Botany, University of Allahabad, Allahabad–211004, India

CONTENTS

ABSTRACT

The fresh leaves of *Mentha piperita* L. (Lamiaceae) were collected, and after proper shade drying, the essential oil was extracted through hydrodistillation using Clevenger's apparatus. The gas chromatography-mass spectroscopy (GC-MS) analysis of the oil showed that menthol (37.20%) and menthone (22.52%) were the major active constituents. In the present work, its antibacterial efficacy was screened with the disc diffusion method, where the maximum zone of inhibition was found against *Salmonella typhi* (17 ± 0.1 mm). Further, the in vitro efficacy of oil was validated against bacterial pathogens such as *Escherichia coli* (ATCC 25922), *Proteus vulgaris* (MTCC 1771), *Salmonella typhi* (MTCC 733), *Klebsiella pneumoniae* (MTCC 109), and *Pseudomonas aeruginosa* (MTCC 103) by using the broth microdilution method (CLSI). The minimum inhibitory concentration (MIC) was recorded as 0.83 mg/mL, 1.19 mg/mL, 0.30 mg/mL, 0.61 mg/mL, and 1.09 mg/mL, respectively. Furthermore, the oil showed toxicity against heavy inoculum density, quick killing activity, broad antimicrobial spectrum, thermostability, and long shelf-life. The in vivo investigations of its active constituents will be needed to perform in order to confirm the mechanism of action for curing human diseases.

22.1 INTRODUCTION

Water and plants are the two main sources on the Earth to continue life. They are directly linked to each other. Ancient data revealed that humans as well as animals live properly settled near the banks of rivers, lakes and other water resources, because they got drinking water easily. Wherever the scarcity of any of them occurs, life becomes difficult. But as the population grew rapidly, the fresh drinking water became polluted and several fatal bacterial, viral, protozoans, nematodal, and fungal diseases emerged. This is due to rapid pace of urbanization, which increased the demand of infrastructure for better livelihood (Kumar, 2011). Presently, bacterial pathogenic diseases are emerging from unsafe drinking water. Further, consumption of this contaminated water can cause several ill effects in human beings such as the very common bacteria like *Escherichia coli, Vibrio cholerae, Salmonella typhi, Salmonella typhimurium, Shigella dysenteriae, Staphylococcus aureus, Pseudomonas aeruginosa, Proteus vulgaris, Klebsiella pneumoniae,* etc., are responsible for a variety of diseases like cholera, typhoid, dysentery, bacillary dysentery, vomiting, urinary tract infection, and other gastrointestinal symptoms (Kumar et al., 2012). Every day, even though the pharmaceutical industries are producing a number of new antimicrobial drugs, the microorganisms have developed resistance against these drugs. It is because the bacteria have the genetic ability to transmit and acquire resistance to drugs used as therapeutic agents (Nascimento et al., 2000). On the contrary, the revival of interest in plant-derived drugs is mainly due to the current widespread belief that "green medicines" are safe and more dependable than the costly synthetic drugs that have adverse side-effects (Nair and Chanda, 2007). India, one of the developing countries, has a gift of nature that the very rich botanical wealth and a large number of diverse types of medicinal plants grow wildly in every corner of country. Hence, different forms of botanicals such as essential oils (EOs) and plant extracts have been used to cure specific ailments from ancient times (Bhattacharjee, 1998; Kumar et al., 2012).

According to the World Health Organization (WHO) report, about 80% of the world's population relies on traditional medicine, and the majority of the traditional therapies involves the use of plant extracts or their active constituents (WHO, 1993). Further, in another report, many species of mint have already been exploited by our forefathers, and peppermint itself has been used for 250 years ago (Hornok, 1992). The genus *Mentha* of family Lamiaceae is composed of 19 geographically widespread species and 13 named hybrids (Chambers and Hummer, 1994). *M. piperita* commonly known as peppermint is a nonnative herbaceous and perennial plant; the plant height reaches up to 80–100 cm (40 inches) with four-sided stem. The leaves are stalked opposite and toothed. The flowering time is from August to October; flowers are hermaphrodite (have both male and female organs) and pollinated by insects. The flowers are irregular in shape; they are pinkish or purplish (Kirtikar and Basu, 1972). Leaves contains about 0.5–6% volatile oil commonly known as peppermint oil which is composed of 50–78% active constituents dominated by monoterpenes, mainly menthol, menthone and their derivatives, e.g., isomenthone, neomenthone, acetylmenthol, pulegone, and menthofuran. Majority of them are widely in use as conventional medicine for antispasmodic, carminative, refrigerant, stimulant and diuretic and antiseptic purposes (Edris et al., 2003); their essential oils are using in chewing gums, alcoholic beverages, cosmetics, perfumes, toothpastes and mouthwashes (Baytop, 1984). It is one of the major constituent. Menthol is used in medicine for stomach disorders and in ointments for headache, whereas infusion of leaves is used in indigestion and rheumatic pains (Nair and Chanda, 2007). Besides, the plants are also in use in salads, spice, and mint herbage for wool dyeing (Leung and Foster, 2003).

Its active constituent menthol is used in medicine for stomach disorders and in ointments for headache. The infusions of leaves are used in indigestion and rheumatic pains. *Mentha piperita* oil, particularly as an inhalant, relieves nausea and respiratory problems and aids digestion. Digestive, carminative, respiratory, anti-inflammatory, cooling and muscle relaxant are peppermint oil properties. The Indian Materia Medica also proved that the various infusions of leaves of *M. piperita* are used in cases of vomiting, gastric colic, cholera, diarrhea, flatulence, weak digestion, hiccup, and palpitation of the heart (Nadkarni, 1976). Since the plant *M. piperita* is well known in India, it was selected for detailed in vitro antibacterial activity against *Escherichia coli, Proteus vulgaris, Pseudomonas aeruginosa, Klebsiella pneumoniae,* and *Salmonella typhi,* the bacterial pathogens causing severe illnesses/diseases in human beings.

22.2   MATERIALS AND METHODS

22.2.1   THE ESSENTIAL OIL

The plant material (fresh leaves) was collected from the Pratapgarh district of Uttar Pradesh, India. The essential oil was extracted from the leaves of *Mentha piperita* through the hydrodistillation method using Clevenger's apparatus (Clevenger, 1928). The extraction of oil was carried out continuously on a heating mantle at the temperature of 30°C–50°C until no further oil was extracted. The excess water content of the oil sample was removed through anhydrous sodium sulfate (Na_2SO_4), and after filtration, it was stored in a dark tightly closed bottle at +4°C for further investigations. The yield of the obtained essential oil was about 0.63% based on the dry weight of plant leaves.

22.2.2 GAS CHROMATOGRAPHY (GC)

The gas chromatography (GC) analysis of the oil was carried out by Per-kin-Elmer Auto XLGC and a Nucon gas chromatograph model 5765, both equipped with an FID using two different stationary phases, PE-5 (60 m × 0.25 mm; 0.25 μm film coating) and BP-20 (coated with a Carbowax 20 M, 30 m × 0.32 mm × 0.25 μm film thickness), fused silica columns, respectively. Hydrogen was the carrier gas at 1.0 mL/min. The column temperature programming was from 70–250°C at 3°C/min (for PE-5) and from 70°C to 230°C at 4°C/min (for BP-20). The injector and detector temperatures were 200 and 230°C on BP-20 and 220 and 300°C on PE-5 column, respectively. The injection volume of the sample was 0.02 μL neat and the split ratio was 1:30.

22.2.3 GAS CHROMATOGRAPHY/MASS SPECTROMETRY (GC/MS) ANALYSIS

The gas chromatography-mass spectrometry (GC–MS) analysis of the oil was performed on a Perkin-Elmer Turbomass Quadrupole mass spectrom-eter fitted with an Equity-5 (Perkin-Elmer) fused silica capillary column (60 m × 0.32 mm; 0.25 μm film coating). The column temperature was pro-grammed 70°C, initial hold time of 2 min, to 250°C at 3°C/min with a final hold time of 3 min; the analysis was carried out using helium as the carrier gas at a flow rate of 1 mL/min. The injector and source temperatures were 250°C. The injection volume was 0.06 μL neat with a split ratio 1:30. The MS readings were taken at 70 eV with an EI source with mass range of m/z 40–400. The separated components were identified tentatively by match-ing with EI-MS results of National Institute of Standards and Technology (NIST and WILEY), 8th edition and NBS mass spectral library data. The quantitative determination was carried out based on peak area integration.

22.2.4 PATHOGENIC BACTERIAL CULTURES

The specific bacterial cultures *Escherichia coli* (ATCC 25922), *Proteus vulgaris* (MTCC 1771), *Pseudomonas aeruginosa* (MTCC 103), *Kleb-*

*siella pneumoniae (*MTCC 109), and *Salmonella typhi* (MTCC 733) procured from the Microbial Type Culture Collection (MTCC), Chandigarh, India, and the American Type Culture Collection (ATCC), Bangalore (local center in India), were used in this study. The fresh culture of each test organisms was periodically maintained on Muller Hinton agar (MHA) slants and kept at $35 \pm 2°C$ in a BOD incubator for further analysis.

22.2.5 IN VITRO ANTIBACTERIAL INVESTIGATION

22.2.5.1 Preparation and Standardization of Inoculum

Standardized inoculum of each bacterium, i.e., 1×10^7 cells CFU/mL (colony forming units) with 0.5 McFarland standard was taken for antimicrobial assay.

22.2.5.2 Disc Diffusion Method (Zone of Inhibition)

For screening of antibacterial properties of essential oil of *M. piperita*, the disc diffusion method was used (Bauer et al., 1966). Sterilized Petri plates were pre-seeded with 15 mL of Muller Hinton Agar medium and 0.1 mL of bacterial suspension containing 1×10^7 CFU mL^{-1} inoculum was uniformly spread over the media to form a lawn of the culture. The stock solution of various concentrations (100 mg, 50 mg, and 25 mg) of 50% essential oil was prepared in dimethyl sulfoxide solvent (DMSO). Sterile Whatman paper discs (10 mm diameter) were soaked in these concentrations in such a manner so that ultimate amount in each disc was 100 μL, 50 μL, and 25 μL, respectively. After 5 minutes of proper soaking, the discs were place over the bacterial lawn and incubated for 24 h at $35 \pm 2°C$. In set of controls, ampicillin 10 μg (Hi-Media disc) was used as negative control, while discs soaked in sterile distilled water were placed on the lawn as a positive control. Further, the inhibited area called as "zone of inhibition" of growth was measured in mm (millimeter) and compared with the standard reference antibiotics. The zone of inhibition was calculated using the following formula:

$$W = \frac{T - D}{2}$$

where: W – diameter of clear zone of inhibition; T – total diameter of including disc and clear zone; D – diameter of the test Disc.

All measurements were calculated in mm.

22.2.5.3 Evaluation of Antibacterial Activity MIC (Minimum Inhibitory Concentration)

Further, the actual determination of antibacterial activity of the essential oil was investigated using broth micro-dilution method CLSI (NCCLS, 2003). Out of various stock solutions, 50 mg/mL was taken for measuring the minimum inhibitory concentration (MIC). The experiment was conducted in sterile 96-well microtiter plates with lids (SPL) using ELISA reader (Spectramax Plus[384], Molecular Devices Corporation, USA). The experiment was started by filling the column 4 to 11 with 180 μL MHB media and 20 μL of essential oil, so that final volume 200 μL can be made. Further, serial dilutions were made from column 4 to 11 and excess broth (100 μL) was discarded from column 11. Now, in each well (from column 4 to 11), 100 μL of inoculum was added. In this way, column 1 (which contains 200 μL media along with antibiotic and inoculum) served as negative control; column 2 (contains 200 μL sterile MHB media) served as media control; column 3 (contains 190 μL MHB media with 10 μL essential oil) served as drug control; and column 12 (which contains 200 μL inoculum but no essential oil) served as a positive control. Furthermore, mixtures of the wells were mixed thoroughly, and the microplates were incubated at 35 ± 2°C for 18–24 h. After the incubation, microbial growth inhibition was evaluated by measuring absorbance at 492 nm. Each experiment was repeated in triplicate. The MIC was defined as the lowest concentration of essential oil, showing no visible bacterial growth after incubation time. However, bactericidal concentration (MBC) was the lowest inhibitory concentration at which the replicates failed to grow, even after re-inoculating on or in the fresh culture media, i.e., MHB/MHA/NA (Kumar et al., 2011, 2012).

22.2.5.4　Effect of Inoculum Density

The effect of inoculum density on MBC of the essential oil against the test pathogens was determined using the method of Shukla et al. (2013). Inoculum of each bacterium was prepared at CFU/mL of 1×10^3, 1×10^5, 1×10^7, and 1×10^9. The culture plates containing essential oil at their respective MBCs were prepared for the treatment set. However, in case of control set, sterilized water was used instead of the essential oil. Further, observations were recorded after the incubation period of 18–24 h.

22.2.5.5　Minimum Killing Time (MKT)

The MKT of the essential oil of *M. piperita* at the respective MICs and MBCs against the test pathogens were also determined using the microtiter plate assay. All the wells were filled with 100 µL culture media (MHB). Again, 80 µL more media was added to the wells of the 4th column. The plant essential oil (20 µL) was added to the wells of the 4th column and serially diluted up to the 11th column using a multichannel micropipette. Then, 100 µL of inoculum was added to all the wells, except the 1st, 2nd, and 3rd columns. The wells corresponding to the MICs, and MBCs were marked. Prior to incubation, the loop was touched over the wells containing MICs, and MBCs essential oil and drag over the MHA/NA plates, gently. The microtiter plate and the MHA/NA plates were kept for incubation. Further, after 5 min, 30 min, 1 h, 6 h, 12 h, 18 h, and 24 h, the same process was repeated. The MHA/NA plates were kept for incubation at $35 \pm 2°C$ up to 18–24 h. Observations were made after the incubation period (Shukla et al., 2013).

22.2.5.6　Broad Antimicrobial Spectrum

To test the range of spectrum of the oil of *M. piperita*, the minimum bactericidal concentration was subjected for investigation against 10 other bacterial pathogens, namely *Bacillus cereus* (MTCC 430), *Bacillus subtilis* (MTCC 441), *E. coli* (MTCC 723), *Vibrio cholerae* (MTCC 3906), *Salmonella enterica* (MTCC 3858), *Lactobacillus acidophilus* (MTCC

447), *Staphylococcus aureus* (MTCC 96), *Shigella flexneri* (MTCC 1457), *Yersinia enterocolitica* (MTCC 859),,and *Vibrio vulnificus* (MTCC 1145), which were available in the laboratory (Kumar et al., 2013; Shukla et al., 2013).

22.2.5.7   Thermostability and Shelf-Life

Effect of some physical factors, viz. temperature (40°C, 60°C, and 80°C) and time of storage on efficacy of the oil, at MBC, was also determined (Kumar et al., 2013; Shukla et al., 2013). Samples of the oil (1.0 mL) in small vials were exposed at 40°C, 60°C, and 80°C in hot water bath for 30 min separately; and their efficacy was observed against their respective bactericidal concentration. Moreover, the essential oil of *M. piperita* showed the efficacy, even after 3 years, against the same test pathogens.

22.2.5.8   Statistical Analysis

The data of zone of inhibition of essential oil obtained against different bacterial pathogens were statistically analyzed with three-way classification and tested by F-test where significance was evaluated at 5% level. Microsoft excel was used to analyze the standard deviation/error, etc.

22.3   RESULTS AND DISCUSSION

22.3.1   *CHEMICAL COMPOSITION OF ESSENTIAL OIL*

The present investigation was performed to determine the various active constituents and antibacterial activity of *M. piperita,* a traditional medicinal plant in India, against some common infectious bacterial pathogens. The percentage of yield of obtained essential oil from the leaves was 0.63%, though the plant has strong aromatic flavor. Its GC/MS analysis showed that the essential oil has 56 constituents, where menthol (37.20%) along with menthol stereoisomer's (+ neomenthol and + isomenthol) and menthone (22.52%) were the major constituents including the isomenthone (4.70%), 1,8-cineole (4.68%), and methyl acetate (4.18% and Pulegone (3.70%) and Limonene (0.8%) as the minor constituents. The other monoterpenes such as β-caryophyllene, piperitone, pinene, eugenol, carvone, linalool, α-phellendrene, ρ-menthane cadinene, and dipentene were observed in trace amounts.

22.3.2 ANTIBACTERIAL ACTIVITY

Later the antibacterial screening was made by using disc diffusion technique, and it was found that the essential oil possessed high antibacterial activity. The higher concentration 100 mg/mL exhibited the maximum zone of inhibition followed by 50 mg/mL and 25 mg/mL conc. Among all the tested bacterial pathogens, the maximum zone of inhibition was recorded against the *S. typhi* (17 ± 0.1 mm); however, the least susceptible was *P. vulgaris* (8 ± 0.12 mm). From the zone of inhibition results, it was found that the efficacy of peppermint oil increased with increasing concentration. On the other hand, *E. coli* (13 ± 0.2 mm) and *K. pneumoniae* (12 ± 0.32 mm) showed more or less very similar inhibition with respect to various concentrations. Further, the standard ampicillin (antibiotic) was compared against the essential oil; it was showed the largest zone of inhibition (31 ± 1.24). There was no growth of bacteria in the positive control (Table 22.1). The same values are also indicated in the form of graph in Figure 22.1.

TABLE 22.1 Zone of Inhibition of Peppermint Oil Against Bacterial Pathogens

Extract Concentration (μg/mL)	Bacterial Pathogens/Diameter of Zone of Inhibition (mm±S.E.)				
	E. coil	*P. vulgaris*	*S. typhi*	*K. pneumoniae*	*P. aeruginosa*
25	3±2.3	1±1.2	6±0.7	2±0.22	1±0.14
50	8±0.1	3±0.5	13±0.31	7±0.02	4±0.56
100	13±0.2	8±0.12	17±0.1	12±0.32	10±1.2
Standard (Ampicillin)	26±0.72	27±2.5	37±0.1	27±3.6	31±1.24
Control (Only extract)	NG	NG	NG	NG	NG

mm – millimeter; S.E. – Standard Error; NG – No growth.

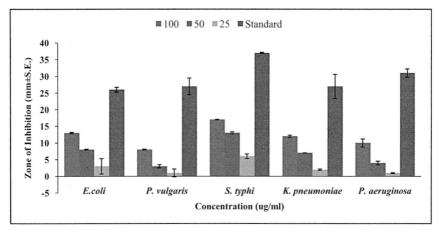

FIGURE 22.1 **(See color insert.)** Effect of essential oil showing zone of inhibition against bacterial pathogens.

22.3.3 DETERMINATION OF MINIMUM INHIBITORY CONCENTRATION (MIC) AND MINIMUM BACTERICIDAL CONCENTRATION (MBC)

The plant essential oils of *M. piperita* results showed a wide variation in the antibacterial properties. The validation of inhibitory action, which was recorded in zones of inhibition at higher concentration in the disc diffusion method, was done through the broth microdilution method. Among all the tested bacterial pathogens such as *Escherichia coli* (ATCC 25922), *Proteus vulgaris* (MTCC 1771), *Pseudomonas aeruginosa* (MTCC 103), *Klebsiella pneumoniae* (MTCC 109), and *Salmonella typhi* (MTCC 733), the most potent inhibitory activity of peppermint oil was found for *S. typhi* with the MIC of 0.30 mg/mL as shown in Table 22.2. It was least effective for *P. vulgaris*; however, for *E. coli* and *K. pneumoniae*, the efficacy was almost similar, which validated the observation of the disc diffusion method. The antibacterial activity was found to progressively increase with increasing concentration of essential oil. The lowest effective dose of the oil was observed at 0.3125 mg/mL. Based on these various concentrations of the essential oil, the minimum inhibitory concentration (MIC) of the oil was recorded as 0.61 mg/mL, 1.19 mg/mL, 0.30 mg/mL, 0.69 mg/mL, and 1.09 mg/mL, respectively (Table 22.2). Besides this, IC_{50} (50%

inhibitory concentration) values of the oil against the test pathogens was also recorded as 0.28 against *Escherichia coli*; 0.54 against *Proteus vulgaris*; 0.14 against *Salmonella typhi*; 0.31 against *Klebsiella pneumoniae*, and 0.58 against *Pseudomonas aeruginosa* (Table 22.2). Further, their average percent (%) growth inhibition was also recorded in Figure 22.2.

22.3.4 BROAD-SPECTRUM PROPERTIES OF ESSENTIAL OIL

Furthermore, the oil's toxicity was also determined against heavy inoculum density as well as other different gram-negative and gram-positive

TABLE 22.2 Antimicrobial Activity of Essential Oil of *M. piperita* (Peppermint Oil)

S. No.	Name of Water Borne Bacterial Pathogens	Antimicrobial Activity		
		IC_{50}	MIC	MBC
1.	*Escherichia coli*	0.28	0.61	0.625
2.	*Proteus vulgaris*	0.54	1.19	1.25
3.	*Salmonella typhi*	0.14	0.30	0.325
4.	*Klebsiella pneumoniae*	0.31	0.69	0.625
5.	*Pseudomonas aeruginosa*	0.58	1.09	1.25

IC_{50} = 50% inhibitory conc.; MIC = minimum inhibitory concentration.

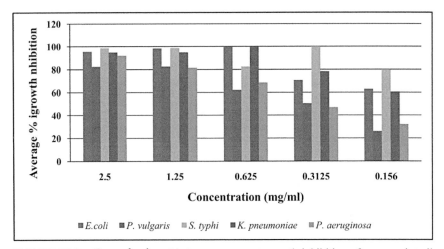

FIGURE 22.2 (**See color insert.**) Average percent growth inhibition of peppermint oil against bacterial pathogens.

bacterial strains. The peppermint oil showed quick killing activity, thermostability, and long shelf-life (Tables 22.3 and 22.4).

The active principles of the medicinal plants are divided chemically into a number of groups, among which are alkaloids, volatile essential oils, phenols and phenolic glycosides, resins, oleosins, steroids, tannins, and terpenes (Habtmaiam et al., 1993). Kazemi et al. (2012) reported that the main components in *M. piperita* oil were menthol, limonene, 1,8-cineole, sabinene, menthyl acetate, and menthone and in *M. spicata* oil were carvone, menthol, limonene, and menthone, which proved the finding of present study. Antibacterial and antifungal activity of these oils and their components were assayed against a variety of human pathogenic bacteria and fungi. In every traditional system of medicine, the peppermint and its oil have been used as an antispasmodic, aromatic, and antiseptic and in the

TABLE 22.3 Detailed In Vitro Investigation of the Peppermint Oil against the Tested Bacterial Pathogens

Properties studied	Tested Bacterial Pathogens				
	E. coil	*P. vulgaris*	*S. typhi*	*K. pneumoniae*	*P. aeruginosa*
Minimum Inhibitory Concentration					
MBC	0.625 mg/mL	1.25 mg/mL	0.3125 mg/mL	0.625 mg/mL	1.25 mg/mL
Minimum Killing Time (MKT)					
Pure oil	35 sec	55 sec	20 sec	40 sec	1 min
MBC	4:00h	5:00h	3:25h	4:25h	5:30h
Inoculum density (CFU/mL)					
$1 \times 10^3 - 1 \times 10^5$	No Growth	No Growth	No Growth	No Growth	No Growth
$1 \times 10^7 - 1 \times 10^9$	No Growth	No Growth	No Growth	No Growth	No Growth
Thermostability (°C)					
50°C–100°C	Effective	Effective	Effective	Effective	Effective
Effect of Storage (in months)					
12 months	+++	+++	+++	+++	+++
24 months	+++	+++	+++	+++	+++
36 months	+++	++	+++	++	++

TABLE 22.4   Toxicological Spectrum of Peppermint Oil against Other Common Bacterial Pathogens

S. No.	Names of Water Borne Bacterial Pathogens	Lethal concentration (1.25 mg/mL)	Hyper lethal concentration (3.2 mg/mL)
1.	*Bacillus cereus*	100^C	100^C
2.	*Bacillus subtilis*	100^C	100^C
3.	*E.coli (MTCC 723)*	100^S	100^C
4.	*Vibrio cholerae*	100^S	100^C
5.	*Salmonella enterica*	100^C	100^C
6.	*Lactobacillus acidophilus*	100^S	100^C
7.	*Staphylococcus aureus*	100^C	100^C
8.	*Shigella flexneri*	100^S	100^C
9.	*Yersinia enterocolitica*	100^S	100^S
10	*Vibrio vulnificus*	100^S	100^C

c – cidal concentration; s – static concentration.

treatment of cancers, colds, cramps, indigestion, nausea, sore throat, and toothaches (Briggs, 1993). In the in vitro experiment, it was seen that the peppermint oil possesses antibacterial activity; moreover, its different commercial preparations exhibit various activities (Lis-Balchin et al., 1997). Therefore, the plants are important source of potentially useful structures for the development of new chemotherapeutic agents. Many reports are available on the antiviral, antibacterial, antifungal, anthelmintic, antimolluscal and anti-inflammatory properties of the plants (Stepanovic et al., 2003). The MIC values observed in the present study showed the correlation with the studies of Fabio et al. (2007), where the MIC values of peppermint oil was reported to range between 1.25 µL/mL (*E. coli*) and 3.125 µL/mL (*K. pneumoniae*); the present study results are also in accordance with finding of Sartoratto et al. (2004), who reported that gram-positive organisms were inhibited by peppermint oil at very low concentrations. In another study, it was seen that *C. albicans* was the most sensitive microorganism (MIC=0.312 mg/mL) to peppermint oil, followed by *S. epidermidis* (MIC=0.625 mg/mL), *B. cereus*, *S. typhimurium*, *E. aerogenes*, *S. aureus*, and *E. coli* (MIC = 1.25 mg/mL). However, *P. aeru-*

ginosa with an MIC value of 5 mg/mL was more resistant that others (Iscan et al., 2002). So, in comparison with experiment of Iscan et al. (2002), the efficacy of oil was found to be more potent in the present study as mentioned in Table 22.3. The peppermint oil is non-toxic and non-irritant in low dilutions, but sensitization may be a problem due to the menthol content, and due to this, it can cause irritation to the skin and mucus membranes and should be kept well away from the eyes. It should also be avoided during pregnancy and should not be used in children below 7 years (German Commission E Monographs (Phytotherapy), 1990). Some reports on peppermint oil also suggested that, in any form, the oil is not recommended for those with hiatal hernia, gallbladder disease, or while pregnant or nursing. Its overdose symptoms are slow breathing, rapid breathing, abdominal pain, diarrhea, nausea, vomiting, blood in urine, no urine production, convulsions, depression, dizziness, twitching, unconsciousness, uncoordinated movement, and flushing (http://www.drugs.com/enc/peppermint-oil-overdose). Although the essential oils have variable composition of components, some of which are actually toxic to human, insects and plants if ingested in large quantities; even though, a large number of active principle responsible for such activities are used in the development of drugs for the therapeutic use in human being (Gubaron, 2000; Talebi et al., 2006). But the difference between antibacterial activities of essential oils may be related to the concentration and nature of contents, the respective composition, the functional groups, the structural configuration of the essential oil against bacterial pathogens, their possible synergistic interaction, and ecological/plant growth factors (Chang et al., 2001). Besides, seasonal variation, especially harvest time, also affects biological activities of oils (Kizil, 2010; Mimica-Dukic et al., 2003). The hydrophobicity, an important characteristic of essential oils which enables them to associate in the lipids of the bacterial cell membrane and mitochondria, disturbing the structures and rendering them more permeable leading to leakage of ions and other cell contents which decreased the intracellular ATP pool of bacterial or fungal cell and also increased extracellular ATP, indicating disruptive action on the cytoplasmic membrane (Kumar et al., 2012; Sikkema et al., 1994). The differences in MIC values of bacteria may be related to differential susceptibility of the bacterial cell wall, which is the functional barrier, to minor differences present in the outer membrane in the cell wall composition (Zhao et al., 2001). According to Jeyakumar et al. (2011), the diffusion of oil was found to be less as compared with the broth microdilution method; this is due to the insufficient penetration of oil in the agar medium, and hence, concentration of oil was required to inhibit the growth of pathogenic bacteria in the agar well/disc diffusion method. Moreover, the phylogenetic variations also play an important role in the differences between the MIC of the oil (Kumar et al., 2011, 2012). A previous study showed the toxic effect of garlic oil was attributed to sulfur and allicin components, which had ability to react with SH groups of enzymes and change their properties (Mitchell, 1980). Literature also revealed that gram-negative bacteria are more resistant to the essential oils present in plants (Kumar et al., 2011; Smith-Palmer et al., 1998). The region behind that the cell wall of gram-negative bacteria essentially contains lipopolysaccharides (LPS), which creates a barrier to accumulate the oil on their cell surfaces. Further, in another study, the active constituent of the oil, menthol, has bactericidal effects and a spicy odor (Hornok, 1992). The antimicrobial activity of peppermint oil is due to the presence of terpenoides menthol, 1-8 cineole, methyl acetate, menthofuran, isomenthone, limonene, β-pinene, germacerene-d, trans-sabinene hydrate, and pulegone (Sartoratto, 2004). Furthermore, some essential oils with the same common name may be derived from different plant species. Secondly, the method used to assess antimicrobial activity, strains, and the choice of the test organisms varies between different studies. Menthol has also been reported to have antimicrobial activity (Iscan et al., 2002). This may be because of phenolic nature of the menthol, and components that are phenolic in nature generally decreased the intracellular ATP pool of bacterial cell and also increased extracellular ATP, indicating disruptive action on the cytoplasmic membrane (Helander et al., 1998; Kumar et al., 2011, 2012). The interaction of peppermint oil and menthol with the antibiotics was studied on the same bacterial strain with the checkerboard method, and the antiplasmid activity of peppermint oil and its main constituent menthol was observed; this implies that menthol-containing substances are potential agents that could eliminate the resistance plasmids of bacteria. The compound preferentially kills the plasmid-containing bacteria due to their increased sensitivity to menthol (Shrivastava, 2009). This was hypothesized through some research data that indicated irritation and toxicity caused by eugenol, menthol, and thymol; hence, the cytotoxicity study of these compounds suggested that gum irritation may be related to membrane lysis and surface activity and that tissue penetration may be related at least partly to membrane affinity and lipid solubility (Manabe et al., 1987). On the other hand, the experiment clearly picturized that, the cells appeared oblong; edges become abnormal, triangular, elongated and incapable of formation of septum in dividing. Cell wall disrupted and exhibited thickened in some parts and breakdown in other which leading to leakage of cytoplasmic materials. Therefore, these variations in the activity between different organisms were observed, and hence, the oil was highly effective against *S. typhi* but was least effective against *P. vulgaris*; however, for *E. coli* and *K. pneumoniae,* the bactericidal efficacy was almost similar at the same concentration (0.625 mg/mL).

22.4 CONCLUSIONS

The use of essential oils of herbs and spices are playing more important role in food and other beverages because of their specific taste. Likewise, the medicinal plant *Mentha piperita* was chosen for the present work, as this plant is traditionally known and has continuously been in practice of use as a therapeutic agent against a variety of diseases. The systematic studies conducted on peppermint oil indicated the immense potential against the infectious bacterial pathogens. Moreover, the present finding provides enough experimental data like heavy inoculum density, quick killing activity, broad antimicrobial spectrum, thermo stability, long-shelf life, etc., through both the techniques. Further in vivo experiments can help to explore the actual mechanism of action of peppermint oil. Therefore, it is recommended to check the toxicity at lower doses for medicinal formulation in the favor of human welfare.

ACKNOWLEDGMENTS

The author is thankful to the authorities of the Mizoram University, Aizawl, as well as the Biological Product Laboratory, Department of Botany, University of Allahabad, for providing the research facilities.

KEYWORDS

- **antibacterial**
- **CLSI**
- **GC-MS**
- **hydrodistillation**
- **thermostability**

REFERENCES

Bauer, A. W., Kirby, W. M. M., Sherris, J. C., & Turch, M., (1966). Antibiotic susceptibility testing by a standardized single disc method. *American Journal of Clinical Pathology, 45,* 494–496.

Baytop, T., (1984). *Therapy with Medicinal Plants in Turkey (Past and Present).* Istanbul University, pp. 520.

Bhattacharjee, S. K., (1998). *Handbook of Medicinal Plants.* Pointer Pub, Jaipur–03, India pp. 1–6.

Briggs, C., (1993). *Peppermint: Medicinal Herb and Flavouring Agent CPJ, 126,* pp. 89–92.

Chambers, H. L., & Hummer, K. E., (1994). *Chromosome Counts in the Mentha Collection at the USDAARS National Germplasm Repository.* Taxon., *43,* 423–432.

Chang, S. T., Chen, P. F., & Chang, S. C., (2001). Antibacterial activity of leaf essential oils and their constituents from *Cinnamomum osmophloeum. J. of Ethnopharma., 77,* 123–127.

Clevenger, J. F., (1928). Apparatus for the determination of volatile oil. *J. Am Pharm. Assoc., 17,* 346–346.

Edris, A. E., Shalaby, A. S., Fadel, H. M., & Abdel-Wahab, M. A., (2003). Evaluation of a chemotype of spearmint (*Mentha spicata* L.) grown in Siwa Oasis, Egypt. *European Food Research and Technology, 218,* 74–78.

El-Meleigy, M. A., Ahmed, M. E., Arafa, R. A., Ebrahim, N. A., & El-Kholany, E. E., (2010). Cytotoxicity of four essential oils on some human and bacterial cells. *Journal of Applied Sciences in Environmental Sanitation., V*(N), 134–150.

Fabio, A., Cermelli, C., Fabio, G., Nicoetti, P., & Quaglio, P., (2007). Screening of the antibacterial effects of a variety of essential oils on microorganisms responsible for respiratory infection. *J Phytother Res., 21,* 374–377.

Gubaron, (2000). Toxicity myths: the actual risks of essential oil use. *Perfumer and Flavorist. 25*(2), 10–28.

Habtmaiam, S., Gray, A. L., & Waterman, R. G., (1993). A new antibacterial sequterpene from *Premma oligotricha. J. of Natural Product, 5*(1), 140–43.

Helander, I. M., Alakomi, H. L., Kala, K. L., Sandholm, T. M., Pol, I., Smid, E. J., Gorris, L. G. M., & Wright, A. V., (1998). Characterization of the action of selected essential oil components on gram-negative bacteria. *J. Agric. Food Chem., 46*, 3590–3595.

Hornok, L., (1992). *The Cultivation of Medicinal Plants,* Cultivation and processing of medicinal plants. John Wiley & Sons, Chichester, pp. 187–196.

http://www.drugs.com/enc/peppermint-oil-overdose.html. As assessed on 25/08/08.

Iscan, G., Kirimer, N., Kurkcuoglu, M., Baser, K. H. C., & Demirci, F., (2002). Antimicrobial screening of *Mentha piperita* essential oils. *J. of Agricultural Food Chemistry, 50,* 3943–3946.

Jeyakumar, E., Lawrence, R., & Pal, T., (2011). Comparative evaluation in the efficacy of peppermint (*Mentha piperita*) oil with standards antibiotics against selected bacterial pathogens. *Asian Pacific Journal of Tropical Biomedicine,* 253–257.

Kazemi, M., Rostami, H., & Shafiei, S., (2012). Antibacterial and antifungal activity of some medicinal plants from Iran. *J. Plant Sci., 7*(2): 55–66.

Kirtikar, K. R., & Basu, B. D., (1972). *Indian Medicinal Plants International Book Distributors,* Allahabad, India, 2nd Ed., vol. I to IV.

Kizil, S., Hasimi, N., Tolan, V., Kilinc, E., & Yuksel, U., (2010). Mineral content, essential oil components and biological activity of two *mentha* species", (*M., piperita* L. *M., spicata* L.), *Turkish Journal of Field Crops*m., *15*(2), 148–153.

Kumar, A., (2011). Effect of selected botanical antimicrobials for enhancing the potability of drinking water from various sources. *PhD., Thesis,* Mizoram University, Aizawl, pp. 1–229.

Kumar, A., Gupta, R., Mishra, R. K., Shukla, A. C., & Dikshit, A., (2012). Pharmacophylogenetic investigation of *micromeria biflora* benth and *citrus reticulata* blanco. *Natl. Acad. Sci. Lett.,* DOI 10.1007/s40009-012-0029-7, 1–7.

Kumar, A., Mishra, R. K., Srivastava, S., Tiwari, A. K., Pandey, A., Shukla, A. C., & Dikshit, A., (2011). Role of phylogenetic analysis for anti-bacterial activity of essential oil of *Trachyspermum ammi* L., against water borne pathogens. *Advances in Environmental Biology, 5*(6), 1271–1278.

Kumar, A., Shukla, A. C., Mishra, R. K., Fanai, L., & Dikshit, A., (2013). Tradition to technology: an approach for the management of water borne bacterial pathogens. *Scie. Tech. J.,* Mizoram University, *1*(2), 48–56.

Leung, A. Y., & Foster, S., (2003). *Encyclopedia of Common Natural Ingredients Used in Food, Drugs and Cosmetics.* New York, John Wiley & Sons, pp. 649.

Lis-Balchin, M., Deans, S. G., & Hart, S., (1997). A study of the variability of commercial peppermint oils using antimicrobial and pharmacological parameters. *Med. Sci. Res., 25*, pp. 151–152.

List of German Commission E. Monographs (*Phytotherapy*), (1990). Peppermint oil (Menthae piperitae aetheroleum) Published March 13, 1986, Revised March 13, 1990, September 1, 1990, and July 14, 1993 available online link: http://www.heilpflanzenwelt.de/buecher/BGA-Commission-E-Monographs/index. htm.

Manabe, A., Nakayama, S., & Sakamoto, K., (1987). Effects of essential oils on erythrocytes and hepatocytes from rats and dipalitoyl phosphatidylcholine-liposomes. *Japanese Journal of Pharmacology, 44,* 77–84.

Mimica-Dukic, N., Bozin, B., Sokovic, M., Mihajlovic, B., & Matavulj, M., (2003). Antimicrobial and antioxidant activities of three *Mentha* species essential oils. *Planta Med., 69,* 413–419.

Mitchell, J. C., (1980). Contact sensitivity to garlic (Allium). *Contact Dermat, 6*(5), 356–357.

Nadkarni, K. M., (1976). *The Indian Materia Medica, Popular Prakashan,* Mumbai, India, vol. I and II, 1st (edn).

Nair, R., & Chanda, S. V., (2007). Antibacterial activities of some medicinal plants of the western region of India. *Turk J. Biol., 31,* 231–236.

Nascimento, G. G. F., Locatelli, J., Freitas, P. C., & Silva, G. L., (2000). Antibacterial activity of plant extracts and phytochemicals on antibiotic-resistant bacteria. *Braz. J. Microbiol., 31,* 247–256.

NCCLS, (2003). *National Committee for Clinical Laboratory Standards.* Methods for dilution antimicrobial susceptibility tests for bacteria that grow aerobically (5th Ed.). Approved standard, M7-A5. Pennsylvania: Wayne.

Sartoratto, A., Machado, M. A. L., Delarmelina, C., Figueira, G. M., Duarte, M. C. T., & Rehder, V. L. G., (2004). Composition and antimicrobial activity of essential oils from aromatic plants used in Brazil. *Braz. J. Microbiol., 35,* 275–280.

Shrivastava, A., (2009). A review on peppermint oil. Asi*an Journal of Pharmaceutical and Clinical Research, 2*(2), 27–33.

Shukla, A. C., Fanai, L., Kumar, A., & Dikshit, A., (2013). *In vitro* antidermatophytic activity of *Lantana camara* L., against *Trichophyton mentagrophytes* and *T., rubrum.* Int. *J. of Current Discovery and Innovations,* 86–91.

Sikkema, J., De Bont, J. A. M., & Poolman, B., (1994). Interactions of cyclic hydrocarbons with biological membranes. *Journal of Biological Chemistry., 269*(11), 8022–8028.

Smith-Palmer, A., Stewart, J., & Fyfe, L., (1998). Antimicrobial properties of plant essential oils and essences against five important food-borne pathogens. *Letters in Food Microbiology, 26,* 118–122.

Stepanovic, S., Ant, N., Dakic, I., & Savabic-blahobit, M., (2003). *In vitro* antimicrobial activity of *propilis* & antimicrobial drug. *Microbial Res., 158,* 353–357.

Talebi, S., Afshari, J. T., Rakhshandeh, H., Seifi, B., & Boskabadi, M. H., (2006). *In vitro* antiproliferative effect of fresh red garlic on human transitional cell carcinoma (TCC-5637) cell line. *International Journal of Agriculture and Biology, 8*(5), 609–614.

World Health Organization, (1993). Summary of WHO guidelines for the assessment of herbal medicines, *Herbal Gram,* pp. 13, 28.

Zhao, W. H., Hu, Z. O., Okubo, S., Hara, Y., & Shimamura, T., (2001). Mechanism of synergy between *Epigalloca Techingallate* and *Lactams* against Methicillin-Resistant *Staphylococcus aureus. Antimicrob Agents Chemother, 45,* 1737–1742.

CHAPTER 23

UNDEREXPLOITED VEGETABLES IN NORTH EASTERN INDIA: A GATEWAY TO FOOD SECURITY

B. LALRAMHLIMI,[1] AKOIJAM RANJITA DEVI,[2] and KHAIDEM NIRJA[3]

[1]Department of Vegetable Crops, [2]Department of Spices and Plantation Crops, Faculty of Horticulture, [3]Department of Agricultural Extension, Faculty of Agriculture, Bidhan Chandra Krishi Viswavidyalaya, Mohanpur–741252, Nadia, West Bengal, India, E-mail: hlimhlimi371@gmail.com

CONTENTS

ABSTRACT

Northeastern states of India are very diverse climatically, agro-ecologically, and ethnically. About 50% of wild edible plants wealth is located in this region, which is also one of the richest reservoirs of different underexploited vegetable crop species (Arora, 1997). Underexploited vegetables play a crucial role in poor people's livelihood and may have a significant potential for commercialization in this region. Moreover, they also possess several desired medicinal properties. Some underexploited vegetables in northeastern regions are rice bean (*Vigna umbellate*) and tree bean (*Parkia javanica* Merr. Syn. *P. Roxburghii*), Manipur Loosestrife/ Kengoi (*Lysimachia obovata*) in Manipur, Chingit/Indian Pepper (*Zanthoxylum rhetsa*) and Anhling/Black nightshade (*Solanum americanum*) in Mizoram, Indian pennywort (*Centella asiatica*), *Zanthoxylum armatum* in Arunachal Pradesh, Chichiri (*Monochoria hastate*) in Tripura, and East Indian Glory Bower (*Clerodendrum colebrookianum*) in Sikkim. Most of them are very rich sources of vitamins, minerals, and other nutrients such as carbohydrates, proteins, fats and phytochemicals that have anticancer and anti-inflammatory properties, which confer many health benefits. They have the potential to contribute to food security, nutrition, health, income generation, and environmental services as they are adapted to marginal soil and diverse climatic condition. The high nutritional qualities indicate that the cultivation and consumption of these crops may be helpful in overcoming the nutritional deficiencies predominant in many rural areas of this region and boost the socioeconomic conditions. Owing to various human activities, there is depletion of this biodiversity. The topic has been taken up for harnessing diversity of these vegetables that have enormous potential with much diversity for exploitation, their conservation, and food security.

23.1 INTRODUCTION

The northeastern (NE) region of India comprising eight states, namely Arunachal Pradesh, Assam, Manipur, Meghalaya, Mizoram, Nagaland, Tripura, and Sikkim, has vast physiographical variations, which have been represented in six agro-climatic zones. The NE region is one of the richest reservoir of genetic variability and diversity of different vegetable crops (Asati and Yadav, 2004). The northeast India is a part of the Himalaya and the Indo-Burma biodiversity hotspots in the world (Moa et al., 2008). Many local vegetables that have the potential to contribute to food security are produced in this area, and their knowledge on distribution of their genetic diversity and use patterns is still largely limited. It is estimated that half of India's floristic diversity of higher plants of about 7150 species out of a total of 15000 species is concentrated in this region. About 50% of wild edible plant wealth is located in this region, which is also one of the richest reservoirs of different underexploited vegetable crop species (Arora, 1997). With increasing population and consequent shortage of food grains and vegetables, the collection and utilization of various types of unutilized crops are considered very essential. The high nutritional qualities indicate that the cultivation and consumption of these crops may be helpful in overcoming the nutritional deficiencies predominant in many rural areas of the country and boost the socioeconomic condition of the society (Chitta Ranjan et al., 2013). The causes of malnutrition and hunger in our country are not the scarcity of food but an inability to access the available food due to poverty and negligence of important local vegetables. Underexploited vegetables play a crucial role in poor people's livelihood in the rural areas as improved varieties of different vegetables do not reach these regions and contribute significantly in the food of rural masses of the people; these vegetables have a significant potential for commercialization in this region. Moreover, they also possess several desirable medicinal properties to curb diseases and malnutrition as they are rich sources of vitamins, minerals, and other nutrients such as carbohydrates, proteins, and fats. These crops require proper studies to know their values and unknown potentials as a means of improving livelihoods, especially in support of the rural poor vegetable farmers of these regions. They have an unimaginable potential to contribute to food security, nutrition, health, income generation, and environmental services as they are adapted to marginal soil and adverse climatic condition. However, the availability of most of these wild crops is now depleting rapidly owing to various factors such as "*Jhum*/shifting cultivation," forest fire, construction of industries, felling of trees for timber and other socio-economic anthropogenic activities in the area. Therefore, it is extremely necessary to tap their potentials and conserve these regions wealth and work out the ways for commercial-scale propagation of these plant species. The lack of improved varieties, information on unutilized crops and unavailability of capital for farmers makes the commercial-scale production difficult. The farmers' security is as important as food security as they are the sole runner in the food production. Chief Minister's Uzhavar Patthukappu Thittam 2011, has been implemented in Tamil Nadu for the security and welfare of the farmers.

23.2   IMPORTANCE FOR PROMOTION OF UNDERUTILIZED VEGETABLES

The country's population is increasing at the rate of 1.548%, and it is projected that it will be around 1256 million up to 2015 and 1331 million in 2020 (Pandey et al., 2014). The food requirement increases day by day. Although India is the second largest producer of vegetables in the world, the available vegetables are not sufficient to meet the requirement of the people especially in rural areas. A large mass of people depend on local vegetables for their livelihood to meet their daily vegetable requirement to sustain life. Some underexploited tuber crops in the NE regions, like Colocasia, Alocasia, Yams, Cassava, and Sweet potato contributes to carbohydrate requirement of many people in this region. The state of Nagaland is a remote tribal state in NE India and is very rich in floristic diversity. In Nagaland, the angiosperm is represented by over 2500 species belonging to 963 genera and 186 families (Deb and Imchen, 2008). Apart from their use as source of food, some are important due to their medicinal properties, vegetables, fibers, construction materials, dyes, etc., (Chitta Ranjan et al., 2013). These plants/parts contribute significantly in the food of rural masses of Nagaland. Fern (*Diplazium esculentum*) is grown wild in this region and is eaten by the local tribals. Indian pennywort (*Centella asiatica*) is an important unutilized leafy vegetable that has innumerable medicinal properties. *Colocasia esculenta*/Taro is a tuber crop grown in tropical areas as a vegetable food for its edible corm and secondarily as a leaf vegetable. The local tribals of Arunachal Pradesh grow a vegetable having red tomato-like fruits that are slightly bitter in taste and belong to the genus *Solanum* (Rai et al., 2004). Bamboo shoot (*Bambusa tulda*) is used to make fermented bamboo shoot which is eaten and cooked with other vegetables and meat. Amaranth (*Amaranthus caudatus, A. viridis, A. lividus, A. retroflexus*, and *A. Spinosus*) which has high vitamins and mineral content is eaten as boiled stuff. Drumstick (*Moringa oleifera*) is a highly nutritious plant whose whole parts are edible. In Manipur, another kind of brinjal, having round fruit and intermediate in appearance between tomato and brinjal, is grown. In the hilly areas, tree tomato (*Cyphomandra betacca*), a perennial shrub producing red tomato-like vegetables, is also grown. Tree tomato is consumed as delicious chutney when raw or after roasting and peeling off the skin (Rai et al., 2004). *Euryale ferox* grown in ponds and other water bodies of Madhubani, Bihar, is considered one of the most viable sources of income of the local people (Pandey et al., 2014). It is also an important source of income of local people in Manipur. Bathua/Jimilsag (*Chenopodium album*) is eaten as boil leafy vegetable and highly nutritious. Rice bean (*Vigna umbellate)* and Manipur Loose-strife /Kengoi (*Lysimachia obovata*) are also very nutritious and important vegetable. Tree bean (*Parkia roxburghii* G. Don.) is one of the most common of multipurpose tree species in the Manipur and Mizoram (Kumar et al., 2002). In Mizoram, Chingit/Indian Pepper (*Zanthoxylum rhetsa*) and Anhling/Black nightshade (*Solanum americanum*) are important local leafy vegetable with many unknown medicinal properties. Sword bean (*Canavalia ensiformis*) and Jack bean (*Canavalia gladiata*) are eaten as salad and cooked with other vegetables. Bitter brinjal (*Solanum torvum)* is sold in the market of Mizoram and is eaten as vegetable (Asati and Yadav, 2004). Climbing wattle/Khanghu/Biswal/Agla bel (*Acacia pennata*), which has a stinky leaf, is a delicious leafy vegetable in its season and can be eaten as fried or boiled. A wide diversity of Rosella (*Hibiscus sabdariffa)* is present in this region, which is eaten as boiled stuff. Toothache plant/Paracress (*Spilanthes acmella*) is marketed by local tribals as a vegetable. Snake gourd (*Trichosanthes cucumerina*) is also an important crop for this region. One of the interesting species of Vigna namely *V. vexillata* is grown by the tribals of Tripura. It is a legume cum tuber crop with much variation in edible tubers (Arora and Pandey, 1996). Chichiri (*Monochoria hastate*) in Tripura is also an important underexploited aquatic vegetable whose whole part is edible. The wild species *Cucumis hardwickii*, the likely progenitor of cultivated cucumber, is found growing in natural habitats in the foothills of Himalayas and NE region, particularly Meghalaya (Asati and Yadav, 2004). Sophlong (*Moghania vestita*) is sold in the market and eaten as raw with salt. Chow-Chow (*Sechium edule*), a native of tropical America, is a very popular vegetable in the region. Commonly called as squash, a wide diversity of this vegetable grows abundantly without much care and attention. Chow-Chow produces large starchy edible roots in addition to fruits (Rai et al., 2004). In Sikkim and many parts of NE regions, East Indian Glory Bower (*Clerodendrum colebrookianum*) is grown wild without proper care but is an important crop in this region. Rhubarb (*Rheum nobile*) native of Himalayas, Sikkim, Nepal is an important vegetable occurring in the alpine zone at 4000–4800 m altitude is also a rare and important crop in India (Anonymous d., 2015). In **Assam** and many other regions, spine gourd (*Momordica dioica*) is a very important underexploited vegetable. Kakrol (*Momordica cochinchinesis*) and kartoli (*M. dioica*) are widely spread in Assam, the Garo hills of Meghalaya (Ram et al., 2002). Many more underexploited vegetables are in these regions that require deeper studies to know their importance, their potentials to contribute to food security, and their immediate need for promotions and improvements.

23.3 POTENTIAL CONTRIBUTION OF UNDEREXPLOITED CROPS

23.3.1 FOOD SECURITY

Food security exists when all people, at all times, have physical and economic access to sufficient, safe and nutritious food that meets their dietary needs and food preferences for an active and healthy life (World Food Summit, 1996). Later definitions added demand and access issues to the definition. The four pillars of food security, namely *food availability, food access, food utilization, and food stability*, were stated later (World Food Summit, 2009). Food availability decreases each year as human population increases and the cultivated area decreases, collection and utilization of various types of unutilized crops are very much important. Most of them are very rich sources of vitamins, minerals and other nutrients such as carbohydrates, proteins and fats (Rai et al., 2005). In our country, where problem of malnutrition is prevailing in general and micronutrient malnutrition in particular, addressing the household nutritional security is indispensable. A recent study indicates that intake of micronutrients in daily diet is far from satisfactory and largely less than 50% recommended dietary allowance (RDA) is consumed by over 70% of Indian population (Pandey et al., 2014). Northeast states comprising eight states, namely Arunachal Pradesh, Assam, Manipur, Meghalaya, Mizoram, Nagaland, Sikkim, and Tripura, has a total population of 75,587,982 (Census of India, 2011). Vitamin A, iron, and zinc deficiency when combined constitute the second largest risk factor in the global burden of diseases; 330,000 child deaths occur every year in India due to vitamin A deficiency; 22,000 people, mainly pregnant women, die every year in India from severe anemia; 6.6 million children are born mentally impaired every year in India due to iodine deficiency; intellectual capacity is reduced by 15% across India due to iodine deficiency; and 200,000 babies are born every year with neural tube defects in India due to folic acid deficiency (Pandey et al., 2014). In northeast India, these vitamins and minerals deficiencies are very prevalent. Considering the above, it has become the need of the hour for local governments along with researchers, farmers, and nutritionists to search for cheap, reliable, and safe plant-based resources to overcome these deficiencies. Exploring underexploited vegetables could be of high significance for food security, meeting nutritional requirements, and agricultural development as well as an efficient means for crop alternates and thus can effectively contribute to overall improvement of a nation's economy.

23.3.2 NUTRITION

The RDA per capita per day recommended by the Indian Council of Medical Research (ICMR) for adult males includes 475 g cereals, 125 g green leafy vegetables, 80 g pulses, 100 g roots and tubers, 30 g fruits, and 75 g other vegetables. The adult females should have the same quantities of vegetables, roots and tubers and fruits as recommended for the adult males but less quantities of cereals (350 g) and pulses (70 g). Human nutrient requirements vary with sex, age, weight, height, and physical activity. The balanced diet should contain adequate energy source (calories) and nutrients like proteins, carbohydrates, fats, vitamins, minerals, and essential amino acids. The minimum amount of calories and nutrients required per capita per day for adult male and adult female, respectively, are listed in the following table.

Nutrients and calories	Adult male	Adult female
Calories	2800 kcal	2200 kcal
Protein	55 g	45 g
Calcium	400–500 mg	400–500 mg
Iron	20 mg	30 mg
Riboflavin	1.5 mg	1.2 mg
Carotene	3000 µg	3000 µg
Thiamine	1.4 mg	1.1 mg
Ascorbic acid	50 mg	50 mg
Folic acid	100 µg	100 µg
Vitamin B$_{12}$	1.0 µg	1.0 µg

Underutilized vegetables have immense potential for contribution to a particular pocket's of food production because they are well adapted to existing as well as adverse environmental conditions and are generally resistant to pests and pathogens. They are a cheap source of carbohydrates, proteins, fats, vitamins, and minerals. They are available locally and moreover have many nutritional and medicinal benefits. To supply the daily requirements of nutrients and minerals, these underutilized crops are an important option. According to Pandey et al. (2014), the free and total folic acid content of Amaranthus is 41.0 µg and 149.0 µg per 100 g edible portion, respectively. Colocasia also contains 16.0 µg free and 94.0 µg total folic acid per 100 g edible portion, and snake gourd contains 7.5 µg free and 15.5 µg total folic acid per 100 g edible portion. Cluster bean (guar) is also a rich source of folic acid with 50.0 µg free and 144.0 µg total per 100 g edible portion. Gupta et al. (2005) have given the mineral contents and trace element contents of different underutilized vegetables. *Trianthema portulacastrum* has mineral contents of ash (2.29 g), calcium (52.0 mg), phosphorus (22.0 mg), potassium (317.0 mg), sodium (16.0 mg), and magnesium (153.0 mg) per 100 g of edible portion. *Centella asiatica* has a high potassium content of 345.0 mg, ash (2.06 g), calcium (174.0 mg), phosphorus (17.0 mg), sodium (107.8 mg), and magnesium (87.0 mg). *Celosia argentea* has high potassium content of 476.0 mg, ash (2.65 g), calcium (188.0 mg), phosphorus (35.0 mg), sodium (240.6 mg), and magnesium (233.0 mg). *Boerhaavia diffusa* has ash (2.91 g), calcium (330.0 mg), phosphorus (27.0 mg), potassium (381.0 mg), and magnesium (167.0 mg). *Digera arvensis* has a high calcium content of 506.0 mg, ash (3.54 g), phosphorus (63.0 mg), potassium (604.0 mg), and magnesium (232 mg). The trace element contents of underutilized leafy greens such as *Trianthema portulacastrum* are iron (4.16 mg), zinc (0.46 mg), copper (0.12 mg), chromium (0.20 mg), and manganese (0.43 mg) per 100 g of edible portion. *Celosia argentea* also contains iron (13.15 mg), zinc (0.49 mg), copper (0.15 mg), chromium (0.153 mg), and manganese (0.27 mg) per 100 g of edible portion. *Boerhaavia diffusa* contains iron (7.83 mg), zinc (0.44 mg), copper (0.22 mg), and chromium (0.040 mg) per 100 g of edible portion. *Centella asiatica* has high iron content of 14.86 mg, zinc (0.97 mg), copper (0.24 mg), and chromium (0.046 mg) per 100 g of edible portion. These crops showed the mineral and nutrient contents of different underutilized crops, which are higher than most of the crops that are commercially grown. Winged bean has excellent nutritional qualities and is particularly very rich in protein (Rao and Dora, 2002). Jack bean seed has been promoted in developing nations as a potential source of affordable and abundant protein. It has 29.0% protein content (Adebowale and Lawal, 2004) (Table 23.1).

TABLE 23.1   Calories and Protein from Some Important Underexploited Vegetables

Vegetables	Calories (kcal per 100g fresh weight)	Protein (grams/ 100grams fresh weight)	Reference
Grain amaranth	371	14	Anonymous b. (2017)
Taro (cooked)	142	0.52	Anonymous a. (2012)
Lotus root	66	1.58	Anonymous c. (2017)
Rice bean	327	20.9	Rajerison, (2006)
Tree bean (mature pod)	426	18.8	Longvah, & Deosthale, (1998).
Fox nut	360	9.7	Shankar, M. (2010)
Sword bean	59	3	Ekanayake, Jansz, & Nair, (2000).

23.3.3 MEDICINAL PROPERTIES OF SOME IMPORTANT UNDERUTILIZED VEGETABLES

Besides being the source of food, these underexploited vegetables possess several desirable medicinal properties. They are a rich source of phytochemicals that have anti-cancer and anti-inflammatory properties which confer many health benefits. Fern (*Diplazium esculentum*) is used in the treatment of diabetes mellitus in an Aboriginal community in Lohit district of the eastern zone of Arunachal Pradesh and Himalaya. The extracts of *Diplazium esculentum* show anti-inflammatory activity and are also good for constipation (Hui et al., 2012). Indian pennywort (*Centella asiatica*) may help protect the brain from damage due to toxic metal exposure of aluminum. It may help protect against skin cancer caused by exposure to ultraviolet radiation. An extract of *Centella asiatica*, Asiaticoside, may help prevent liver cancer. *Oxalis corniculata* is known to cure dysentery, diarrhea, and skin diseases (Raghavendra et al., 2006). Taro (*Colocasia esculenta, Colocasia affinis*) leaf is eaten for fever and respiratory disorder. It has a medicinal use for leprosy and tuberculosis, earache, alopecia or hair loss and being styptic, it stops the flow of blood; this action helps to arrest arterial disorder (Anonymous a., 2012). The fruit of *Solanum kurzii*, one of the *Solanum* spp, is used as an anti-allergic agent by the Mao Naga tribe of Manipur. The fruit is crushed and applied to the allergic area of the body. This is very effective for any types of allergies. The seed is also edible and used by the tribals of the region as vegetables. The fruit is used for cough and worm infestation (Mao et al., 2009). *Solanum myriacanthum* is a perennial shrub that is used in the folk medicine of Tangkhul Naga tribe of India for treating intestinal worms (Yadav and Tangpu, 2012). *Solanum nigrum* is considered good for cooling hot inflammation, ringworms, ulcers, testicular swellings, gout and ear pain. The Arabs used the bruised fresh leaves to alleviate pain and reduce inflammation. *Solanum torvum* leaf juice is taken orally to reduce body pain (Muthu et al., 2006). According to Asati and Yadav (2004), there are around 12 cultivated species of *Solanum* in NE India, viz, *Solanum macrocarpon* (introduced in the NE region), *Solanum mammosum* (possibly introduced, ornamental with high solasodine percentage), *Solanum xanthcarpum* (used as vegetable and medicinal purpose), *Solanum indicum* (domesticated, used as vegetable and medicine), *Solanum khasianum* (wild and cultivated for solasodine alkaloid), *Solanum torvum* (wild, sold in the market in Mizoram), *Solanum berbisetum* (ripe fruits are eaten), *Solanum ferox* (wild, leaves are used medicinally), *Solanum spirale* (wild but domesticated for medicinal use in Arunachal Pradesh), *Solanum sisymbrifolium* (native of Africa, wildly grown in Meghalaya), *Solanum kurzii* (endemic in Garo hills, Meghalaya), and *Solanum gilo* (introduced in the NE region as a vegetable). For bamboo shoot (*Bambusa tulda*), the fermented bamboo shoot juice has preservative property similar to vinegar, and meat, fish, or vegetables are cooked with it have longer shelf-life. Methanol extracts of *Amaranthus caudatus*, *Amaranthus spinosus*, and *Amaranthus viridis* showed significant anti-diabetic and anti-cholesterolemic activity, which provides the scientific proof for their traditional claims (Girija et al., 2011). High magnesium content in *Amaranthus cruentus* (2.53 mg/100 g), *T. triangulare* (2.22 mg/100 g), *Celosia* (1.41 mg/100 g), and *G. latifolium* (1.32 mg/100 g) may explain their blood pressure lowering properties (Mensah et al., 2008). *Amaranthus cruentus* is used as tapeworm expellant and causes relief of respiratory disease (Mensah et al., 2008). *Amaranthus spinosus* stem and leaf are useful to treat dysentery. *Amaranthus virides* stem and leaf are useful against small pox. Red amaranth is used for the treatment of skin problems, diarrhea, sores, aching, and bleeding gums. Fox nut/ Thangjing is recommended for the treatment of diseases of respiratory, circulatory, digestive, excretory, and reproductive systems (Shankar et al., 2010). Bathua/Jimilsag (*Chenopodium album*) in Ayurveda is used to treat diseases of blood, heart, spleen, and eye; in biliousness conditions, cough, abdominal pain, pulmonary obstruction; and in nervous ailments (Sikawar et al., 2013). Chingit/Indian Pepper (*Zanthoxylum rhetsa*) fruit and stem bark are aromatic, stimulant, astringent, stomachic and digestive; they are prescribed in urinary diseases, dyspepsia, diarrhea and with honey for rheumatism. Fruits are appetizer; useful for treating cholera, asthma, bronchitis, heart troubles, piles and toothache, and relief from hiccup. The carpels yield an essential oil, which is given for treating cholera. The seed oil is antiseptic and disinfectant and is applied on inflammatory dermatosis. The seed oil is used in dry eczema and dandruff of children in Jointiapur of Sylhet. The root barks have cholinergic, hypoglycaemic and spasmolytic activity. Most of the secondary metabolites present in the plant possess antimicrobial and cytotoxic properties. In addition, the fruits, seeds, stem bark and heart wood contain furoquinoline and indolequinazoline alkaloids and terpenoids (Ghani, 2003). *Zanthoxylum acanthopodium* seeds and bark are used in the treatment of dyspepsia, fever, and cholera. *Zanthoxylum armatum* exhibits spasmolytic effects, mediated possibly through Ca(++) antagonist mechanism, which provides pharmacological base for its medicinal use in the gastrointestinal, respiratory, and cardiovascular disorders (Gilani et al., 2010). *Zanthoxylum oxyphyllum* fruit is used for stomach disorder. In China, Anhling/American black nightshade (*S. americanum*) tea from the whole plant is used to treat cancer of the cervix. Extracts from *S. americanum* were found to have selective antiviral activity against the herpes simplex type-1 virus (HSV-1). Methanol extracts of *S. americanum* have high antimicrobial activity against *Escherichia coli, Pseudomonas aeruginosa, Staphylococcus aureus,* and *Aspergillus niger*. In Nigeria, jack bean seed is used as an antibiotic and antiseptic (Olowokudejo et al., 2008). There is also pharmaceutical interest in the use of *C. ensiformis* as a source for the anticancer agents trigonelline and canavanine (Morris, 1999). The bark of Climbing wattle/ Khanghu/Biswal/Agla bel (*Acacia pennata*) is used as antiasthmatic and antibilious. The leaf is used as stomachic, styptic (for bleeding gum), and antiseptic (for scalding of urine), and a decoction of young leaves is taken for headache, fever, and body pain (Khare, 2008). In folk medicine, the calyx extracts of Rosella (*Hibiscus sabdariffa)* are used for the treatment of several complaints, including high blood pressure, liver diseases, and fever. In healthy men, consumption of *H. sabdariffa* showed significant decreases in the urinary concentrations of creatinine, uric acid, citrate, tartrate, calcium, sodium, potassium and phosphate, but not oxalate (Ali et al., 2005). Toothache plant/Paracress (*Spilanthes acmella*) is used for toothache in local areas. *Spilanthes paniculata* young stem and leaf are taken for deworming. Fresh flower is used for relieving toothache. Zombi pea (*Vigna vexillate),* is grown by the tribals of Tripura. The whole plant part is used for medicine. It is effective for joint disorders, arthritis, and swellings in joints. As a hemostatic, it checks hemorrhaging and thus prolongs life in individuals suffering from internal bleeding while building their strength with its nutritive action. In Rice bean (*Vigna umbellate*),

Catechin-7-O-glucoside can be found in the seed. In vitro, this compound has an antioxidant activity leading to a cytoprotective effect. Catechin-7-O-β-d-glucopyranoside scavenges free radicals and protects human B lymphoma BJAB cells on H_2O_2-mediated oxidative stress (Baek Jin-A et al., 2011). Loosestrife /Kengoi (*Lysimachia obovata*) is a herb used as traditional herbal medicine for the treatment of diabetes (Devi et al., 2011). The green portion of the fruit Tree bean (*Parkia roxburghii*) is mixed with little amount of water and applied to wounds and scabies. The fruit of young shoot is eaten for curing diarrhea, dysentery, and food poisoning (Bhardwaj and Gakhar, 2005). It has a very nutritious pod. Sophlong (*Moghania vestita*) is an anthelmintic (Mali and Mehta, 2008). East Indian Glory Bower (*Clerodendrum inerme*) leaf is grounded in water, and the juice is taken orally to treat fever (Muthu et al., 2006) and boiled leaves are taken to get relief from high blood pressure in local areas. Rhubarb (*Rheum nobile*) root is used as an anticholesterol, antiseptic, antispasmodic, antitumor, astringent, demulcent, diuretic, laxative, purgative, stomachic and tonic. Rhubarb roots contain anthraquinones which have a purgative effect and anti-cancer properties (Qing Huang et al., 2006). Spine gourd (*Momordica dioica*) is a protein-rich vegetable. Fruits are used in ulcers, piles, sores, and obstruction of the liver and spleen. It possesses several medicinal properties and is said to be good for those suffering from cough, bile, and other digestive problems. The seeds are used for chest problems and simulate urinary discharge (Ram et al., 2004). Moringa/Drumstick leaf is taken as food, and it reduces body heat and to treat indigestion and eye diseases. Flower is taken as food, and it gives cools eyes and increases sperm production in men (Muthu et al., 2006).

23.3.4   INCOME GENERATION

These underexploited vegetables play an important role that constitutes the daily vegetable requirement of the people. They are locally available wild in different parts of these regions. The tribals collect and sell these vegetables to earn income that could support their daily needs and to earn small amount of income. These crops require proper studies to know their values and unknown potentials as a means of improving livelihoods, especially in support of the rural poor vegetable farmers of these regions. Thus, if this hidden wealth of novel leaves, fruits and its medicinal compound are explored without further delay, these NE states, which are a rich source of genetic biodiversity, will be in a position to occupy a sizeable share in the national and international market for vegetables and herbal medicine.

23.4   THE DEPLETION OF BIO-DIVERSITY OF UNDERUTILIZED VEGETABLES: A MAJOR CONCERN

According to a study of Food and Agricultural Organization (FAO), crop genetic resources are being wiped out at the rate of 1.2% per annum. Tropical forests are falling at a rate of just under 1% per annum or 29 hectares per minute. From 1980 to 1990, this is equivalent to an area the size of Ecuador and Peru combined (Shand, 2000). However, the availability of most of these wild crops are now depleting rapidly owing to various factors such as "*Jhum*/shifting cultivation," forest fire, construction of industries, felling of trees for timber, and other socio-economic anthropogenic activities in the area. Therefore, it is a need of the hour to tap their potentials and conserve these regions' wealth and devise the ways for commercial-scale propagation of these plant species. If these underutilized plants are not conserved, research studies and commercialization of these crops are far beyond the reach of the researchers and the industries.

23.5   CONCLUSION AND RECOMMENDATION

Besides their commercial utility by the local tribes as food, these under-exploited vegetables have immeasurable potential to contribute to food security and have desirable medicinal properties, which require further studies to exploit their potential. The possible reasons for the low utilization of underutilized vegetables in spite of their recognized importance are the lack of seeds, lack of information about their performance and input requirements, lack of information on how they can fit into production systems, and nonviability of indigenous vegetable production like the major cultivated species of vegetables such as tomato, pepper, eggplant, cauliflowers, cabbage, etc., whose improvement and seed production are taken care by the private sector as well as government institutions, while the underutilized vegetables are a neglected lot (Pandey et al., 2014). However, there is still limited research on these underutilized crops and commercialization is still impossible due to negligence. Therefore, there is an immediate need to tap their potentials and conserve these regions' wealth, to harnessing the diversity of these crops, and to popularize and use them as crop alternatives as they are important sources of nutrients with innumerable medicinal properties.

KEYWORDS

- **food security**
- **north-eastern India**
- **phytochemicals**
- **underexploited vegetables**

REFERENCES

Ali, B. H., Wabel, N. A., & Blunden, G., (2005). Phytochemical, pharmacological and toxicological aspects of *Hibiscus sabdariffa* L.: a review. *Phytotherapy Research, 19*(5), 369–375.

Anonymous, a., (2012). Scientific Name: *Colocasia Esculenta Linn.* retrieved from https://sites.google.com/site/medicinalplantshealing/list-of-plants/taro.

Anonymous, b., (2017). Amaranth grain retrieved from *https://en.wikipedia.org/wiki/Amaranth_grain.*

Anonymous, c., (2017). *Nelumbo nucifera* retrieved from *https://en.wikipedia.org/wiki/Nelumbo_nucifera.*

Anonymous, d., (2015). *Rheum nobile* retrieved from *https://en.wikipedia.org/wiki/Rheum_nobile.*

Arora, R. K., (1997). Plant genetic resources of Northeastern region: diversity, domestication trends, conservation and uses. *Proceedings-Indian National Science Academy Part B, 63,* 175–186.

Asati, B. S., & Yadav, D. S., (2004). Diversity of horticultural crops in north eastern region. *ENVIS Bulletin: Himalayan Ecology, 12*(1), 1.

Baek, J. A., Son, Y. O., Fang, M., Lee, Y. J., Cho, H. K., Whang, W. K., & Lee, J. C., (2011). Catechin-7-O-β-d-glucopyranoside scavenges free radicals and protects human B lymphoma BJAB cells on H_2O_2-mediated oxidative stress. *Food Science and Biotechnology, 20*(1), 151–158.

Bhardwaj, S., & Gakhar, S. K., (2005). Ethnomedicinal plants used by the tribals of Mizoram to cure cuts & wounds. *Indian Journal of Traditional Knowledge, 4*(1), 75–80.

Deb, C. R., & Imchen, T., (2008). *Orchid Diversity of Nagaland.* SciChem Publishing House.

Deb, C. R., Jamir, N. S., & Ozukum, S., (2013). A study on the survey and documentation of underutilized crops of three districts of Nagaland, India. *Journal of Global Bioscience, 2*(3), 67–70.

Ekanayake, S., Jansz, E. R., & Nair, B. M., (2000). Literature review of an underutilized legume: *Canavalia gladiata* L., *Plant Foods for Human Nutrition, 55*(4), 305–321.

Gilani, S. N., Khan, A. U., & Gilani, A. H., (2010). Pharmacological basis for the medicinal use of *Zanthoxylum armatum* in gut, airways and cardiovascular disorders. *Phytotherapy Research, 24*(4), 553–558.

Girija, K., Lakshman, K., Udaya, C., Sachi, G. S., & Divya, T., (2011). Anti–diabetic and anti–cholesterolemic activity of methanol extracts of three species of Amaranthus. *Asian Pacific Journal of Tropical Biomedicine, 1*(2), 133–138.

Kar, A., & Borthakur, S. K., (2008). Wild vegetables of Karbi-anglong district, Assam. *Natural product Radiance, 7*(5), 448–460.

Khare, C. P., (2008). *Indian Medicinal Plants: An Illustrated Dictionary.* Springer Science & Business Media.

Longvah, T., & Deosthale, Y. G., (1998). Nutrient composition and food potential of *Parkia roxburghii*, a less known tree legume from northeast India. *Food Chemistry, 62*(4), 477–481.

Mali, R. G., & Mehta, A. A., (2008). A review on anthelmintic plants. *Natural Product Radiance, 7*(5), 466–475.

Mao, A. A., Hynniewta, T. M., & Sanjappa, M., (2009). Plant wealth of Northeast India with reference to ethnobotany. *Indian Journal of Traditional Knowledge, 8*(1), 96–103.

Mensah, J. K., Okoli, R. I., Ohaju-Obodo, J. O., & Eifediyi, K., (2008). Phytochemical, nutritional and medical properties of some leafy vegetables consumed by Edo people of Nigeria. *African Journal of Biotechnology, 7*(14).

Muthu, C., Ayyanar, M., Raja, N., & Ignacimuthu, S., (2006). Medicinal plants used by traditional healers in Kancheepuram District of Tamil Nadu, India. *Journal of Ethnobiology and Ethnomedicine, 2*(1), 43.

Pandey, A. K., Dubey, R. K., Singh, V., & Vida, E., (2012). Addressing the problem of micronutrient malnutrition in NEH region–underutilized vegetables as a source of food. *International Journal of Food and Nutritional Sciences, 3*(3).

Rahman, M. S., Khan, M. M. H., & Jamal, M. A. H. M., (2010). Anti-bacterial evaluation and minimum inhibitory concentration analysis of *Oxalis corniculata* and *Ocimum santum* against bacterial pathogens. *Biotechnology, 9*(4), 533–536.

Rai, N., Asati, B. S., & Yadav, D. S., (2004). Conservation and genetic enhancement of underutilized vegetable crop species in North Eastern region of India. *Low External Input Sustainable Agriculture* (LEISA).

Rajerison, R., (2006). *Vigna Umbellata* (Thunb.) Ohwi and H., Ohashi. In: Brink, M., & Belay, G. (Editors). PROTA (Plant Resources of Tropical Africa / Ressources végétales de l'Afrique tropicale), Wageningen, Netherlands.

Ram, D., Kumar, S., Verma, A., & Rai, M., (2004). Variability analysis of underutilized nutritive vegetable Kartoli: Indian collection. *Report-Cucurbit Genetics Cooperative, 27,* 66.

Shankar, M., (2010). A review on gorgon nut. *International Journal of Pharmaceutical and Biological Archive, 1*(2).

Sheikh, Y., Maibam, B. C., Biswas, D., Laisharm, S., Deb, L., Talukdar, N. C., & Borah, J. C., (2015). Anti-diabetic potential of selected ethno-medicinal plants of north east India. *Journal of Ethnopharmacology, 171,* 37–41.

Sikarwar, I., Wanjari, M., Baghel, S. S., & Vashishtha, P., (2013). A review on phytopharmacological studies on *Chenopodium album* Linn. *American Journal of Pharm Research, 3*(4), 3089–3098.

Tag, H., Kalita, P., Dwivedi, P., Das, A. K., & Namsa, N. D., (2012). Herbal medicines used in the treatment of diabetes mellitus in Arunachal Himalaya, northeast, India. *Journal of Ethnopharmacology, 141*(3), 786–795.

United States Department of Agriculture, Jack bean *Canavalia ensiformis* retrieved from https://www. google. co. in/url?sa=t&rct=j&q=&esrc=s&source=web&cd=8&cad=r ja&uact=8&ved=0ahUKEwjZ9_bU2NjSAhXLt48KHY5cDQ8QFghIMAc&url=htt ps%3A%2F%2Fplants. usda. gov%2Fplantguide%2Fdoc%2Fpg_caen4. docx&usg =AFQjCNHfDKXvFhcaHZH9gTRxo-832wyfHQ&sig2=xw2ZgWDkktVNwpyHV mia8g&bvm=bv.149397726, d. c2.

Yadav, A. K., & Tangpu, V., (2012). Anthelmintic activity of ripe fruit extract of *Solanum myriacanthum* Dunal (Solanaceae) against experimentally induced *Hymenolepis diminuta* (Cestoda) infections in rats. *Parasitology Research, 110*(2), 1047–1053.

CHAPTER 24

ESSENTIAL OIL OF *THYMUS SATUREJOIDES* COSS. IN THE HIGH ATLAS OF MOROCCO: FROM TRADITIONAL MEDICINE TO COMMUNITY NATURAL PRODUCT DEVELOPMENT

BERNADETTE MONTANARI[1*] and AMRITESH C. SHUKLA[2]

[1]Department of Geography, University of Urbana Champaign, USA (Social Dimensions of Environmental Policy (SDEP)), Tel: +12173774102, E-mail: bernadettemontanari@hotmail.com

[2]Department of Horticulture, Aromatic and Medicinal Plants, Mizoram University, Aizawl–796004, India

CONTENTS

ABSTRACT

This article provides an insight into the phytochemistry of the essential oil of *Thymus saturejoides* Coss. (thyme) and its potential for natural product development in Megz, a village of the Agoundis valley, in the High Atlas mountains of Morocco. The aromatic plant is widely harvested by the local population as an important source of income and used in traditional Berber medicine. To assess the commercial viability of the plant, the essential oil was distilled, analyzed, and compared with thyme oil distillates from Asni Moulay Brahim situated at 55 km, and a recent study of thyme samples collected in the lower part of the Agoundis valley. The gas chromatography–mass spectrometry (GC-MS) analysis of the thyme oil from Megz revealed borneol (32.89%), carvacrol (18.05%), and thymol (0.46%) as the active constituents. These have already been reported for their antimicrobial, antiseptic, antioxidant, and other pharmaceutical properties. We suggest that the added value of the essential oil of thyme from Megz may therefore be destined for the aromatherapy market, and for more specific indications in the pharmaceutical and sanitary industries.

24.1 INTRODUCTION

People have been resorting to traditional medicine since ancient times often in the lack of conventional medicine and proximity to medical facilities. An estimated 70,000 plant species have medicinal value and are employed in traditional medicine worldwide (Lange, 2006; Schippmann, et al., 2002). Medicinal and aromatic plants (MAPs) are also gathered to generate income and to enhance livelihood (Ticktin, 2004). While many have condemned the practices of wild harvesting as unsustainable, because of the enormous pressure exerted at local, national, and environmental levels, and which may result in species extinction (Ticktin, 2004; Sheldon, et al., 1997), wild harvesting of medicinal plants does, nonetheless, allow isolated rural populations to maintain vital subsistence (Robins, 2000; Schipmann, et al., 2005; Ticktin, et al., 2002). Medicinal plants used in the traditional systems of medicine hold the potential to add value to the socioeconomic welfare of communities and to contribute to conservation strategies (Hamilton, 2004).

Morocco is the second most diverse country of species in the Mediterranean basin in terms of biological resources after Turkey. The country offers a rich flora and high endemism, 41 ecosystems, and 7000 vegetal species of which 4500 are vascular plants. Six hundred species, which have been reported as endemic, are listed as having medicinal and aromatic uses and harvested from the wild or cultivated (Vasisht, Kumar, 2004). Within the Mediterranean region, Morocco stands as the ninth largest exporter of MAPs on the global scale (Lange, 2004; Ozhatay, et al., 1997). These are used mainly commercially in the pharmaceutical, cosmetic, culinary, and food industries (USAID, 2006) The country's export of pharmaceutical plants alone has increased from 5510 metric tons (MT) to 12133 MT between 1993 and 2007, which represents more than US$ 34 million of MAPs and an estimated US$ 18 million worth for essential oils (CFC, 2012). The main species harvested and processed in various parts of the country are *R. officinalis* L., (rosemary) (1 million ha of the plant for the production of rosemary essential oil, with an estimated annual yield of 60 MT), thyme, lavender, and *Artemisia* species, *M. pulegium* L., (pennyroyal), *O. vulgare* L., (oregano), and *C. sativum* L., (coriander). Prices for essential oils vary widely from US$ 2 to US$ 10 for 10 mL and potentially can reach US$ 200 to 600 per kg, especially for specialized production used in the food, cosmetic, and pharmaceutical industries.

The present study was conducted between 2007 and 2009 in Megz, a village of the Agoundis valley in the High Atlas mountains of Morocco (Figure 24.1). As one of the seven valleys originating from the Toubkal, the Agoun-

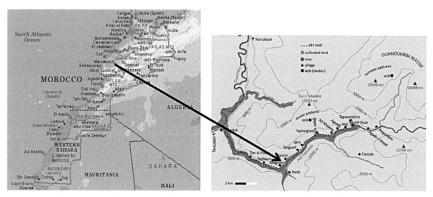

FIGURE 24.1 **(See color insert.)** Map of Morocco and the study area.

dis valley benefits from streams descending in altitude that carry snowmelt water across various altitudinal zones and ecosystems. Temperatures may vary throughout the year from 5°C to 6°C during the winter season and over 30°C in the months of July and August. Rain fall also varies considerably according to the seasons and may range from a low of 0.92 mm in July to a high 67.36 mm in December and 70.13 mm in January. Snow fall usually occurs from October until May, more or less abundant at 1200 m; however, it becomes substantial at 2000 m (Benaboubou, 2004). Owing to its topographic and geographic position, Agoundis is one of the narrowest and enclave valleys of the High Atlas, enclosed between abrupt forested slopes, offering very little cultivable space. The duality of this spatial structure produces noticeable differences in the landscape and in the availability of resources. The strong declivity of the slopes favors the streaming and erosion of the ground, thus necessitating the construction of terraces. Because of the altitude ranges, local families have traditionally diversified livelihood strategies according to the seasons. Millennia of human modification have shaped the typicality and diversity of these landscapes to control erosion and to promote agriculture. The Agoundis valley has, therefore, access to a remarkable human-shaped landscape (Gerbati, 2004).

Although the environment is biologically rich, especially in MAPs, the natural resources of the region are overall declining owing to overharvesting in the face of the increasing demand for phytoaromatic products and the needs of a growing population. In the mountains during the summer months, local people harvest the aromatic plants thyme, *Salvia aucheri Bentham var. canescens* Boiss and Heldr (sage), and *Lavandula dentata* L. (lavender) for both herbal medicine and for trade (Montanari, 2004), the most lucrative being thyme. These plants are not only one of the few sources of income, but their utilization and collection also represent an important aspect of the transmission of herbal plant knowledge within the community. The plants are then traded down the valley via several middlemen to urban markets in Marrakech and beyond. The trade follows two commodity chains, one official (legal) and one unofficial (illegal) (Montanari, 2013). This income, which varies in terms of the amount of plant material collected, represents a significant contribution to the household economy.

Although this huge demand for plant products for domestic and commercial use puts enormous pressure at the local and regional levels, these figures suggest that the essential oil sector has potential for adding value to otherwise fragile and marginal landscapes and to provide employment, especially in isolated rural communities. Some communities have indeed managed to overcome these obstacles and to achieve positive outcomes in the plant trade. This is the case, for example for *P. Africana* Hook.f., Kalkman *(red stinkwood)* in Cameroon where local communities signed agreements with external companies to ensure sustainable revenues and practices (Ndam and Marcelin, 2004). Similarly, in Madagascar, middle-men buy dried *Centella asiatica* L., Urban (gotu kola) from harvesters and are responsible for packaging (Rasoanaivo, 2009). Exporters in Namibia pay a percentage to the harvesters for good harvesting practices of *Harpagophytum procumbens* Burch., (devil's claw) (Cole, 2009; Tonye Mahop, 2009), as is the case for the minor millets in the Kolli Hills of Tamil Nadu, India (Gruère, et al., 2007). These cases have in common the full and active integration of local people in either self-help groups or small-scale enterprises, and agreements signed directly between external companies and the communities. Such arrangements have potential for positive financial outcomes at all levels, i.e., village, local, and national.

Thyme (Figure 24.2) for instance is a promising source of antibacterial and anti-inflammatory products according to the European Pharmacopeia (WHO, 2002). The essential oils, which are mainly found in the flowering stems of the plant (Hmamouchi, 2001), are rich in borneol (with high antimicrobial activity), flavonoids (derived from apigenol and luteolol),

FIGURE 24.2 *T. saturejoides* Coss.

phenolic acids (particularly cafeic and rosmarinic acids), tannins, resins, and other chemical compounds responsible for the majority of these pharmacological effects (Garcia-Martin, et al., 1974; Grieve, 1974; Guenther, 1955). This article therefore seeks to demonstrate that the essential oil of thyme could significantly increase the potential of a natural product development on a commercial scale operated by the local people in the valley, with the aim of returning benefits to the local population.

24.2   MATERIALS AND METHODS

24.2.1   STUDY AREA

The study was conducted between 2007 and 2009 in Megz, a village of the Agoundis valley in the High Atlas mountans of Morocco. The village is situated at an altitude of 1300 m, 8 km from the main rural commune of Ijoukak, 100 km from Marrakech in Al Haouz province, close to the Toubkal National Park, and has a rich biodiversity.

24.2.2   PLANT DATA COLLECTION WITH THE INHABITANTS OF MEGZ

The data regarding the use of thyme were randomly collected with a total of 60 female informants aged between 20 and 65 years and a total of 50 male informants aged between 20 and 67 years. This coverage allowed a wide age spectrum of individuals involved (or with a history of involvement) of herbal knowledge and collection, but excluding minors. All interviews were conducted using an interpreter assisting with the translation of the Berber language back to French. The interviews sought to identify the ethnomedicinal uses of the plant, the used parts for treating specific ailments as well as the locations where the plant was collected. Plant information/collections were made in various locations, i.e., with the women while they were collecting vegetables, weeding the gardens, or doing laundry down by the river. The men were accompanied while they were working in the gardens, ploughing the land, when they went to the mountains to collect the aromatic plants and firewood, or while they were building a dam in the river. The identification of the collected plant thyme (voucher specimen n° BML07) was compared with botanical plates from "Medicinal plants of North Africa" (Boulos, 1983), and their therapeutic properties were confirmed with the available literatures, viz., "La pharmacopéemarocainetraditionnelle," "Les plantesmédicinales du Maroc" (Bellakhdar, 1997; Sijelmassi, 2003). A plant taxonomist at the Laboratory of Vegetal Ecology, University Cadi Ayyad, Marrakech, assisted with the identification of plant voucher specimens. These were then deposited at the Laboratory of Vegetal Ecology at the Faculty of Sciences, University Cadi Ayyad in Marrakech.

24.2.3 THYME ESSENTIAL OIL DISTILLATE

Flowering tops of thyme were randomly collected from four different locations on Tanammirt, Tissilu, Wankrim, and Wijdane Mountains around Megz in the Agoundis valley. However, for the purpose of this study, only the sample collected from Wijdane Mountain was analyzed. The fresh aerial parts of the plant, when just coming into flower (discarding the lower portion of the stem, together with any yellow or brown leaves), were used for steam distillation and for estimation of the bioactive constituents. The fresh aerial parts (1 kg) were steam distilled for 24 h in a 20-liter distillation flask fitted with an oil estimator. Light amber colored oil (2.73 mL, w/ dry weight) was obtained. This was done at the Faculty of Sciences, Cadi Ayyad University in Marrakech.

Staff at the "Laboratoire de Biotechnologies végétalesappliquées aux plantesaromatiques et médicinales" at Jean Monnet University in St Etienne, France, performed the gas chromatography (GC) and mass spectrometry (MS) analyses. The essential oil sample was diluted 1/50 with hexane and injected with a 1:200 split. The GC analysis of the oil was performed on a Perkin-Elmer GC 8500, using a fused silica capillary column (25 m × 0.55 mm, film thickness 0.245 μm), coated with dimethyl siloxane (BP-1). The oven temperature was programmed from 60°C to 220°C at 5°C/min, then held isothermally at 220°C; detector temperature, 300°C; carrier gas-nitrogen at an inlet pressure of psi; split, 1: 80. GC-MS data were obtained on a Shimadzu QP-2000 Mass Spectrometer instrument at

70 eV and 250°C. GC column: Ulbon HR-1 (equivalent to OV –1), fused silica capillary column (0.25 mm × 50 m, film thickness 0.25 μm). The initial temperature was 100°C for 7 mm, and then heated at 5°C/min to 250°C. The carrier gas helium was used at a flow rate of 2 mL/min. The results of the GC-MS analysis were then compared with studies of thyme distillate conducted by Jafaari et al. (2007) and by Ghalbane et al. (2011) in the lower part of the Agoundis valley. Moreover, data were analyzed using statistical method of randomized block design (Gomez and Gomez, 1983) in Table 24.2.

24.3   RESULTS AND DISCUSSION

24.3.1   HERBAL MEDICINE USE

The Agoundis valley is botanically rich; many species are endemic, while others are cultivated in gardens. In the absence of conventional medicine, the villagers resort to medicinal plants that are widely used in ethno-medicine, and most families have harvested supplies of plants stored in their houses. It is women who usually prepare and administer plant mixtures that are consumed on a daily basis in tea, coffee, infusion and other more complex herbal preparations for treating and bringing relief from common ailments (Montanari, 2014). Thyme is used profusely in herbal medicine. It is taken regularly, more or less on a daily basis, as a fresh herbal tea infusion during the harvest season or outside the harvesting season in the dried herb form. Because thyme contains antioxidant, anti-infectious, antispasmodic, anti-microbial, and anti-inflammatory properties (Jafaari, et al., 2007), the dried herb is used to relieve common ailments in traditional folk medicine throughout Morocco (Bellakhdar, 1997). In Megz, women take regularly the powdered herb for painful menses, to relieve gastric disorders (stomachache, bile complaints, indigestion, and intestinal trouble). It is also administered for respiratory disorders such as colds, coughs, chills, and headache. People, however, have cautioned that thyme should not be taken over long periods of time as it will damage teeth and gums (Montanari, 2014). It has another important use as a preservative in the confection of Smen. Smen is the term applied to preserved butter that is prepared with the addition of salt and thyme. The addition of the leaves and stems allows the mixture to be preserved for a year due to its antimicrobial properties (Banqour, 1985; Gutierrez, et al., 2008).

24.4 RESULTS OF THE ESSENTIAL OIL DISTILLATE OF THYME FROM MEGZ

According to the literatures, the chemical composition of thyme oil has also been studied by various researchers (Table 24.1).

It is carvacrol and thymol that gives thyme its antioxidant properties, a phenolic structure present in various concentrations of thyme essential oil (El Bouzidi, et al., 2013). Studies conducted on the essential oil of thyme reveal important radical scavenger actions and their potential antibacterial, antifungal, antiviral, and antioxidant properties (Ghalbane, et al., 2011; Alaoui, et al., 2012). However, the borneol contained in the essential oil of thyme contributes to its antimicrobial and anti-inflammatory properties, indicated in respiratory viral or bacterial chronic infections arthritis, rheumatism, deep physical and sexual asthenia, candida infections, cystitis, leucorrhoea, and externally for acne, infected wounds, dermatitis, and other skin problems (Laboratoire de Combe d'As, 2014; Zhiri and Baudoux, 2008). Further research on the aqueous extract of thyme has shown that its analgesic action is more potent than acetyl salicylic acid (ASA), acting through an opioid-mediated mechanism (Elhabazi, et al., 2008).

From the studies, carvacrol, thymol and borneol stand as the major constituents of interest of thyme from these three mountain locations and are typically of the carvacrol-thymol chemotype. However, based on the observations of Tables 24.1 and 24.2, the carvacrol from Megz is lower than in the thyme species samples collected from Asni Moulay Brahim in the High Atlas (35.90%), and it is higher than the thyme species samples collected in the lower part of the Agoundis valley (9.1%). On the other hand, the thymol content (0.46%) is lower than that in both samples collected by Jafaari et al. (2007) and the lower part of the Agoundis valley by Ghalbane et al. (2011), that is, 0.94% and 0.1%, respectively. However, the borneol content of thyme collected from Megz (32.89%) is predominantly higher than that of the other thyme analyzed from the two locations (Table 24.2). This phytochemical variation shows that thyme species and

TABLE 24.1 Chemical Composition of Thyme Oil: An Overview

Authors	Chemical Constituents	Remarks
Rovesti, 1971	camphene-pinene (0.15%), p-cymene (15–50%), linalool (3–13%), linalyl acetate (0–6%), bornol(2–8%), carvacrol (0–20%), thymol (5–60%).	Reported as additions in the chemical composition from various Italian essential oils of thyme
Garcia-Martin, et al., 1974	α-thujene (0.5%), β-pinene (4.6–4.7%), myrcene (0.4–0.9%), α-phellandrene(0.1–0.2%), limonene + 1,8-cineole 35.7–44.4%, γ-terpinene 0.3%, trans-linalool oxide 0.5%, cis-linalool oxide 1.0–1.1%, camphor 11.6–16.3%, β-terpinelol 0.6–0.9%, α-terpineol + borneol + bornyl acetate 7.8- 8.9%, α-terpinyl acetate 0.7–1.4%, geranyl acetate 0–0.5%, geraniol 0.1–0.2%.	Reported as additions in the chemical composition of Spanish essential oils of thyme
Richards, et al., 1975	a-terpinene 1.0%, trans–sabinene hydrate 0.3%, methyl carvacrol 3.0%, terpinen-4-ol 0.3% and caryophylln 0.9%.	Reported as additions in the chemical composition of French essential oils of thyme
Lawrence, 1984	sabinene trace, verbenene trace, 1-octen-3-ol 0.1%, methyl thymol 1.5%, verbenone 0.2%, α-muurolene 0.3%, p-cymen-8-ol 0.1%.	Reported as additions in the chemical composition of Spanish essential oils of thyme
Jafaari, et al., 2007	camphene 0.58%, β-pinene trace, α-terpinene 0.37%, p-cymene 2.17%, g-terpinene 0.37%, linalool 30.03%, camphor 0.48%, borneol 30.03%, bornyl acetate 1.73%, thymol 0.94%, carvacrol 35.90%, α-copaene trace, caryophyllene 0.16%.	Reported as additions in the chemical composition of essential oil of thyme from AsniMoulayBrahim
Ghalbane, et al., 2011	α-Tricyclene 0.1%, Tricyclene 0.1%, α-Thujene 0.1%, α-Pinene 1.6%, Thuja-2,4(10)-diene trace, Camphene 4.0%, Verbenene trace, Octen-3-ol trace, Sabinene trace, β-Pinene 0.5%, Myrcene 0.2%, 3-δ-carene trace, α-terpinene 0.1%, p-cymene 5.1%, Limonene 0.5%, γ-Terpinene 0.1%, Dehydro-p-cymene trace, (E)-Sabinene hydrate 0.2%, Linalol 5.8%, Thujone trace,	Reported as additions in the chemical composition of essential oils of thyme from the lower part of the Agoundis valley

TABLE 24.1 (continued)

Authors	Chemical Constituents	Remarks
	p-Menth-2-en-1-ol 0.1%, Camphor 1.0%, Pinocarveol0.3%, (E)-Verbenol 0.2%, Isoborneol 0.1%, Borneol 29.5%, Terpinen-4-ol 2.1%, Dihydrocarvone 1 1..0%, α-Terpineol 6.5%, (Z)-dihydrocarvone trace, carveol trace, bornylformate 1.1%, Methyl thymol 0.6%, Thymol 0.1%, Bornyl acetate 5.5%, Carvacrol 9.1%, α-Cubébène 0.1%, Copaène 0.9%, β-Bourbonene 0.4%, β-Patchoulene 0.1%, α-Gurjunene 0.4%, β-caryophyllene 8.2%, β-Cubebene 0.2%, α-Panasisen trace, Aromadendrene trace, α-Humulene 0.4%, allo-Aromadendrene 0.3%, γ-Muurolene 0.3%, β-Gurjunene 0.2%, Ledene 0.3%, γ-Cadinene 1.0%, δ-Cadinene 0.7%, β-Ionone 0.1%, Guaiazulene 0.3%, Caryophyllene oxide 2.5%, Isoaromadendreneepoxyde 0.2%, GeranylLinalol 0.2%.	
Present study	gtricyclene 0.27%, a-thujene 0.30%, a-pinene 5.62%, camphene 11.00%, β-pinene 0.71%, myrcene 0.23%, a-terpinene 0.31%, p-cymene 4.15%, limonene 0.41%, g-terpinene 1.74%, linalool 3.65%, camphor 0.31%, borneol 32.89%, a-terpineol 7.07%, methyl carvacrol 0.82%, bornyl acetate 2.39%, thymol 0.46%, carvacrol 18.05%, a-copaene 0.23%, caryophyllene 5.13%, a-humulene 0.09%, g-cadinene 0.17%, d-cadinene 0.46%, caryophyllene oxide 0.15%.	Reported as additions in the chemical composition of essential oil of thyme from Megz

TABLE 24.2 Comparison of the Three Main Compounds of Interest Found in Thyme Essential Oil in the Three Different Locations of the High Atlas

Average of bioactive compounds	*T. saturejoides, Coss., (*Megz, High Atlas*)* Area%	*T. saturejoides, Coss.,* (Asni, High Atlas*)* Area%	*T. saturejoidesCoss., (Agoundis valley, High Atlas)* Area%
Borneol	32.89	30.03	29.5
Carvacrol	18.05	35.90	9.1
Thymol	0.46	0.94	0.1

their chemical composition may vary according to soil variation, vegetative cycle, seasonal variations, and extreme climatic conditions that may affect their chemotyped identity (Thompson, et al., 2013). Therefore, in concordance with Ghalbane et al. (2011) and Jafaari et al. (2007), these results suggest that the essential oil of thyme from these three locations has powerful antibacterial and antioxidant activities and can make an important contribution to aromatherapy, natural preservation of food, and as antibacterial ingredients for food and pharmaceutical industry.

24.5 CONCLUSION

Molecules from natural products will in the future continue to play a preponderant role as active substances in the discovery and validation of new drugs. Of approximately 420,000 species, less than 5% have gone through screening for one or several biological actions and the vast majority of antibacterials or 78% of new chemicals are derived from natural product molecules. With the advance in technology, especially the use of high-performance liquid chromatography (HPLC), and spectrometry for the rapid characterization of extracts and molecules, phytochemical analysis,has a major impact in research on natural products. Natural compounds obtained from indigenous drugs therefore hold the potential to be utilized not only in the discovery of new drugs but also to benefit local communities from which the substances are extracted. The study presented here has shown that thyme species and their chemical compositions are largely altered by climate condition, soil variation, vegetative cycle, and seasonal variation. However, the phytochemical analysis of thyme collected from Megz as well as the significantly varied bioactive compound, viz., borneol, carvacrol, and thymol contents of thyme collected from the other two agro-climatic locations are worthy of attention. The quality of the phytochemical content of thyme from Megz therefore contributes to the European markets for aromatherapy use. While the aim is not to distill the essential oil of thyme from Megz on an industrial scale, it is rather to target a niche market production, an emblem of community-produced essential oil. This would necessitate contractual agreements with fair trade companies to ensure that ethical practices are respected to return benefits to the local population.

ACKNOWLEDGMENTS

The authors are thankful to the communities of Megz (Agoundis valley, High Atlas of Morocco) for sharing their valuable information, the funding organizations (The Gen Foundation, FfWG (Funds for Women Graduates), Radcliffe-Brown & Firth Trust, Funds (Royal Anthropology Institute), John Ray Trust) who have permitted fieldwork, and the authorities of Mizoram University, Aizawl (India), for providing various types of support during the compilation of the paper.

KEYWORDS

- **aromatic plant**
- **community development**
- **essential oil**
- **morocco**
- **natural product**
- **thyme**

REFERENCES

Alaoui, J. C., El Bouzidi, L., Bekkouche, K., Hassani, L., Markouk, M., Wohlmuth, H., et al., (2012). Chemical composition and antioxidant and anticandidal activities of essential oils from different wild moroccan *thymus* species. *Chemistry & Biodiversity, 9,* 1188–1197.

Banqour, N., (1985). Conservation artisanale des aliments par les plantes aromatiques dans Marrakech et sa region: Essai de conservation du Smen par le thym et son huile essentielle. Thèse. Faculté des Sciences Semlalia, University Cadi Ayyad, Marrakech.

Bellakhdar, J., (1997). *La Pharmacopée Marocaine Traditionnelle.* Ibiss Press, Casablanca.

Benaboubou, M., (2004). Etude ethnobotanique du Haut Atlas. Diplôme des études supérieures spécialisées. Gestion de l'environnement et développement durable. Faculté des lettres et sciences humaines université, Mohamed, V., Agdal Rabat.

Boulos, L., (1983). *Medicinal Plants of North Africa.* Reference Publication Inc. Algonac, Michigan, pp. 286.

Cole, D., (2009). Sustainable harvesting of devils claw in Namibia. In: Philips, L. D., (Ed.), *Plants, People and Nature.* Benefit sharing in practice. AAMPS Publishing, Mauritius, pp. 77–100.

Common Fund for Commodities (CFC), (2012). *International Workshop on Potentials and Constraints for Aromatic Plants and Essential Oils Production and Marketing in Africa,* Workshop Report. Project CFC/FIGSTF/30.03–05 May, Rabat, Morocco.

El Bouzidi, L., Alaoui, J. C., Bekkouche, K., Hassani, L., Wohlmuth, H., et al., (2013). Chemical composition, antioxidant and antimicrobial activities of essential oils obtained from wild and cultivated Moroccan Thymus species. *Industrial Crops and Products, 43,* 450–456.

Elhabazi, K., Ouacherif, A., Laroubi, A., Aboufatima, R., Abbad, A., Benharref, A., Zyad Chait, A., & Dalal, A., (2008). Analgesic activity of three thyme species, *Thymussatureioides, Thymus maroccanus* and *Thymusleptobotrys. African Journal of Microbiology Research, 2,* 262–267.

Garcia-Martin, D., Fernandez-Vega, F. I., Lopez de Bustamante, F. M., & Garcia-Vallejo, C., (1974). *Ministry Agriculture Instituto Nacional de Investigaciones* Agarias, Madrid.

Gerbati, F., (2004). La mobilisation territoriale des acteurs du développement local dans le Haut Atlas de Marrakech. Thèse pour l'obtention du Docteur de l'Université Joseph Fournier, Grenoble, France.

Ghalbane, I., Belaqziz, R., Ait Said, L., Oufdou, K., Romane, A., & El Messoussi, S., (2011). Chemical composition, antibacterial and antioxidant activities of the essential oils from *Thymus satureioides* and *Thymus pallidus. Natural Product Communications,* vol. *6*(0), 1–3.

Gomez, K. A., & Gomez, A. A., (1983). *Statistical Procedures for Agricultural Research* (2nd ed.) John Wiley and Sons. New York, NY. U.,SA.

Grieve, M., (1974). *A Modern Herbal.* Jonathan Cape, Thirty Bedford Square, London, pp. 808–812.

Gruère, G., Nagarajan, L., & Oliver King, E. D. I., (2007). Marketing underutilised plant species for the poor: A case study of minor millets in Kolli Hills, Tamil Nadu, India. *Global Facilitation Unit for Underutilized Species (GFU).*

Guenther, E., (1955). *The Essential Oils*. D., Van Nostrand Co. Inc. New York, vol. *3*, pp. 744.

Gutierrez, J., Barry-Ryan, C., & Bourke, P., (2008). The antimicrobial efficacy of plant essential oil combinations and interactions with food ingredients. Interna*tional Journal of Food Microbiology. vol. 124*(1), 91–97.

Hamilton, A. C., (2004). Medicinal plants, conservation and livelihoods. *Biodiversity and Conservation, 13*, 1477–1517.

Hmamouchi, M., (2001). Les plantes médicinales et aromatiques marocaines (2nd Ed.) Rabat.

http://www.combedase.com/aromatherapie.html (Retrieved 09/01/2015).

Jafaari, A., Ait Mouse, H., Rakib, M., Ait Mbarek, L., Tilaoui, M., Benbakhta, C., Boulli, A., Abbad, A., & Zyad, A., (2007). Chemical composition and anti-tumour activity of different wild varieties of Moroccan thyme. *Brazilian Journal of Pharmacognosy, 17*(4), 477–491.

Laboratoire de Combe d'As. 2014.

Lange, D., (2004). Medicinal and Aromatic Plants: Trade, Production, and Management of Botanical Resources. In: Craker, L. E., et al., (Eds.), Proc. XXVI IHC – Future for Medicinal and Aromatic Plants. *Acta. Hort., 629*, ISHS.

Lange, D., (2006). *International Trade in Medicinal and Aromatic Plants*: Actors, *Volumes and Commodities*. In: Bogers, R. J., Cracker, L. E., Lange, D. (Eds.), Medicinal and aromatic plants. Agricultural, Commercial, Ecological, Legal, Pharmacological and Social Aspects, Series: Wageningen UR Frontis Series vol. *17*, Springer, The Netherlands, pp. 155–170.

Lawrence, B. M., (1984). Progress in essential oils. *Perfumer Flavorist, 9*, 31.

Montanari, B., (2004). Alternative paths to commercialising medicinal and aromatic plants in a community of the High Atlas Mountains, Morocco. *MSc Dissertation*, University of Kent, Canterbury, UK.

Montanari, B., (2013). The future of agriculture in the high atlas mountains of Morocco: The Need to integrate traditional ecological knowledge. In: Mann, S. (Ed.), *The Future of Mountain Agriculture,* Springer Geography, Springer-Verlag Berlin Heidelberg, pp. 51–72.

Montanari, B., (2014). Aromatic, medicinal plants and vulnerability of traditional herbal knowledge in a berber community of the high atlas mountains of Morocco. Chinese Academy of Sciences, *Plant Diversity and Resources, 36*, 3. DOI: 10.7677/ynzwyj201413160 .

Ndam, N., & Marcelin, M. T., (2004). 'Chop but no broke pot': the case of Prunus Africana in Mount Cameroon. In: Sunderland, T., Ndoye, O., (Eds.), *Forest products, Livelihood and Conservation*, vol. 2, Africa. Centre for International Forestry Research, Jakarta, pp. 37–52.

Ozhatay, N., Koyuncu, M., Atay, S., & Byfield, A. J., (1997). The wild medicinal plant trade in Turkey. *TRAFFIC Turkey*. Do_alHayatıKorumaDerne_i (DHKD). Istanbul.

Rasoanaivo, P., (2009). The work of IMRA-SOAMADINA. In: Phillips, L. D., (Ed.), P*lants, People and Nature*. Benefit sharing in practice. AAMPS. Mauritius, pp. 45–50.

Richards, H. M. J., Miquel, J. D., & Sandret, F. G., (1975). Huiles essentielles de thym du Maroc et de thym de Provence. Compositions comparées inter analytique. *Parfum. Cosmet. Aromes., 6*, 69–78.

Robins, C. S., (2000). Comparative analysis of management regimes and medicinal plant trade monitoring mechanisms for American Ginseng and Goldenseal. *Conservation Biology, 14*, 1422–1434.

Rovesti, P., (1971). Incidences écologiques sur la composition des huiles. Des variétés chemo taxonomiques des thyms spontanés en Italie. *Perfum. Cosmet. Savons, 1*, 139–47.

Schipmann, U., Leaman, D. J., Cunningham, A. B., & Walter, S., (2005). Impact of cultivation and collection on the conservation of medicinal plants: Global trends and issues. III WOCMAP Congress on medicinal and aromatic plants. Conservation, cultivation and sustainable use of medicinal and aromatic plants. *ISHS Acta. Horticulturae*, vol. 2, 676.

Schippmann, U., Leaman, D. J., & Cunningham, A. B., (2002). Impact of Cultivation and Gathering of Medicinal Plants on Biodiversity: Global Trends and Issues. In: *FAO: Biodiversity and the Ecosystem Approach in Agriculture, Forestry and Fisheries*. Satellite event on the occasion of the Ninth Regular Session of the Commission on Genetic Resources for Food and Agriculture, Rome, 12–13 October. Inter-Departmental Working Group on Biological Diversity for Food and Agriculture.

Sheldon, J. W., Balick, M. J., & Laird, S. A., (1997). Medicinal plants: can utilization and conservation coexist? *Advances in Economic Botany, 12*, 1–104.

Sijelmassi, A., (2003). Les plantes médicinales du Maroc. *Editions Le Fennec*, Casablanca.

Thompson, J., Charpentier, A., Bouguet, G., Charmasson, F., Roset, S., Buatois, B., et al., (2013). Evolution of a genetic polymorphism with climate change in a Mediterranean landscape. *Proceeding of the National Academy of Sciences (PNAS) vol. 110*(8), 2893–2897.

Ticktin, T., (2004). The ecological implications of harvesting non-timber forest products. *Journal of Applied Ecology, 41*(1), 11–21.

Ticktin, T., Nantel, P., Ramirez, F., & Johns, T., (2002). Effects of variation on harvest limits for non-timber products species in Mexico. *Conservation Biology, 16*, 691–705.

Tonye Mahop, M., (2009). *Prunus Africana* harvesting in Cameroon: An ABS success story? In: Phillips, L. D., (Ed.). *Plants, People and Nature*. Benefit sharing in practice. AAMS Publishing. Mauritius, pp. 51–54.

United State Agency for International Development (USAID), (2006). Projet filière des plantes aromatiques et médicinales. Mission de l'USAID au Maroc. Agence américaine pour le développement humain. Washington, D. C.,

Vasisht, K., & Kumar, V., (2004). *Compendium of Medicinal and Aromatic Plants, Africa.*

World Health Organization (WHO), (2002). *Traditional Medicine Strategy 2002–2005*. World Health Organization, Geneva.

Zhiri, A., & Baudoux, D., (2008). *Chemotyped Essential Oils and their Synergies*. Publisher: Edition Inspir Development. Luxembourg.

CHAPTER 25

POTENT NUTRIMENTAL AND ETHNOMEDICINAL HORTICULTURAL FLORA FROM NORTH CENTRAL TERAI FORESTS OF U.P., INDIA

S. C. TRIPATHI and T. P. MALL

Postgraduate Department of Botany, Kisan PG College, Bahraich, U.P., India, Tel: +91-5252-235113, 9450259294, E-mail: dr_sctripathi@yahoo.co.in

CONTENTS

ABSTRACT

Bahraich, Uttar Pradesh, India, is blessed with diversified flora of more than 1200 plant species, and has Tharu tribals inside as well as around the forests. The tribals have strong belief in magicotherapeutic properties of plants for treatment of their ailments. The vegetation of the area is mainly characterized by large member of herbaceous plants growing on variety of habitats along with scattered occurrence of many indigenous and exotic species of trees and shrubs in open areas or cultivated in gardens and along road sides. The North Central Terai belt in which Bahraich is situated is next to North East and Western Ghats, which represents one of the 18 hot spots of the world mega biodiversity. Despite this richness, the wealth of traditional knowledge is being lost, as the traditional culture is gradually disappearing; hence, we have taken this project to document and describe the potent, nutrimental, ethnomedicinal, and horticultural flora of the studied area.

25.1 INTRODUCTION

Bahraich district is one of the Terai districts of eastern Uttar Pradesh, situated in Upper Gangetic Plane. It lies between 27°43' and 28°51' North Latitude and 81°8', and 82°10', East longitude, with a total area of about 6944 sq km. Botanically, the area is very interesting. In north, the Himalayas rise as a virtual wall beyond the snow line. Above the alluvial plain lies the Terai strip, a seasonally marshy zone of sand and clay soils.

This north Terai region has higher rain fall than the plains, and the downward rushing rivers of the Himalayas slow down and spread out in the flatter Terai zone, depositing fertile silt and reproductive means during the monsoon season and receding in the dry season. Terai, as a result, has higher water level and is characterized by moist sub-tropical condition and a luxuriant turnover of green vegetation all the year around.

The study area is blessed with several floras by nature; it is referred as natural paradise, and it is very rich in ethnic and floristic diversity. The Tharu tribes are endowed with vast knowledge of medicinal plant and have strong belief in magicotheropeutic properties of plants for the treatment of various ailments. The district is having good population of tribals, mostly Tharus residing in villages, viz., Phakeerpuri, Amba, Balai gaon, and Ramapur of Mihinpurwa block; their knowledge regarding plants has descended from one generation to another as a domestic practice (Brahman, 2000). Due to the vast area of natural forests, the Bahraich is also known as City of Forests.

The land surface is a level tract sloping gently from North West to South East. A remarkable feature fills landscape is the total absence of any hill or hillocks. The soil is composed of Gangetic alluvium. Because much of the ground is liable to inundation, the particles deposited are very fine. Bahraich enjoys monsoon type of climate, very much influenced by the Himalaya being nearer to the region. The climate is markedly periodic and is divided into three seasons, i.e., rainy, winter, and summer season. The general temperature range between 3°C and 43°C. The general vegetation of the area is tropical deciduous type. However, some of the trees are evergreen and semi-evergreen. The forests are only restricted to northern portion of the district bordering up to foothills of Nepal. The middle and southern part of the area are under the influence of human and their domestic animals. Thus, the vegetation of this area is being damaged by intense grazing, fire, and cutting down of plants for fodder, fuel, and for various developmental projects. A vast area is also under cultivation. The vegetation of these areas is mainly characterized by a large number of herbaceous plants growing on variety of habitat along with scattered occurrence of many indigenous and exotic species of trees and shrubs in open areas or cultivated in gardens and along road sides.

Plants have a significant contribution toward the wealth of a country. During recent years, the exploration of our plant wealth and its economic utilization have rightly been given due importance. The contribution on the economic aspects of our plants are scattered over numerous literatures. The revision of the information based on modern collection and field observation has been advocated by Rao (1958). Gupta (1967) emphasized that the information we already possess on the economic aspect of plants should be revised thoroughly based on personal enquiries and experimentations. India presents colorful mosaic of about 563 tribal communities that have acquired considerable knowledge on the use of plants for their livelihood, healthcare and other purposes through their long association with the forests, inheritance, practices, and experiences. Plants with medicinal properties enjoyed the highest reputation in the indigenous system of medicines all over the world. India has one of the oldest, richest, and most diverse cultural traditions called folk tradition associated with the use of medicinal plants. Traditional folk medicine is the application of indigenous beliefs, knowledge, skills, and cultural practices concerned with human health.

The ethnic people have provided several miraculous plants of medicinal value for modern civilization. Both the ayurvedic and Siddha system of medicine originated more than 300 years ago and are prevalent in North and South India (Lgnacimuthu et al., 2006). The traditional definition of medicinal plant is given in Ashtasane Hrdays 2006 AD Sutra sihana ch 9, Vrse 10 as "There is nothing in this universe which is non-medicinal which cannot be used for many purposes and by many modes" (Shanker et al., 2000). India represents one of the 12 mega biodiversity centers of the world, and has two of the world's 18 biodiversity hot spots. North East and Western Ghats ranks first followed by the north central forests of Terai region. This Terai belt, well blessed and inhabited by tribal community inside the forest as well as around the forest area, is a natural paradise for ethnobotanical, mycological, plant pathological as well as work related with wildlife alone or interdisciplinary work. The World Health Organization (WHO) has also recognized the role of traditional system of medicine and considers it as a part of strategy to provide healthcare to masses. Folk medicine is gaining importance. Much of this wealth of knowledge is being lost as traditional culture is gradually disappearing (Hamilton, 1995). Tribal people throughout the world have developed their own culture, customs, cults, religious rites, myths, folk tales and songs, foods, medicinal practices, etc. Numerous wild as well as cultivated plants play a very important and vital role among these cultures, and this interrelationship has evolved over generations of experience and practices (Maheswari, 1983).

25.2 MATERIALS AND METHODS

The survey of the Bahraich districti was carried out during 2010–2014. Rapport was established with local elderly persons and the vaids (Ayurvedic physicians), Hakims of the locality as well as Tharu tribes of the surveyed area. Inquiries were made on the plant material used for curing different ailments. Elderly men and women folk were interviewed by the questionnaire method, which resulted in heterogeneity of information. Participation in their feasts, festivals, and other social events, etc., was of great use in collecting information on flora and their use. The plant was collected in flowering and fruiting stage so as to facilitate identification. It was identified by authentic literatures and floras, viz Hooker (1872–1997); Sharma et al. (1993a,b); Dubey (2004); Saini et al. (2010); and Saini (2006). The herbarium of flora was prepared according to the method described by Jain and Rao (1976) and Rao (1958) and deposited in Herbarium of the Department for the record and reference.

25.3 RESULT AND DISCUSSION

25.3.1 ENUMERATION

The study area has eight floras of family Apocynaceae (dicot) and three floras of family Liliaceae m(Monocot).

25.3.2 FAMILY APOCYANACEAE

1. *Carissa carendas*—A horticultural plant used in cure of anemia, acid reflux, analgesic for headache and migrane anorexia, analgesic for Aphthous ulcers, antihelminthic, and antihistamine. It contains alkaloids and is sour in taste due to presence of vitamin C. It is antiscorbutic. Parts used: leaves, fruits, roots, and dried stem bark. Center of origin is India.

2. *Carrisa edulis*—Roots contain an active ingredient, carissin, that is proved useful in the treatment of cancer; the twigs contain quebrachytol and cardioglycosides that are useful as an antihelminthic against tapeworm. The boiled leaves and poultice is applied to relieve toothache. Root bark is mixed with spices and used as an enema for lumbago and other pain. Root scrapings are used for glandular inflammation; ground up roots is used as a remedy for venereal diseases, to restore vitality and to treat gastric ulcers, cause abortion, and as an expectorant. An infusion of roots along with other medicinal plants is used for treating chest pains, and a root decoction is also used for treating malaria. *Carrisa edulis* is an attractive ornamental tree and is suitable for planting in amenity

areas; the abundant branching naturet and the presence of spines make the plant suitable for planting as a protective hedge.

3. ***Catharanthus roseus***—Flower extracts have wound healing activity, it is used to treat diabetes, it has anticancer and antimicrobial properties.

4. ***Holarrheria antidysentrica***—Stem, root bark and seeds are used, it is primarily used for the treatment of dysentery, bleeding disorder menorrhea, hemorrhoids, edema, tumors, abscesses aches, branchites, colic disorders diarrhea, spinitis and as a vermifuge, laxative and astringent.

5. ***Nerium indicum***—The whole plant is used; it's constituents are oleandrin, tanin, nerrin, phytosterine and L—strophnathinrosaginin, nerlin, volatile oil, fixed oil, neriodorine and nerriodorein. Leaves & flowers are thought to have action as cardiotonic, diaphoretic, diuretic, emetic, expectorant and used to treat malaria and in termination of embryo. Tincture or decoction is used to reduce swelling and scabes, Root powder is used as an external remedy for hemorrhoids and ulcers around genitals. It also has anticancer properties.

6. ***Rauwolfia serpentina***—It is used in high blood pressure, nervous disorders insomnia, and mental disorder such as agitated psychosis and in insanity. The juice of the root is given internally to treat snake bite. The decoction of leaf is invariably used in typhoid, malaria, and other fevers. It is anti-inflammatory, antidiurectic, and anticholinergic. It is one of significant medicinal herb used by the people as indigenous medicine.

7. ***Rauwolfia tetraphylla***—Roots are sedative tonic febrifuge; it is a valuable remedy in high blood pressure and is used in insomnia, mental disorders, and painful ailments of the bowels, hypochondria and irritating condition of the central nervous system. Roots are used in anxiety, excitement, schizophrenia, and epilepsy. Root extracts are also used a valuable remedy in diarrhea, dysentery, cholera, colic, and fever, Root paste along with orange peel is used to treat fever by the tribal of Mihinpurwa Block. The root or leaf juice is used against piles and as a remedy for sterility in women.

8. ***Thevetia peruviana***—Bark or leaf decoction is taken to loosen the bowels as an emetic and is said to be an effective cure in intermittent fevers. In Senegal, water in which leaves and bark are macerated is taken to cure amenorrhea.

25.3.3   FAMILY – LILIACEAE

1. ***Aloe barbadensis***—It is used to treat wounds, heals skin infections, cure chopping, decrease hair loss, and eliminate hemorrhoids. It is also used to cure eczema, sun burns, amputation, stump ulcers, lacerations, colds, tuberculosis, gonorrhea, asthma, dysentery, and headache. It is also used as insect repellent and as a laxative.
2. ***Asparagus racemosus***—The roots are used following a regime of processing and drying. It is used as uterine tonic and as a galactogogue (to improve production of breast milk); it is also used in hyperacidity and as a best general health tonic.
3. ***Glorisa superba***—It is also known as flame lily and has a wide variety of uses within traditional medicine. It contains alkaloid colchicines, which has been used effectively to treat gout, intestinal worms, infertility, wounds and other skin problems. The roots and leaves are used as an antidote for snake bite, as a laxative, and to induce abortion. It has proven useful in the treatment of chronic ulcers, arthritis, cholera, colic, and kidney problems. Glory lily extract is useful to rectify many respiratory disorders. The sap from the leaf tip is used for pimples and skin eruptions. Glorisa paste can be applied for crump inflammation like wounds, piles, and skin-related problems. Its powder is also helpful in relieving from menstrual disturbances.

25.4   CONCLUSION

The rural and tribal repository of the studied area contains many medicines for the treatment of various ailments. It is hoped that this effort will not only provide additional support to the earlier findings but also provide clues for new materials having traditional potentiality for the benefits of mankind.

ACKNOWLEDGMENTS

The authors are thankful to all the local knowledge holders who helped in one way or the other. Thanks are also due to The Principal, Kisan P.G. College, Bahraich, for his permission to conduct this project and for facilities and to The Chief Wildlife Warden, Uttar Pradesh Government, Lucknow, for due permission and facilities. Thanks to Prof. S. K. Singh, Retired Professor and Head, Department of Botany, DDU University of Gorakhpur and Late Dr. D.C. Saini, Scientist Grade F, Birbal Sahni Paleobotany Research Institute, Lucknow, for identification of certain plants and for their admirable help and encouragements.

KEYWORDS

- **ethnomedicinal**
- **horticultural**
- **mega biodiversity**
- **nutrimental**
- **potent**
- **tribals**

REFERENCES

Brahman, M., (2000). Some Ethnomedicinal plants of Akola and Sanganer talukes of Ahmadnagar. *J. Indian Bot. Soc.*, *81*, 213–215.

Dubey, N. K., (2004). *Flora of BHU Campus*, Printed and Published by BHU Publication Cell, 1–180.

Gupta, R., (1967). *Seasonal Flowers of the Indian Summer Resorts*, Mussoorie Hills, New Delhi.

Hamilton, A., (1995). The people and plants initiative, In: *Ethnobotany: Methods and Manual,* by Martin, G. J., WWW International, Chapman and Hall, London, pp. 10–11.

Hooker, J. D., (1872–1897). *The Flora of British India,* 1–7, (London).

Ignacimuthu, S., Ayyangar, M., & Shanker, S. K., (2006). Ehhnobolanical investigations among tribes in Madurai district of Tamil Nadu (India). *J. Ethnobiol Ethnomedi, 2,* 25.

Jain, S. K., & Rao, R. R., (1976). *Hand Book of Field and Herbarium Methods,* Today and Tomorrow Printers and Publishers, New Delhi, 33–58.

Maheswari, J. K., (1983). *Developments in Ethnobotany. J. Econ. Tax. Bot., 4*(1), 1–5.

Rao, R. S., (1958). History and importance of Indian herbaria. *J. Ind. Bot. Soc., 3,* 152–159.

Saini, D. C., (2006). Flora of Bahraich District, Uttar Pradesh, *J. Eco Taxon Bot., 29*(V), 843–886.

Shanker, D., Ved, D. K., & Geeta, V. G. A., (2000). *Green Pharmacy Indian Health Traditions,* The Hindu Special issue with the Sunday Magazine, 1–2.

Sharma, B. D., & Samjappa, M., (1993). With assistance from Bal Krishnan, N. P., Ed. *Flora of India,* vol. 3, BSI. Calcutta Deep Printers New Delhi.

CHAPTER 26

ELEMENTAL DETERMINATION OF TWO MEDICINAL PLANTS OF MIZORAM USING EDXRF

R. LAWMZUALI,[1] K. BIRLA SINGH,[2] and N. MOHONDAS SINGH[1]

[1]Department of Chemistry, School of Physical Sciences, Mizoram University, Aizawl–796004, India

[2]Department of Zoology, Pachhunga University College, Aizawl–796001, India, E-mail: nmdas08@rediffmail.com

CONTENTS

ABSTRACT

Medicinal plants have been used by humans for centuries in folklore medicine. Medicinal plants are also incorporated into the historical medicine of virtually all human cultures. The present work discusses the elemental determination of elements in two medicinal plants *Solanum nigrum linn* and *Spilanthes acmella* by using energy dispersive X-ray fluorescence (ED-XRF). *S. nigrum linn* belongs to the Solanaceae family and is used in the treatment of cardiac, skin disease, and inflammation of kidney, etc. *S. acmella* belongs to the family Compositae and is called the toothache plant and it is also used as an anti-inflammatory and analgesic. Twelve elements were determined in this study.

26.1 INTRODUCTION

Mizoram is a state rich in plant life, and traditional medicines procured from plants are employed for the treatment of numerous ailments since time memorial. Such knowledge has been handed down through generations, and although modern medicines have seen a huge breakthrough, these traditional medicines are still being used to a certain degree. Among these, two of the most commonly used medicinal plants are *Solanum nigrum linn* and *Spilanthes acmella* (Figure 26.1).

Solanum nigrum linn commonly known as Black Nightshade is a dicot weed in the Solanaceae family. It is an annual branched herb of up to 90 cm high, with dull dark green leaves, juicy, ovate or lanceolate, and toothless to slightly toothed on the margins. Flowers are small and white with

Solanum nigrum linn   *Spilanthes acmella*

FIGURE 26.1   Medicinal plants: (a) *Solanum nigrum* linn and (b) *Spilanthes acmella*.

a short pedicellate and five widely spread petals. Fruits are small, black when ripe. The herb is antiseptic, antidysenteric, and antidiuretic and is used in the treatment of cardiac, skin disease, psoriasis, herpesvirus infection, and inflammation of the kidney. The root bark is laxative and is useful in the treatment of ulcers on the neck, burning of throat, inflammation of the liver, and chronic fever. Berries are bitter and pungent useful for treating heart disease, piles, and dysentery.

Spilanthes acmella is an indigenous herb belonging to the family Compositae. The plant has yellow/red gumdrop-shaped flowers. The leaves are arranged opposite to one another and are 2.5 cm to 5 cm long. It is an important medicinal plant and is commonly known as Akkalkara plant with rich source of therapeutic constituents. It is called the toothache plant because by chewing the leaves or flowers, it produces a numbing effect on the tongue and gums. The flower heads of *S. acmella* can be used to relieve toothache and also has anti-inflammatory and analgesic effects.

However, complete data regarding the elemental composition of these plants are not known. The World Health Organization (WHO) has estimated that 80% of the population of developing countries relies on plant-based traditional medicines to maintain their primary healthcare needs. High treatment cost and side effects along with drug resistance are the major problems associated with synthetic drugs. Because traditional medicine is not only easily accessible but also affordable, there is an increased emphasis on the the use of plants to treat human diseases. Therefore, the global markets are turning to plants as a potential and realistic source of ingredients for healthcare products.

26.2 METHODOLOGY

26.2.1 SAMPLE PREPARATION

The two plants were thoroughly washed with triple distilled water to eliminate contamination due to dust and environmental pollution, air-dried, and then oven dried at 60°C. After drying, they were grounded into fine powder using mortar and pestle. The powdered samples were then formed into pellets.

26.2.2   ANALYSIS USING ED-XRF SPECTROMETER

The elemental analysis of plant samples was carried out using a Xenematrix Ex-3600 energy dispersive X-ray fluorescence (ED-XRF) spectrometer, which consists of an oil-cooled Rh anode X-ray tube (maximum voltage 50 kV, current 1 mA). The measurements were carried out in vacuum using different filters (between the source and sample) for the optimum detection of elements. For example, for P, S, Cl, K, and Ca, no filter was used, and a voltage of 6 kV and current of 200 mA were used, and samples were run for 200 sec. A 0.05-mm-thick Ti filter was used in front of the source for Mn, Fe, Cu, and Zn, with an applied voltage of 14 kV and a current 900 mA, and samples were run for 400 sec. For higher Z elements such as Se, Br, Rb and Sr, Fe filter of 0.05 mm thickness was used at a voltage of 23 kV and 200 mA current, and samples were run for 600 sec. The X-rays were detected using a liquid nitrogen-cooled 12.5 mm^2 Si (Li) semiconductor detector (resolution 150 eV at 5.9 keV). The X-ray fluorescence spectra were quantitatively analyzed by the software integrated with the system. This software uses the fundamental parameter method approach, which combines a theoretical basis of X-ray emission and absorption with experimental measurements for unknown sample analyses. Here, all matrix corrections, etc., are taken into account. The experimental results were subjected to statistical analysis using Excel 2007 and SPSS package v.17.0. Values are presented as standard error of mean (SEM).

26.3   RESULT AND DISCUSSION

The elemental determination of the two medicinal plants, viz,. *S. nigrum linn* and *S. acmella* using ED-XRF yielded 12 different elements that are given in Table 26.1.

From the result, we can see that both contain high concentration of potassium (K), calcium (Ca), phosphorus (P), and sulfur (S) (Figure 26.2).

K is one of the essential elements of human diet and play an important role in vital cellular mechanisms. As a cofactor, it catalyzes the conversion of ADP to ATP, and K deficiency can create resistance in fat and muscle

TABLE 26.1 Elemental Concentration (mg/L) of the Selected Medicinal Plants of Mizoram. Values are Mean ± SEM of 6 Observations Each

	Solanum nigrum linn	*Spilanthes acmella*
Calcium (Ca)	16659.7±329.5	12542.9±346.2
Potassium (K)	42180.9±789.3	46224.4±690.6
Phosphorus (P)	9816.7±125.1	12310.1±312.8
Sulphur (S)	7577.9±333.1	5608.2±116.1
Iron (Fe)	577.2±110.6	736.1±90.8
Manganese (Mn)	89.4±6.5	140±3.9
Zinc (Zn)	59.1±9.9	152.5±8.9
Copper (Cu)	11.3±0.1	10.2±0.7
Selenium (Se)	0.23	13.9±0.1
Strontium (Sr)	91.5±5.2	58.1±2.4
Rubidium (Rb)	20.7±0.7	16.2±3.0
Bromine (Br)	15.3±1.4	110.4±2.2

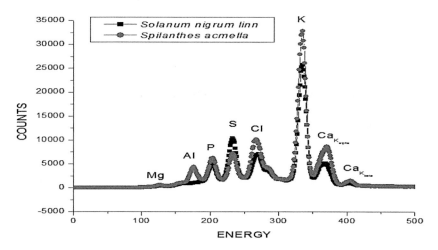

FIGURE 26.2 **(See color insert.)** EDXRF spectra of major elements in *Solanum nigrum* linn and *Spilanthes acmella.*

cells, etc., from insulin, increase serum triglycerides, may lower HDL and blood supply to the vital organs, and increase the chances of stroke. Ca is also an important element that plays a pivotal role in the physiology and biochemistry of the cells in humans. It has an important role in signal transduction pathways, where it acts as a secondary messenger, in neurotransmitter release from neurons, contraction of all muscle cell types, and fertilization. P is an element that makes up 1% of a person's total body weight and is present in every cell of the body. The main function of P in the human body is in the formation of bones and teeth. As a macroelement of the body, the role of S is to act as an integral part of many important compounds found in all body cells which are indispensable for life.

The presence of trace elements like iron (Fe), manganese (Mn), zinc (Zn), copper (Cu), selenium (Se), strontium (Sr), rubidium (Rb), and bromine (Br) in notable concentrations explained to a certain degree their medicinal properties (Figures 26.3 and 26.4).

Trace element plays an important role in human health because they participate in biological functions that contribute to growth and good health. Fe is a necessary nutrient element and is a core component of RBC. It is needed for healthy immune system and for energy production. Mn is one of the important essential elements required in carbohydrate metabolism as well as an antioxidant in SOD enzymes. As a contaminant, no maximum permissible limit (MPL) has been fixed for Mn in vegetables. The upper tolerable limit of Mn for humans is 2–11 mg/day.

Zn is an important trace element involved in numerous aspects of cellular metabolism and required for the catalytic activity of more than 200 enzymes. Zn also plays an important role in immune function, wound healing, protein synthesis, DNA synthesis, and cell division. Cu is known

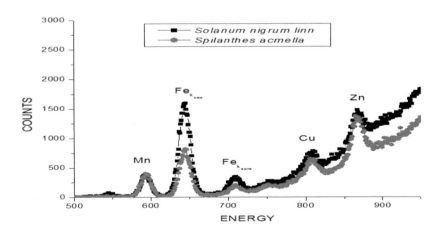

FIGURE 26.3   (See color insert.) ED-XRF spectra of minor elements in *Solanum nigrum* linn and *Spilanthes acmella*.

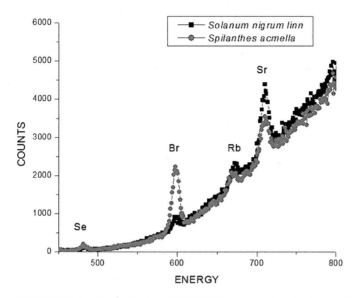

FIGURE 26.4 **(See color insert.)** ED-XRF spectra of earth elements in *Solanum nigrum linn* and *Spilanthes acmella*.

to play an important role in human metabolism, largely because it allows many critical enzymes to function properly. As an antioxidant, Cu scavenges or neutralize free radicals and may reduce or help prevent some of the damage they cause. The concentration of Cu in plants varied much with dependent nearby factors like proximity industries and use of fertilizers and Cu-based fungicides. Se is an element which behaves as both an antioxidant and anti-inflammatory agent. As a contaminant, there is no limit for Se. Rb ions are utilized by human body in a manner similar toK ions, being actively taken up by plant and animal cells.

26.4 CONCLUSION

The chemical characterization of the two medicinal plants, viz., *Solanum nigrum linn* and *Spilanthes acmella* by ED-XRF reveals the presence of variable amounts of different types of major elements (Ca, P, K, and S), minor elements (Zn, Fe, Cu, Mn, and Se), and earth elements (Sr, Rb, and Br).

ACKNOWLEDGMENT

The authors are thankful to the UGC-DAE Consortium for Scientific Research, Kolkata Center, for aiding in the research work.

KEYWORDS

- **EDXRF**
- **elements**
- **medicinal plants**

REFERENCES

Acharya, E., & Pokhrel, B., (2006). Ethno-medicinal, Plants used by Bantar of Bhaudaha, Morang, Nepal. *Our Nature, 4,* 96–103.

Chakraborty, A., et al., (2004). Preliminary studies on anti-inflammatory and analgesic activities of *Spilanthes acmella* in experimental animal models. *Indian. J. Pharmacology, 36*(3), 148–150.

Chopra, R. N., Nayara, S. L., & Chopra, I. C., (1956). *Glossary of Indian Medicinal Plants,* Council of Scientific & Industrial Research, New Delhi, 168–169.

Cooper, M. R., & Johnson, A. W., (1984). *Poisonous Plants in Britain and Other Effects on Animals and Man.* Ministry of Agriculture, Fisheries Food, *161,* 219–220.

Gokhle, V. G., et al., (1945). Chemical investigation of *Spilanthes acmella* (Murr.). *J. Chemical Society, 1,* 250–252.

Hsiu-chen, et al., (2010). Chemical composition of *Solanum nigrum* Linn extract and induction of autophagy by leaf extract and its major flavonoids, *Journal of Agricultural and food Chemistry, 58*(15), 8699–8708.

Kritikar, K. R., & Basu, B. D., (1935). *Indian Medicinal Plants.* M/S Bishen Singh Mahendrapal, New Delhi, India, pp. 267–268.

Pronob Gagoi, & Islam, M., (2012). Phytochemical screening of *Solanum nigrum* L and *S. myriacanthus* dunal from districts of Upper Assam, India, *IOSR Journal of Pharmacy, 2*(3), 455–459.

Zakaria, Z. A., Gopalan, H. K., Zainal, H., et al., (2006). Antinociceptive, anti-inflammatory and antipyretic effects of *Solanum nigrum* chloroform extract in animal models. *Yakugaku Zasshi, 126,* 1171–1178.

INDEX